Generic Drug
Product Development

Generic Drug Product Development

Specialty Dosage Forms

Edited by

Leon Shargel
Applied Biopharmaceutics, LLC
Raleigh, North Carolina, USA

Isadore Kanfer
Rhodes University
Grahamstown, South Africa

CRC Press
Taylor & Francis Group
Boca Raton London New York

CRC Press is an imprint of the
Taylor & Francis Group, an **informa** business

CRC Press
Taylor & Francis Group
6000 Broken Sound Parkway NW, Suite 300
Boca Raton, FL 33487-2742

First issued in paperback 2019

© 2010 by Taylor & Francis Group, LLC
CRC Press is an imprint of Taylor & Francis Group, an Informa business

No claim to original U.S. Government works

ISBN-13: 978-0-8493-7786-0 (hbk)
ISBN-13: 978-0-367-38439-5 (pbk)

This book contains information obtained from authentic and highly regarded sources. While all reasonable efforts have been made to publish reliable data and information, neither the author[s] nor the publisher can accept any legal responsibility or liability for any errors or omissions that may be made. The publishers wish to make clear that any views or opinions expressed in this book by individual editors, authors or contributors are personal to them and do not necessarily reflect the views/opinions of the publishers. The information or guidance contained in this book is intended for use by medical, scientific or health-care professionals and is provided strictly as a supplement to the medical or other professional's own judgement, their knowledge of the patient's medical history, relevant manufacturer's instructions and the appropriate best practice guidelines. Because of the rapid advances in medical science, any information or advice on dosages, procedures or diagnoses should be independently verified. The reader is strongly urged to consult the relevant national drug formulary and the drug companies' and device or material manufacturers' printed instructions, and their websites, before administering or utilizing any of the drugs, devices or materials mentioned in this book. This book does not indicate whether a particular treatment is appropriate or suitable for a particular individual. Ultimately it is the sole responsibility of the medical professional to make his or her own professional judgements, so as to advise and treat patients appropriately. The authors and publishers have also attempted to trace the copyright holders of all material reproduced in this publication and apologize to copyright holders if permission to publish in this form has not been obtained. If any copyright material has not been acknowledged please write and let us know so we may rectify in any future reprint.

Library of Congress Cataloging-in-Publication Data

Generic drug product development : specialty dosage forms / edited by Leon Shargel, Isadore Kanfer.
     p. ; cm. – (Drugs and the pharmaceutical sciences ; 204)
    Includes bibliographical references and index.
    ISBN-13: 978-0-8493-7786-0 (hardcover : alk. paper)
    ISBN-10: 0-8493-7786-2 (hardcover : alk. paper)   1. Generic drugs.
2. Drug development. 3. Drugs–Dosage forms. 4. Drugs–Therapeutic equivalency. I. Shargel, Leon, 1941– II. Kanfer, Isadore. III. Series: Drugs and the pharmaceutical sciences ; v. 204.
    [DNLM: 1. Dosage Forms. 2. Drugs, Generic. 3. Biopharmaceutics–methods. 4. Therapeutic Equivalency. W1 DR893B v.204 2010 /
QV 785 G326 2010]
    RS55.2G456 2010
    615'.19–dc22
                                                2009045723

Visit the Taylor & Francis Web site at
http://www.taylorandfrancis.com

and the CRC Press Web site at
http://www.crcpress.com

Preface

This book is an additional volume in a series of books on Generic Drug Product Development. The objectives of the first book, *Generic Drug Product Development—Solid Oral Dosage Forms*, described from concept to market approval, the development of therapeutic equivalent generic drug products including scientific, regulatory, and legal challenges. The second volume, *Generic Drug Product Development—Bioequivalence Issues*, focused on current problems concerning the scientific demonstration of bioequivalence of pharmaceutical dosage forms, a test, and a reference product that would lead to regulatory approval of the generic product. This book, *Generic Drug Product Development—Specialty Dosage Forms*, focuses on the development of drug products other than solid oral dosage forms for systemic drug absorption. These specialty dosage forms may be given by various routes of administration to deliver drug for systemic or local activity.

Specialty drug products are difficult to define. Many of these products require special routes of administration, for example, inhalation, transdermal, parenteral (e.g., intramuscular depot injection), topical application, or medicated stent. In these cases, the drug delivery is performed with a device such as a metered dose inhaler, transdermal patch, or sterile injection needle, and the device is involved in drug delivery as well as the release of the drug at the site of administration. Specialty drug products may be designed for systemic drug delivery or for local drug delivery, where the active ingredient is not intended for systemic absorption. Although, drug products intended for local activity may be systemically absorbed, the systemically absorbed drug does not reflect the activity of the drug at the application site.

Bioequivalence studies are expensive, time-consuming and have a possibility of failure. Failure to demonstrate bioequivalence of a proposed generic dosage form results not only in a loss of money and time, but also may lead to a management decision not to pursue further development of that product.

Bioequivalence is more easily established for a large number of oral drug products intended for systemic drug absorption in which the drug and/or metabolites can be measured in the plasma. For these drug products, regulatory agencies and the scientific community are in general agreement as to the design of a bioequivalence study and the statistical analyses of the results. For specialty drug products, such as drug products given by a non-oral route of administration or for drugs not intended to be absorbed into the systemic circulation, pharmaceutical equivalence and bioequivalence can be difficult to demonstrate. Alternate approaches to the design and evaluation of bioequivalence studies are usually needed for these drug products.

The development of a therapeutic equivalent, generic specialty drug product is generally product specific. Each specialty drug product has its own unique issues in the determination of pharmaceutical equivalence and bioequivalence. This volume discusses the problems in the development of a generic, therapeutic equivalent specialty drug product.

Chapter 1 introduces the subject of specialty drug products and provides a general background for the development of a therapeutic equivalent, generic drug product. Nonsystematically absorbed drug products that are given orally are discussed in Chapter 2. These oral drug products are intended for local activity and may have negligible or in some cases, measurable systemic drug absorption from the gastrointestinal tract. Currently, the U.S. Food and Drug Administration (FDA) discourages the use of systemic drug concentrations for drug products that are applied locally or intended only for local action. Therefore, alternate approaches for the determination of bioequivalence must be used for these products.

Chapters 3 and 4 discuss topical dosage forms from two approaches. Chapter 3 reviews manufacturing and regulatory issues, whereas Chapter 4 discusses various approaches for the assessment of bioequivalence of topical dermatological dosage forms. Suppositories are unique semisolid drug products that can be given rectally or vaginally for either local or systemic activity. The development of suppository drug products is discussed in Chapter 5. Chapters 6 and 7 discuss nasal and inhalation drug products. Chapter 6 provides a manufacturing perspective, whereas Chapter 7 gives regulatory and bioequivalence perspectives. Transdermal drug delivery is increasing in popularity and is discussed in Chapter 8. This chapter reviews the various approaches that are now available for providing systemic drug absorption from a dosage form applied onto the skin. Chapter 9 discusses development of generic modified release parenterals such as microspheres, liposomes, and other carrier-mediated systems from both regulatory as well as scientific perspectives. Application of principles of Bioequivalence (BE), Quality by Design (QBD), and In Vitro In Vivo Correlation (IVIVC) to the pharmaceutical development of these systems is detailed.

The development of a generic, therapeutic equivalent biotechnology-derived drug product such as recombinant, cell-derived protein is both a scientific and regulatory challenge. These products are sometimes known as biosimilars or follow-on biologic drugs. The demonstration of pharmaceutical equivalence for biotechnology-derived drugs such as protein drugs is a major scientific challenge. Moreover, regulatory agencies such as the U.S. FDA have not as yet formalized a pathway for review and approval of these products. Chapter 10 provides some insight on the manufacture of these biotechnology-derived drug products.

The audience for this volume includes members of the pharmaceutical community including academia, pharmaceutical scientists and regulatory health agencies who are interested in the development and regulation of interchangeable, multisource, generic drug products. The emphasis of this volume focuses on issues pertaining to developing therapeutic equivalent specialty drug products.

Leon Shargel
Isadore Kanfer

Contents

Contributors

Mohammed N. AliChisty G & W Laboratories, Inc., South Plainfield, New Jersey, U.S.A.

Wai Ling Au Faculty of Pharmacy, Rhodes University, Grahamstown, South Africa

John Bell Stewart Erl Associates, Loughborough, U.K.

Diane J. Burgess Department of Pharmaceutical Sciences, University of Connecticut, Storrs, Connecticut, U.S.A.

Gary W. Cleary Corium International, Inc., Menlo Park, California, U.S.A.

Candis Edwards St. John's University, College of Pharmacy & Allied Health Professions, Queens, New York, U.S.A.

Mary Beth G. Erstad Paddock Laboratories, Inc., Minneapolis, Minnesota, U.S.A.

Paul Flanders Mylan, London, U.K.

Mario A. González P'Kinetics International, Inc., Pembroke Pines, Florida, U.S.A.

Betsy Hughes-Formella bioskin GmbH, Hamburg, Germany

Isadore Kanfer Faculty of Pharmacy, Rhodes University, Grahamstown, South Africa

Marc Lefebvre Algorithme Pharma Inc., Montreal, Quebec, Canada

Siddhesh D. Patil Millennium: The Takeda Oncology Company, Parenteral Formulation Sciences, Cambridge, Massachusetts, U.S.A.

Suzanne M. Sensabaugh HartmannWillner LLC, Columbia, Maryland, U.S.A.

Leon Shargel Applied Biopharmaceutics, LLC, Raleigh, North Carolina, U.S.A.

Gur Jai Pal Singh Division of Bioequivalence, U.S. FDA, Rockville, Maryland, U.S.A.

Ralph Nii Okai Tettey-Amlalo Faculty of Pharmacy, Rhodes University, Grahamstown, South Africa

K. Rosh Vora G & W Laboratories, Inc., South Plainfield, New Jersey, U.S.A.

1 Introduction

Leon Shargel

Applied Biopharmaceutics, LLC, Raleigh, North Carolina, U.S.A.

Specialty drug products discussed in this textbook are defined as nonorally absorbed drug products for which pharmaceutical equivalence is not easily demonstrated and in many instances bioequivalence cannot be determined using only the measurement of drug concentrations in plasma. For many specialty drug products, the demonstration of pharmaceutical equivalence and bioequivalence is more difficult compared to solid oral drug products intended for systemic activity, consequently, the requirements and procedures for bioequivalence have have not been agreed upon by both the scientific and regulatory communities. When the usual assessments of bioequivalence and pharmaceutical equivalence are not applicable, scientific and regulatory challenges become barriers to the development and approval of generic drug products. Moreover, legal challenges often arise when there is no general agreement on these issues.

MULTISOURCE GENERIC DRUG PRODUCTS

Multisource drug products are marketed by more than one manufacturer and contain the same active pharmaceutical ingredient (API) or drug substance in the same dosage form and are given by the same route of administration. Many multisource drug products meet various compendial[a] monograph standards of strength, quality, purity, and identity. However, drug substances and drug products that solely meet the same compendia monograph standards should not be automatically considered as interchangeable, therapeutically equivalent generic drug products. The approval of interchangeable, therapeutic equivalent, multisource, generic drug products is dependent on the review of an application [e.g., Abbreviated New Drug Application (ANDA)] by an appropriate regulatory agency such as the Food and Drug Administration (FDA).

The term *therapeutic equivalent* is defined by the U.S. Food and Drug Administration (FDA) (1)

> Drug products are considered to be therapeutic equivalents only if they are pharmaceutical equivalents and if they can be expected to have the same clinical effect and safety profile when administered to patients under the conditions specified in the labeling.
>
> FDA classifies as therapeutically equivalent those products that meet the following general criteria:
>
> 1. They are approved as safe and effective.
> 2. They are pharmaceutical equivalents in that they

---

[a] Compendial monographs are found in the United States Pharmacopeia-National Formulary (USP), European Pharmacopeia (EP), Japanese Pharmacopeial (JP), and others.

a. contain identical amounts of the same active drug ingredient in the same dosage form and route of administration, and
b. meet compendial or other applicable standards of strength, quality, purity, and identity.
3. They are bioequivalent in that
    a. they do not present a known or potential bioequivalence problem, and they meet an acceptable in vitro standard, or
    b. if they do present such a known or potential problem, they are shown to meet an appropriate bioequivalence standard.
4. They are adequately labeled.
5. They are manufactured in compliance with Current Good Manufacturing Practice regulations.

Multisource drug products must be approved as a therapeutic equivalent by an appropriate regulatory agency, such as the FDA, to be marketed as interchangeable drug products. These multisource drug products must be pharmaceutical equivalent, bioequivalent, and meet with other regulatory requirements such as manufactured by current good manufacturing practices, cGMP.

The approach for the demonstration of bioequivalence of two pharmaceutically equivalent drug products containing drugs that are systemically absorbed is widely accepted by the scientific and regulatory agencies (2). Generally, the active drug substance and/or active metabolites are quantitatively determined in plasma after the administration of the drug product to normal, healthy subjects by using a crossover design. The basis of bioequivalence is statistically determined by comparison of the C_{max} and AUC values as discussed earlier (2).

DRUG PRODUCT PERFORMANCE

Drug product performance may be defined as the release of the API from the dosage form, leading to systemic availability of the API to achieve a desired therapeutic response (3). Drug product performance for drugs that are intended for local pharmacodynamic activity considers the release of API at the site of action. Drug product performance must be considered in the design and manufacture of any finished dosage form (drug product). An approved multisource generic drug product must demonstrate equivalent drug product performance.

Drug product release from a dosage form may be measured using either in vivo or in vitro approaches. The in vivo release is measured by a comparative bioavailability (bioequivalence) study, whereas the in vitro release is most often measured by a comparison of dissolution and/or drug release profiles under various conditions.

DRUG PRODUCT QUALITY AND PERFORMANCE ATTRIBUTES

Quality standards are important attributes that must be built into the drug product. Quality cannot be tested into drug products. Quality should be built in or should be designed and confirmed by testing. The United States Pharmacopeia–National Formulary, USP–NF, and other official compendia contain monographs for a substance or preparation that includes applicable standards of strength, quality, purity, and identity also including the article's definition—packaging, storage, and other requirements and specifications. It cannot be automatically assumed that if two drug products meet the specifications of quality in a compendial monograph, the two products will have equal performance in vivo.

TABLE 1 Drug Product Quality and Performance Attributes

Product quality
- Chemistry, manufacturing, and controls (CMC)
- Microbiology
- Identity, strength, quality, purity, and potency of drug product
- Validation of manufacturing process and identification of critical quality attributes

Product performance
- In vivo: Bioavailability and bioequivalence
- In vitro: Drug release/drug dissolution

With a greater understanding of the drug product and its manufacturing process, regulatory agencies and pharmaceutical manufacturers are developing a systematic approach to achieve quality and drug product performance. Table 1 distinguishes between drug product quality and performance attributes.

Drug Product Quality and Drug Product Performance, In Vitro

Drug product release in vitro as measured by dissolution and/or drug release is a major tool in product development. Other uses of dissolution/drug release studies are (*i*) to establish specifications, possibly through the use of an in vitro–in vivo correlation (IVIVC), (*ii*) to demonstrate batch-to-batch quality, and (*iii*) to demonstrate stability of the drug product.

The difference between dissolution as a quality control, QC test, and dissolution as an in vitro performance test is that the in vitro performance test involves dissolution profile comparisons in pHs of 1.2, 4.5, and 6.8 between the Test drug product and the Reference drug product, whereas, the QC control test may involve limited sampling in only one medium. For example, many immediate-release dosage forms require only a single-point dissolution test under a single set of conditions. The QC dissolution test can only be used after the original

TABLE 2 Level of Changes Under SUPAC

Change level	Example	Comment
Level 1	Deletion or partial deletion of an ingredient to affect the color or flavor of the drug product	Level 1 changes are those that are unlikely to have any detectable impact on formulation quality and performance
Level 2	Quantitative change in excipients greater than allowed in a Level 1 change	Level 2 changes are those that could have a significant impact on formulation quality and performance
Level 3	Qualitative change in excipients	Level 3 changes are those that are likely to have a significant impact on formulation quality and performance. A Level 3 change may require in vivo bioequivalence testing

Source: Adapted from Refs. 4, 5.

batches have been validated for their performance by having a complete set of data that may include dissolution profiles, IVIVC, and BE measures.

Drug Product Quality and Drug Product Performance, In Vivo

Drug product performance in vivo as measured by a bioequivalence study is the most important and definitive test for drug product performance and is often considered as the "gold standard." For many generic drug products, comparative in vitro dissolution/drug release profiles between Test and Reference drug products are different, often due to differences in the manufacture, formulation, and/or drug release mechanism for each product. However, the differences in dissolution profiles may not necessarily be reflected in the bioavailability of the drug. Thus, it is possible that the in vivo bioequivalence study may show that the Test and Reference drug products are bioequivalent even with dissimilar in vitro drug release profiles.

SPECIALTY DRUG PRODUCTS AND DRUG PRODUCT PERFORMANCE

Unlike general oral drug products containing drugs for systemic absorption, drug delivery of some specialty drug products such as inhalation products, transdermal products, and medicated stents are complicated by the device in which the drug is contained. The drug delivery device such as a metered-dose inhaler can affect the drug delivery rate, particle size, plume geometry, etc., of the drug solution or suspension that is dispensed to a patient. This drug delivery will then impact on the therapeutic efficacy, that is, drug product performance.

Pharmaceutical Equivalents

As discussed above, two drug products (including specialty drug products) must be pharmaceutical equivalents to comply with the definition for therapeutic equivalents. The FDA defines therapeutic equivalence as drug products that "contain the same active ingredient(s), are of the same dosage form and route of administration, and are identical in strength or concentration (e.g., chlordiazepoxide hydrochloride, 5 mg capsules). Pharmaceutically equivalent drug products are formulated to contain the same amount of active ingredient in the same dosage form and to meet the same or compendial or other applicable standards (i.e., strength, quality, purity, and identity), but they may differ in characteristics such as shape, scoring configuration, release mechanisms, packaging, excipients (including colors, flavors, preservatives), expiration time, and, within certain limits, labeling" (1).

The determination of pharmaceutical equivalents for all drug products is made by the FDA. Because of patents and different approaches to drug delivery, multisource drug products may be formulated differently and have different drug release mechanisms. For example, two generic fentanyl transdermal systems (patch) have been approved. One fentanyl transdermal patch uses a reservoir system, whereas, the other fentanyl transdermal patch uses a matrix system. Thus, the regulatory agencies, such as FDA, make the determination or pharmaceutical equivalence or *sameness* but this does not infer that such products are bioequivalent and thus must be tested against the RLD. Once the generic product that is a pharmaceutical equivalent is found to be bioequivalent to the RLD, then such a generic dosage form is considered to be therapeutically equivalent and *interchangeable*.

Pharmaceutical Alternatives

The U.S. FDA does not extend pharmaceutical equivalence to capsule and tablet formulations of the same drug even if both products demonstrate bioequivalence with respect to the drug. According to the FDA, "Data are generally not available for FDA to make the determination of tablet to capsule bioequivalence" (1). Currently, FDA does not consider capsule and tablet dosage forms to be therapeutic equivalents nor interchangeable even if these products are bioequivalent. Some countries, however, consider bioequivalent tablet and capsule drug products of the same drug as interchangeable.

FDA defines pharmaceutical alternatives as drug products that "contain the same therapeutic moiety, but are different salts, esters, or complexes of that moiety, or are different dosage forms or strengths (e.g., tetracycline hydrochloride, 250 mg capsules vs. tetracycline phosphate complex, 250 mg capsules; quinidine sulfate, 200 mg tablets vs. quinidine sulfate, 200 mg capsules). Different dosage forms and strengths within a product line by a single manufacturer are thus pharmaceutical alternatives, as are extended-release products when compared with immediate-release or standard-release formulations of the same active ingredient.

Establishment of Pharmaceutical Equivalence

As discussed above, the establishment of pharmaceutical equivalence for some specialty drug products is difficult. For example, most topical drug products are considered semisolid dosage forms such as creams, lotions, gels, and ointments. These products are intermediate between liquid and solid, and depending on the measurement, their properties are a mixture of solid and liquid behavior.

The FDA has identified three in vitro methods to assess structural similarity of topical products:

Q1: Qualitative similarity in composition
Q2: Quantitative similarity in composition
Q3: Structural similarity
- Refers to the physical attributes and state of the product.
- Reflects changes in manufacturing or physical state of starting materials.

Q1 and Q2 similarity can be demonstrated by accepted scientific methods. However, Q3 similarity is more difficult to demonstrate. Several questions need to be answered:

- What type of data is needed to demonstrate that two products are Q3 equivalent? Some data might include structural similarity, arrangement of matter, and state of aggregation.
- How should the Q3 concept be validated or demonstrated?
- Can Q3 be used to detect changes in manufacturing processes?
- What does Q3 similarity imply about bioequivalence?

Some approaches to Q3 have considered physical state of starting materials and manufacturing methods.

For nasal aerosols and nasal sprays (6), tests may include

- single actuation content (SAC) through container life,
- droplet size distribution by laser diffraction,
- particle/droplet size distribution by cascade impactor,
- particle/droplet size distribution,
- spray pattern,
- plume geometry, and
- priming and repriming.

Whether bioequivalence or drug product performance is affected by small changes in Q3 has not been established. The assumption is that if a drug product meets Q1, Q2, and Q3, the drug products might be considered as pharmaceutical equivalents. However, an in vivo bioequivalence study using a clinical or pharmacodynamic endpoint is still most important for establishing therapeutic equivalence.

Scale-up and Postapproval Changes

Drug product performance must also be considered whenever there is a change in the formula or manufacturing process. Often, a pharmaceutical manufacturer will make manufacturing changes after market approval. These changes are termed scale-up and postapproval changes (SUPAC) (4,5). SUPAC guidances[b] are published by the U.S. FDA for manufacturers of approved drug products who want to make a manufacturing change including

- a component and composition of the drug product,
- the batch size,
- the manufacturing site,
- the manufacturing process or equipment, and
- packaging.

The SUPAC guidances describe various levels of postapproval changes according to whether the change is likely to impact on the quality and performance of the drug product. The level of change is classified by the FDA as the likelihood that a change in the drug product might affect the quality of the product. Each SUPAC change may be expected to have a minimal (Level 1), moderate (Level 2), or major (Level 3) change in the performance of the drug product. If the change is considered a Level 3 change, then an in vivo measure drug product performance such as a bioequivalence study must be performed on the original drug product and the "changed" drug product (Table 1).

FDA SUPAC guidances defines (*i*) levels of change; (*ii*) recommended chemistry, manufacturing, and controls (CMC) tests for each level of change; (*iii*) recommended in vitro dissolution tests and/or in vivo bioequivalence tests for each level of change; and (*iv*) documentation that should support the change. The guidance also specifies application information that should be provided to the FDA to ensure continuing product quality and performance characteristics.

SUPAC is an issue for all drug products. As a dosage form becomes more complex in terms of manufacture and drug release, the more difficult it is to

---

[b] http://www.fda.gov/cder/guidance/index.htm.

predict how a SUPAC change for this product will affect drug product performance, in vivo. For example, FDA requires a new in vivo bioequivalence study if the manufacture does a slight change for a modified release oral tablet that has already been approved.

SUMMARY

Quality standards are an important component in the manufacture of drug products. Multisource drug products that meet compendia standards for quality may not have the same performance, in vivo. Therapeutic equivalence and generic drug substitution are established by a regulatory agency such as the FDA. The demonstration of therapeutic equivalence and pharmaceutical equivalence for oral drug products intended for systemic drug absorption is well established. The establishment of pharmaceutical equivalence and therapeutic equivalence for many specialty drug products, particularly those products containing drugs intended for local action, is not fully accepted by the scientific community and the regulatory agencies. The relationship of qualitative, quantitative, and structural similarity (Q1, Q2, and Q3) for the drug product and its in vivo performance cannot always be predicted. Thus, postapproval changes (SUPAC) must be very carefully considered as to how any change in the product will affect product performance and ultimately safety and efficacy of the drug. SUPAC is an issue for all drug products.

REFERENCES

1. Approved Drug Products with Therapeutic Equivalence Evaluations, "Orange Book," U.S. FDA. www.fda.gov/cder/orange/obannual.pdf.
2. Shargel L, Kanfer I. Generic Drug Product Development—Solid Oral Dosage Forms. New York: Informa Healthcare, 2005.
3. Shargel L. Drug product performance and interchangeability of multisource drug substances and drug products. Pharm Forum 2009; 35(3):744–749.
4. FDA Guidance for Industry: Immediate Release Solid Oral Dosage Forms Scale-Up and Postapproval Changes: Chemistry, Manufacturing, and Controls, In Vitro Dissolution Testing, and In Vivo Bioequivalence Documentation, November 1995.
5. FDA Guidance for Industry: SUPAC-MR: Modified Release Solid Oral Dosage Forms Scale-Up and Post-Approval Changes: Chemistry, Manufacturing and Controls; In Vitro Dissolution Testing and In Vivo Bioequivalence Documentation, October 1997.
6. FDA Guidance for Industry: Bioavailability and Bioequivalence Studies for Nasal Aerosols and Nasal Sprays for Local Action, Draft Guidance, April, 2003.

2 Nonsystemically Absorbed Oral Drug Products

Marc Lefebvre

Algorithme Pharma Inc., Montreal, Quebec, Canada

INTRODUCTION

For most formulations used by the patient population, the oral form will always be preferred considering the relative ease of self-administration of the drug product. In general, the oral formulation follows similar "digestive" processes, starting with degradation/dissolution of the drug product down to the absorption of the active moieties. However, as the absorption process of oral drugs is limited or influenced by intrinsic factors, some of them may not go through drug release and dissolution steps and probably not reach the systemic circulation. In consequence, the absorption process of the parent drug and/or its related metabolites will differ between drug products as well as between different formulations of the same drug (e.g., immediate-release vs. extended-release).

According to the Code of Federal Regulations (CFR) for bioavailability and bioequivalence requirements (21 CFR Part 320), the approval of a subsequent entry drug market is to be based on a comparative bioavailability study using a typical crossover design (1). For two drug products to be bioequivalent, similar pharmacokinetic profiles, using C_{max} (maximum concentration) and AUC (area under the curve) parameters, must be observed to ensure therapeutic equivalence (2–4). These two biomarkers are usually acceptable when the active ingredient is intended for systemic activities; however, if the therapeutic and/or the toxic activity of the drug is mostly local or if the pharmacokinetics of the drug cannot be determined, it is usually recommended to assess bioequivalence based on pharmacodynamic or clinical endpoints.

In order to assess, in vivo, the bioequivalence of two drug products, regulatory agencies usually recommend the evaluation of bioequivalence on a priority basis, as per 21 CFR Part 320.24 (1):

1. Bioequivalence based on the comparison of the serum/plasma/blood pharmacokinetic measurements.
   - This approach is particularly applicable to dosage forms intended to deliver the active moiety to the bloodstream for systemic distribution within the body.
2. Bioequivalence based on the comparison of the urine pharmacokinetic measurements.
   - This approach is applicable when the drug cannot be quantified in serum/plasma/blood and when it is excreted mainly in urine.

3. Bioequivalence based on the comparison of the pharmacodynamic measurements.
   - This approach is recommended when appropriate analytical methods are not available for measurement of the concentration of the moiety, but a method is available for the measurement of an appropriate acute pharmacological effect.
   - This approach may be particularly applicable to dosage forms that are not intended to deliver the active moiety to the bloodstream for systemic distribution.
   - One of the critical steps of pharmacodynamic evaluation is to select the most appropriate dose of the drug to be administered (5).
4. Bioequivalence based on the comparison of the clinical endpoints.
   - This approach is the least accurate, sensitive, and reproducible among the general approaches for determining bioavailability or bioequivalence, but may be considered sufficiently accurate for determining the bioavailability or bioequivalence of dosage forms intended to deliver the active moiety locally.
   - Although the least sensitive approach, the choice of the most accurate endpoint is a critical step for this design.

Other regulatory agencies like the Therapeutic Products Directorate (TPD; Canada) and the European Medicines Agency (EMEA) follow a similar structure for the design recommendations for bioequivalence studies.

The objective of this chapter is not to provide or confirm any particular study design for specific drug products, but mainly to explore the bioequivalence requirements of poorly absorbed oral drugs intended for localized gastrointestinal activities. Some of these may have significant to very insignificant systemic absorption, but it may be difficult to differentiate between "systemically or nonsystemically absorbed oral drugs." Different examples representative of different groups are discussed and presented in the last section of this chapter. Hopefully, this section will stimulate discussion of possible designs for the different categories of oral formulations.

DESIGN OF CHOICE

Although the use of pharmacodynamic or clinical endpoints would be recommended for locally acting drugs, the appropriate study design for the assessment of bioequivalence for some oral (or rectal) drug products may be debatable. This is in contrast with most nonoral, nonsystemically absorbed drug products that are typically evaluated from clinical endpoints studies.

Most agencies agree that the assessment of bioequivalence from the plasma concentration–time profile should be limited to drugs intended for systemic action (2–4). It is also understood that pharmacokinetic studies of locally acting drugs could be better related to safety. Although usually true, pharmacokinetic data may provide useful and significant information mainly for locally acting drugs that target the gastrointestinal tract, as they are well connected to the performance of the formulation.

As shown in Figure 1, the systemic activity of a drug administered orally can be related to its pharmacokinetic profile. Any difference in formulation

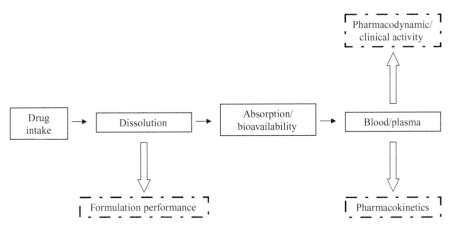

FIGURE 1 Schematic profile of an oral drug product intended for systemic activities.

performance should influence the pharmacokinetics of the drug and explain any possible difference in product efficacy/safety.

Interestingly, the relation between the performance of the formulation and its pharmacokinetics is schematically the same for locally acting drugs (Fig. 2). However, in this case, blood/plasma level cannot be directly connected to the pharmacodynamic and/or the clinical activity, although they are still related to the performance of the formulation.

Considering those facts, it is unknown if the pharmacokinetics of the locally acting drugs are adequate biomarkers for the bioequivalence assessment of two formulations. In some cases, it would be acceptable, while in other cases, the pharmacodynamic or clinical endpoints study would probably be a better marker of the comparative performance. The use of in vitro dissolution testing may also be recommended.

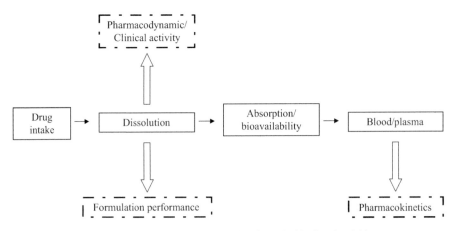

FIGURE 2 Schematic profile of an oral drug product intended for local activities.

Impact of New Analytical Methods

Over the years, the development of highly sensitive analytical methods has helped to determine some previously elusive pharmacokinetic profiles (parent drug vs. metabolite), but has also raised controversies for specific already-approved formulations as established concentration–time profiles have been deemed inadequate or inaccurate. Therefore, it is not surprising to see the use of pharmacokinetic endpoints for the evaluation of many oral formulations, and it would not be surprising to see, in a near future, the use of other pharmacokinetic endpoints for locally acting drugs. However, for some drug products, the pharmacodynamic/clinical endpoints study will always be required for their approval.

The bioequivalence criteria for the comparison of some oral formulations intended to act locally within the gastrointestinal tract and of some prodrugs are probably the most controversial classes of drug products. Almost all these drugs are absorbed, even if poorly, but a portion of their activities may occur locally or systemically, or conversely, the clinical outcome may be systemic even if the drug is acting locally within the gastrointestinal tract. The dilemma really started with the new analytical methods that now allow the detection of very low concentrations of different moieties in biological fluids.

As presented in Table 1, most drugs or prodrugs showing low bioavailability usually go through significant presystemic metabolism, but some of them may still exhibit reliable pharmacokinetic profiles. In other cases, the low percentage of absorption is not related to presystemic metabolism but simply to a poor absorption of the drug because of its low solubility/permeability and/or its poor dissolution. As it can be noted in Table 1, most of the oral drugs acting locally are usually not well absorbed as they are expected to act within the gastrointestinal tract.

Consequently, when localized activity is reported, additional or different data analysis should be provided, including clinical endpoints.

As described in the next sections, some requirements may be different between agencies or for similar drugs within the same agency. Interestingly, most regulatory agencies appear to be open for discussion on other options knowing

TABLE 1 Examples of Drugs/Prodrugs That Show Low Bioavailability or Poor Absorption Following an Oral Administration

Drug Products				
Drugs / Prodrugs		Drugs		
Partially bioavailable	Poorly bioavailable	Partially absorbed	Poorly absorbed	Not absorbed
Trimebutine[a]	Clopidogrel	Mesalamine[b]	Orlistat[b]	Cholestyramine
Ramipril[a]	Misoprostol[c]	Ezetimibe[d]	Rifaximin[b]	Vancomycin
Selegiline[a]	Fenofibrate[c]	Trospium	Acarbose[b]	Sucralfate
Valacyclovir[a]	Leflunomide[c]	Balsalazide[d]	Alendronate	Sevelamer

[a]Bioequivalence usually assessed based on the parent drug or its main metabolite profile (differs between agencies).
[b]Pharmacokinetic profile may or may not be reliable or acceptable for agencies.
[c]Bioequivalence usually assessed based on the main metabolite profile.
[d]Bioequivalence may be assessed based on both the parent drug and its main metabolite profile.

that new evaluations could be more sensitive than pharmacodynamic/clinical endpoints studies to assess the bioequivalence of two formulations.

PRODRUGS OR POORLY ABSORBED DRUGS

Although previously impossible to obtain adequate concentration–time profiles of some parent drugs (for example, due to extensive metabolism or poor absorption), it is now possible to determine and compare their pharmacokinetic profiles. Drugs traditionally considered to be nonsystemically absorbed can now be detected in biological fluids.

The active metabolite was previously the main comparative marker for some oral formulations like selegiline and trimebutine (6,7), but even more so when dealing with prodrugs such as enalapril (8). As shown below for trimebutine, significant differences in plasma concentrations of the parent drug versus its metabolite (*N*-desmethyl-trimebutine) may explain the decision to use the metabolite profile (Fig. 3) (9). Although acceptable, the assessment of bioequivalence from the pharmacokinetic profile of the main active metabolite may not be considered as precise as that of the parent drug, even if it is inactive. In this situation, the performance of the formulation is not only assessed from the systemic exposure of the metabolite but also from the systemic metabolism of the parent drug. In a way, those highly metabolized drug products may be classified as nonsystemically absorbed oral drugs.

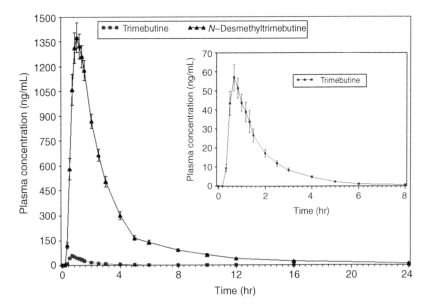

FIGURE 3 Mean (±SD) plasma concentration–time profile of trimebutine and *N*-desmethyl-trimebutine following a single 200 mg oral dose administration of trimebutine to healthy volunteers.

Parent Drug or Metabolite?

Is it better to use the concentration–time profile of a metabolite versus the parent drug? The role of metabolites in bioequivalence studies has been questioned for many years. Although it is widely recognized that measurement of metabolite concentrations is crucial to understanding the clinical pharmacology characteristics of a new molecular entity, a clear consensus on the role of metabolites in the assessment of bioequivalence has never been achieved (10). One school of thought argues that the parent drug alone is sensitive enough to decipher formulation differences, whereas another school of thought believes that establishing bioequivalence criteria on all the species that contribute to safety and efficacy is the only way to ensure the interchangeability of two products (10).

Interestingly, the comparison of prodrugs may be in contrast with the primary reason to use pharmacokinetic parameters as biomarkers, that being the systemic activity (11). Some may complain that although the corresponding metabolites are intended for systemic activities, there is no direct relationship between the pharmacokinetics of the prodrug and its "pharmacological action." However, an extrapolation may be performed between the pharmacokinetics of the parent drug and its own metabolite.

The conclusion of bioequivalence from the pharmacokinetic parameters of the parent drug does not guarantee it would be the same for the active metabolite, and vice-versa (10), but there is probably a better chance to meet bioequivalence on each moiety when the parent drug is judged to be bioequivalent. One of the reasons could be related to the number of volunteers included in the study: as prodrugs are usually more variable for C_{max} and AUC parameters than their metabolites, a higher number of volunteers is usually included in the study. This would not be necessarily the case when the objective of the study is to assess bioequivalence of two drug products from the metabolite pharmacokinetic profile.

The issue of inactive parent drugs is even more debatable considering that EMEA Member States (MS) officially approve bioequivalence comparisons using the profile of the "active" compound even if it refers to the main metabolite. This seems to be the case for many prodrugs like valacyclovir, which is biotransformed to its active form (acyclovir) (12) and ramipril, for which the bioequivalence is to be assessed through the pharmacokinetic profile of its main active metabolite profile, ramiprilat (13).

Nevertheless, the EMEA guidance still recommends the comparison of the pharmacokinetics of the parent drug, and to use the active metabolite only if the profile of the parent compound cannot be adequately determined (4,14). Also, the assessment of bioequivalence from the parent drug and/or the active metabolite is required when the pharmacokinetic system is nonlinear (14). The relevant current U.S. FDA policy and guidelines related to bioavailability and bioequivalence recommend the assessment of the parent drug, when possible, for all drugs, including prodrugs (4). However, it is also recommended to provide supportive information on the main active metabolite profile. In Canada, TPD recommends only the assay of the parent drug regardless of its activity (15).

Overall, although the answer is not evident, it is probably best in the interest of safety for the patient that bioequivalence decision-making should be based on the parent drug, whenever possible (16).

Would it be possible to find similarities between prodrugs and poorly absorbed drugs intended for gastrointestinal activity? Would a metabolite profile be appropriate to compare and confirm similarity between two formulations? This could probably be acceptable on a case-by-case basis. It remains that the debate is still open when dealing with a poorly absorbed oral drug.

DIFFERENT ABSORPTION/BIOAVAILABILITY PROFILES

As mentioned previously, most oral formulations produce a drug concentration–time profile and most of them generate systemic activities, even in cases of low bioavailability. However, when an oral (or rectal) drug is intended for gastrointestinal action, it is mainly developed to limit its absorption: for local gastrointestinal action, absorption of the drug into systemic circulation is usually unnecessary for efficacy and may be detrimental to its safety. Indeed, low bioavailability would be an advantageous property since it would limit systemic adverse events.

On the other hand, although poorly absorbed, this does not mean that there is no pharmacokinetic profile, that there is no relationship between pharmacokinetics and the local presence of the drug, or that the absorption of the drug is not relevant as it may be modified by disease or age. Although the mechanism of action is mainly local, the systemic presence of the drug may also bring about adverse events. Besides drugs intended for local gastrointestinal action, other drugs are also poorly absorbed but are nevertheless intended for systemic action, or for a mix of local and systemic activities.

The definition of nonsystemically absorbed drugs is certainly not the same today as it was few years ago: most drugs can now be detected in biological fluids and their pharmacokinetics determined. In order to compare clinical study designs of poorly absorbed drugs, different examples are presented and classified into four categories. The description of the next categories is not related to any particular approved guidance or guideline, but appears to be adequate for the discussion of the present chapter:

I. Drug partially absorbed or bioavailable and intended for systemic and/or local activity.
II. Drug poorly absorbed and intended for systemic activity.
III. Drug poorly absorbed and intended for local activity.
IV. Drug not absorbed or sparingly absorbed and intended for local activity.

Although pharmacodynamic or clinical endpoints studies could be required for some of these categories because of the mechanism of action of the drug (17), bioequivalence may also be assessed by the systemic exposure profile through the standard comparative bioavailability design. However, this would usually not be the case for the fourth category, even if other alternatives to the clinical endpoints might be suggested.

DRUG PARTIALLY ABSORBED OR BIOAVAILABLE AND INTENDED FOR SYSTEMIC AND/OR LOCAL ACTIVITY

Ezetimibe

Ezetimibe belongs to a new class of lipid-lowering compounds that selectively inhibit the intestinal absorption of cholesterol, without necessarily affecting the absorption of fatty acids, bile acids, or the fat-soluble vitamins A, D, and E (18).

Following oral administration, ezetimibe is rapidly absorbed and extensively conjugated to a phenolic glucuronide (ezetimibe-glucuronide) that appears to be more active than the parent drug (19,20). However, there is no information available on the absolute bioavailability of ezetimibe, as it is practically insoluble in water (18). Interestingly, the mean ezetimibe peak plasma concentration is attained within 4 to 12 hours of administration, while its main metabolite, ezetimibe-glucuronide, reaches its peak exposure only 1 to 2 hours after administration (18).

Glucuronidation is primarily a phase-II enzymatic reaction leading to a conjugated, polar, and usually inactive compound that is rapidly excreted in urine (21). On the other hand, those highly polar glucuronidated molecules may also be excreted in bile and consequently may be secreted back in the intestine, leading to a possible enterohepatic recycling, after being hydrolyzed in the intestine (21).

As the peak time of ezetimibe-glucuronide is attained shortly after ezetimibe administration, it may be related to a localized enzymatic reaction or to a rapid systemic transformation of the drug. However, it seems that a large portion of ezetimibe is metabolized at the brush border of the small intestine (18). Additionally, as the peak time of ezetimibe is highly variable with a concentration–time profile showing multiple peaks, this would indicate the presence of an enterohepatic cycling (19). In this way, ezetimibe may be considered as an active prodrug.

According to the U.S. FDA (22), the recommended design to assess bioequivalence of two ezetimibe products would be a standard crossover comparative study evaluating pharmacokinetics of both free- and total-ezetimibe after a single-dose administration in the fasting and fed states. Considering that ezetimibe-glucuronide is the active metabolite, is the main moiety detected in plasma, and is generated before its absorption, it is required, as per guidance, to apply bioequivalence criteria on both moieties (2,3). Interestingly, although ezetimibe and its metabolite are intended for local gastrointestinal action, their clinical outcomes are to limit systemic absorption of cholesterol (18). Then, this drug does not exert any action on any physical property of the intestine nor is it used for the treatment of any disease of the gastrointestinal tract.

The recommendation to assess bioequivalence of ezetimibe products by the usual C_{max} and AUC parameters of free- and total-ezetimibe would presume that their plasma concentration–time profiles are related to their activity even if the mechanical process responsible for inhibition of cholesterol absorption is still not well known. This could be questioned if the comparison was based only on the parent drug. Considering that ezetimibe-glucuronide is secreted back in the intestine from the bile, then following its absorption, this systemic process improves the inhibition of the absorption of cholesterol. It has to be noted that approximately 78% of a dose of ezetimibe is excreted in the feces predominantly as ezetimibe, with the balance found in the urine mainly as ezetimibe-glucuronide (23).

Therefore, the measurement of the active metabolite is certainly providing additional information on the possible equivalence of the two formulations: the combination of these two markers is probably even more sensitive than a clinical endpoints study. In addition, as the plasma concentration–time profile of ezetimibe is in part dependant of the enterohepatic recycling, then occurring after its systemic absorption, plasma concentration–time profiles of both free- and total-ezetimibe are judged to be related to their local activity.

Then, the pharmacokinetic profile of free-ezetimibe could be linked to the comparison of the performance of the formulation, while that of total-ezetimibe is used as a biomarker of the local presence of the drug. These conditions would be difficult to find for nonoral formulations, but it shows that locally gastrointestinal activity may be assessed and compared from usual pharmacokinetic parameters.

Anti-inflammatory Drugs

This section reviews specific information on two local anti-inflammatory agents, mesalamine and balsalazide, and raises issues applicable to the other formulations of mesalamine as they are all expected to use the same mechanism of action.

Mesalamine or 5-Aminosalicylic Acid (5-ASA)

Sulfasalazine has been the standard treatment for acute inflammatory bowel disease and for maintaining remission over a long period (24). However, up to one-third of patients receive the drug experienced adverse reactions that limited its use (25–28). As mesalamine is the active component of sulfasalazine and as its carrier (sulphapyridine) appears to be responsible for most of the numerous adverse events, several new formulations containing only mesalamine have been developed in order to reduce the number of adverse events and to improve a therapeutic benefit.

Of all drugs under discussion in this chapter, mesalamine is probably the most controversial. In fact, although the mechanism of action seems to be localized in the intestine, the drug is also absorbed systemically. The implication of the systemic presence of mesalamine on its anti-inflammatory activity is still unknown, although it seems evident that there is a significant topical action (25–27).

The actual formulations available can be administered orally or rectally (enema and suppository). Although the present chapter focuses on oral formulations, the comparison between the oral and rectal formulations of mesalamine is important, as they are all interrelated and all intended for the same (or similar) intestinal action.

For all three formulations, the bioavailability of mesalamine is low, ranging from 10% to 15% for a rectal dose (26,29) up to 20% to 25% after an oral dose (27,28). This bioavailability seems, however, to be increased in patients where it ranges between 35% and 50% for all formulations, possibly due to the increased permeability of the inflamed mucosa (29–32). It is unknown whether this increased bioavailability improves the efficacy of the drug and/or increases the occurrence of adverse events.

The main difference between these formulations does not seem to be related to the mechanism of action but more to the delivery site of the drug (25–28). In fact, none of those oral and rectal formulations can be classified as bioequivalent or therapeutically equivalent: no comparative trials have been conducted with equivalent mesalamine doses to determine if any of these formulations are superior in the treatment of ulcerative colitis, despite that they are all effective (33). They only differ with regard to the release of mesalamine in the intestinal tract, which likely influences the clinical outcome in some patients (33).

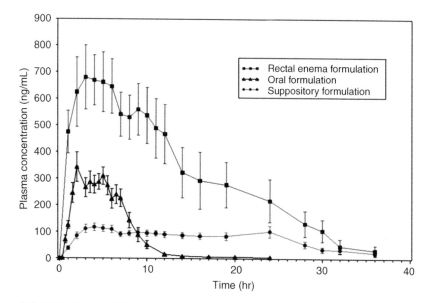

FIGURE 4 Mean (±SD) plasma concentration–time profiles of mesalamine (5-ASA) following a single-dose administration of the rectal enema (4 g), the suppository (500 mg) and the extended-release capsule oral formulation (500 mg; Pentasa®) of mesalamine to healthy volunteers.

On the other hand, it is also stated that it may not be possible to differentiate the clinical performance of the different oral formulations of mesalamine (32). Interestingly, this raises the possibility that clinical endpoint studies would not detect differences between two formulations of mesalamine even with possible different pharmacokinetic profiles.

When comparing the plasma concentration–time profile following different single-dose administrations of each formulation (Fig. 4), the absorption of mesalamine from the two rectal formulations appears to last longer than that observed with the oral one. This may be logical given the relative ease of absorption through the small intestine compared to the colon mucosa (34). It is true that it may be difficult to correlate the plasma concentration–time profiles of mesalamine to its local presence in the intestine. On the other hand, the pharmacokinetic profile can show differences in the performance of those formulations.

If pharmacokinetics may provide useful information on the behavior of the formulation, and considering the local activity of mesalamine, would the evaluation of the local distribution be considered in the bioequivalence comparison of mesalamine? Gamma scintigraphy, in addition to the assessment of the pharmacokinetic profiles, would certainly provide adequate information to conclude to bioequivalence (35) and that it could be more accurate than a clinical endpoint study.

Up to now, clinical endpoints studies are usually required by different regulatory agencies to assess bioequivalence of the different formulations of mesalamine. However, as recommended by the U.S. FDA, different scenarios can be questioned to assess their bioequivalence:

Rectal suppository

For the suppository formulation, two different clinical trials are recommended for each dosage strength product (500 and 1000 mg): (*i*) A bioequivalence study with clinical endpoints; and (*ii*) a bioequivalence study comparing the pharmacokinetic profile of mesalamine (36).

As the efficacy of mesalamine seems to be more strongly related to its local presence in the gastrointestinal tract, while its toxicity might be better related to its systemic absorption, it seems logical to think that the typical bioequivalence trial based on pharmacokinetic profile together with the clinical endpoints trial are required.

However, considering the possible mechanism of action of mesalamine and the administration of the suppository formulation directly at the site of action, is the conduct of a clinical endpoints study absolutely required? There is no doubt that the conclusion of a clinical endpoints study would confirm the therapeutic equivalence of the studied formulations, but could it be shown by other study designs assessing the local distribution of mesalamine? As per the U.S. FDA recommendation (36), this is already the case for the rectal enema formulation of mesalamine (see next section) where only the bioequivalence comparison based on its pharmacokinetic profile is required. If we take into account that the suppository and the rectal enema are different formulations, are they sufficiently different to justify a clinical endpoints study for the suppository one?

For instance, in a study assessing the distribution of mesalamine following a multiple dose administration of the suppository formulation (27), rectal biopsies were performed at the end of the treatment to determine the local presence of mesalamine and its main (mostly inactive) metabolite. This "chronic" treatment was judged necessary to provide enough time for mesalamine distribution into rectal tissues. The ratio of mesalamine/metabolite in the rectal biopsies was about 4:1, while in urine, the ratio was about 1:100 (27). These results show that the local presence of unchanged mesalamine in the gastrointestinal tract can be determined and is probably related to its local efficacy. Interestingly, similar amounts of mesalamine were detected locally in rectal biopsies following the repeated administration of the enema and the suppository formulation (27,30).

In contrast to rectal biopsies, less-invasive technique like gamma scintigraphy would provide other way to validate mesalamine distribution as it allows localizing drug release at the site of action (35). This, however, needs to be accurately validated and approved by agencies.

As confirmed by regulatory agencies, the clinical endpoints study is the least sensitive method to assess bioequivalence. As mesalamine appears to act mainly locally, it would be logical to recommend a clinical endpoints study but regulatory agencies would certainly welcome scientific discussion to develop better comparison of these two formulations.

Rectal enema

As described above, the rectal enema of mesalamine is mainly indicated for the treatment of ulcerative colitis. However, in contrast to the suppository formulation indicated for the treatment of proctitis (26), the bioequivalence of two enema formulations, as per the U.S. FDA recommendation (36), can be assessed only through the comparison of the standard pharmacokinetic parameters (C_{max} and AUC).

Interestingly, besides the dose administered, the pharmacokinetic profile of mesalamine following a single-dose administration of the rectal enema appears to be similar to the suppository formulation (Fig. 4) (34). In both cases, the absorption of mesalamine appears to last over 12 to 24 hours after drug administration and the plasma concentrations appear to be proportional to the doses administered (4 g vs. 500 mg).

In contrast to the suppository formulation, the rectal enema is designed to reach the distal segment of the colon in order to exert its anti-inflammatory action. As for the suppository formulation, although the assessment of the systemic pharmacokinetic profile is judged acceptable, the evaluation of the local distribution of mesalamine could be a feasible alternative.

Oral formulations

For the bioequivalence assessment of the different oral modified-release formulations of mesalamine, only a clinical endpoints study is required by the U.S. FDA agency (36). This recommendation seems to be related to the high variability in the absorption of the drug and to the absence of a relationship between plasma concentrations of mesalamine and the efficacy and/or toxicity events.

In contrast to the suppository formulation, the oral formulation was developed in order to reach, unchanged, the distal segment of the intestine, or the colon. In fact, with the presence of delayed and extended-release formulations, some oral formulations of mesalamine start to release the drug in the upper segment of the intestine, while others are preferred when the disease is confined to the large bowel (32).

For example, as shown in Figure 4, the pharmacokinetic profile of mesalamine following a single-dose administration of an extended-release capsule (Pentasa®) to healthy volunteers appears to be different than that observed with the suppository or the rectal enema formulations (34). Although the absorption rate seems to be similar for the oral capsule and the enema, the period of absorption and elimination seems to be much shorter for the oral delayed-release formulation. This confirms that pharmacokinetic endpoints may show differences in the performance of different formulations.

As the absorption of oral formulations of mesalamine is occurring in part in the small intestine, it could be more complicated to find any correlation between the plasma concentrations of mesalamine and the efficacy or toxicity of the drug in the colon (34). On the other hand, the plasma pharmacokinetic profile may detect true differences in the time of release of the drug in the intestine, and this may help to differentiate properties of marketed oral formulations (33). Although the efficacy might not correlate well with the pharmacokinetics of mesalamine, it remains that the different oral formulations display differences in their pharmacokinetics. This is also a reason why it is not recommended to interchange different oral formulations like Asacol®, Pentasa®, and Salofalk®, as differences in pharmacokinetics may correlate with possible differences in clinical efficacy (33).

As about 70% of the oral dose is not absorbed and reaches the colon intact and as the formulation is expected to release mesalamine at pH 7 or higher (27,28,37), it is likely that the pharmacokinetic profile of mesalamine does not reflect all parts of drug's pharmacological activity. This may be in contrast to the rectal formulations where the drug quickly reaches its site of action.

To conclude, the clinical endpoints study may be better justified for the oral formulations than for the suppository formulation. This remains true until an approved well-defined methodology can be used to determine the local distribution of mesalamine in the intestine or colon, thus providing better evidence of adequate tissue concentrations at the site of action.

Balsalazide

Balsalazide is a prodrug that is enzymatically cleaved by bacterial azoreduction in the colon to release equimolar quantities of mesalamine. Importantly, this enzymatic activity does not occur in the small intestine but seems to be specific to (or to mainly occur in) the colon. Each daily dose of balsalazide (6.75 g) is equivalent to 2.4 g of mesalamine (38). As described above, the mechanism of action of mesalamine is unknown, but appears to be topical rather than systemic.

In healthy individuals, the systemic absorption of the prodrug balsalazide is very low and variable, but occurs approximately 1 to 2 hours after a single oral dose administration (38). As noted in Figure 5, balsalazide is absorbed and detected rapidly, while the peak absorption time of its main active metabolite (mesalamine) occurs about 5 to 10 hours after a single dose (39).

In contrast to the oral formulation of mesalamine, the assessment of bioequivalence of two balsalazide products based on the comparison of C_{max} and AUC parameters is judged acceptable by the U.S. FDA agency (40). However, this comparison has to be based on the pharmacokinetic profile of both balsalazide and mesalamine, the active metabolite (mesalamine) being used as a surrogate marker of the clinical endpoints study.

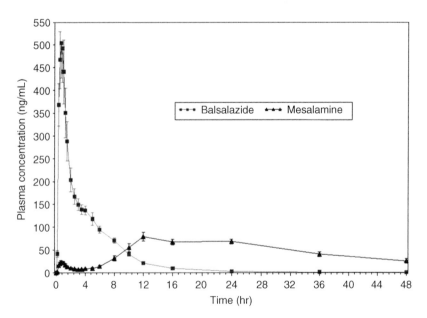

FIGURE 5 Mean ($\pm$SD) plasma concentration–time profile of balsalazide and mesalamine (5-ASA) following a single 2.5 g oral dose administration of balsalazide to healthy volunteers.

As for other products, when the parent drug is biotransformed to an active metabolite before its systemic absorption, it is usually requested to assess bioequivalence using the pharmacokinetic profile of each moiety (2,3). As for ezetimibe, the presence of the active metabolite of balsalazide is certainly critical, as it is also used as a marker for the local presence of the studied formulation. In fact, as the azoreduction activity is occurring in the colon, it is directly related to the amount of balsalazide brought to this site. Therefore, the presence of mesalamine in plasma is probably well correlated to the local presence of the drug in the colon.

Therefore, in this case, balsalazide plasma concentration–time profile provides information on the performance of the formulation, while that of mesalamine is used to compare the therapeutic equivalence of the drug product.

DRUG POORLY ABSORBED AND INTENDED FOR SYSTEMIC ACTIVITY

Alendronate

Bisphosphonates are drugs exhibiting extremely low bioavailability. This poor bioavailability is not related to an extensive metabolism of the drugs, but to a very poor absorption (17). Following their absorption, bisphosphonates are rapidly distributed into the bone and excreted in urine.

In contrast to other poorly absorbed drug products, all bisphosphonates are clearly intended for systemic action. Whereas the urine excretion profile must be used to compare the behavior of alendronate formulations, plasma pharmacokinetic profiles of other bisphosphonates (risedronate and ibandronate) can be reliably estimated.

Alendronate has a bioavailability that is usually lower than 1% (41). Of the portion absorbed, about 50% is rapidly excreted in urine, while the other 50% is rapidly distributed into the bone (41). This affinity to the bone is so intense that the exact period of time of elimination of alendronate is unknown. As per the product monograph, the half-life of alendronate is estimated to be about 10 years (41,42).

Due to the rapid excretion in urine and to the absence of pre-systemic metabolism, it is acceptable to compare the behavior of two alendronate formulations using their urine excretion profiles. The intra-subject variability of alendronate pharmacokinetic parameters (as well as for the other bisphosphonates) is relatively high (usually over 40%) given its poor bioavailability (43); therefore, a high number of healthy subjects (usually more than 100) are required to assess bioequivalence of two alendronate formulations. The bioequivalence criteria are based on the 90% confidence intervals for the exponential of the difference between the Test and the Reference product for the ln-transformed parameters R_{max} (maximal rate of urinary excretion over all time intervals) and Ae_{0-x} (amount excreted over a x-hour interval) that should be within the 80.00% to 125.00% acceptance range.

Considering that alendronate as well as other bisphosphonates are poorly systemically absorbed, but acting systemically, the assessment of bioequivalence from their pharmacokinetic profiles is then judged acceptable by all agencies.

DRUG POORLY ABSORBED AND INTENDED FOR LOCAL ACTIVITY

Acarbose

Acarbose is a complex oligosaccharide that inhibits α-glucosidase activity in the brush border membrane of the small intestine. This delays the digestion of ingested carbohydrates, thereby resulting in a lowering of blood glucose concentration following meals (postprandial) (44).

The mean systemic bioavailability of acarbose following oral administration is between 0.7% and 1.6% (45,46). Because acarbose acts locally within the gastrointestinal tract, this low systemic bioavailability is therapeutically desired. Acarbose is metabolized exclusively in the gastrointestinal tract, principally by intestinal bacteria, but also by digestive enzymes (44).

Considering the mechanism of action of acarbose and that the active moiety (although poorly available) does not have the potential to cause toxicity, there is clearly no relationship between the plasma concentration of the drug and its pharmacological activity. The list of adverse events for acarbose (44) includes mainly events related to the gastrointestinal tract (e.g. flatulence, diarrhea, abdominal pain, nausea, vomiting and constipation) which confirms the local action of the drug. As shown in human clinical trials, acarbose reduces postprandial hyperglycemia and plasma insulin concentrations (45). Therefore, glycemia is judged to be an adequate marker for the bioequivalence comparison of two oral formulations of acarbose.

As described previously, the use of a pharmacodynamic endpoint has to be linked to any possible variation of the plasma concentration-time profile of the drug (3). Therefore, the dose of the drug must be selected in order to detect any possible difference in bioavailability of the drug products under study. For acarbose, it is usually recommended to conduct pilot studies with ascending doses (25 mg and more) in order to select the most appropriate one for pivotal study. The dose is chosen from the pharmacodynamic glucose response observed following the administration of a solution containing 75 g of glucose.

Once the dose is selected, the two formulations are to be compared using a standard, two-way, crossover design, including a baseline evaluation. Bioequivalence evaluation is to be based on the reduction of serum glucose levels following single administration of acarbose with 75 g of sucrose relative to the serum glucose levels observed at baseline (46). The pharmacodynamic endpoints are based on the C_{max} and AUC of the serum glucose level using the standard 90% confidence interval which should be within the 80.00% to 125.00% acceptance range.

In the present case, a pharmacodynamic validated methodology was available and is surely preferable to a clinical endpoint study. The variation in glucose response is sufficiently sensitive to detect any possible difference in the activity of two different formulations of acarbose.

Orlistat

Orlistat is an inhibitor of gastric and pancreatic lipase for obesity management that acts by inhibiting the absorption of dietary fats (47,48). Systemic absorption of the drug is therefore not needed for activity (47,48).

Following the oral administration of orlistat, the drug and its main metabolites are mainly excreted in feces with about 83% in the unchanged form (47). However, a very small portion is absorbed through gastrointestinal tract, but plasma concentrations of orlistat are usually extremely low (49). Per instance, C_{max} is about 0.8 ng/mL following a single 120 mg dose, and then it would be extremely difficult to obtain a full reliable pharmacokinetic profile following the administration of an adequate dose (47,49). On the other hand, although feasible, it would probably be not acceptable to assess bioequivalence based on the pharmacokinetic profile of orlistat due to its local gastrointestinal activity.

As per literature, bioequivalence of orlistat could be assessed from pharmacodynamic or clinical endpoints. One possible pharmacodynamic endpoint included the measurement of fecal fat excretion (50) but, in contrast to glycemia used for acarbose, this method may be judged dubious, although acceptable for others. Therefore, as no pharmacodynamic evaluation appears to be presently fully validated, and also as recommended by US FDA (51), a clinical endpoint (probably comparing body weight loss and/or serum lipid profile in obese patients) would be required to determine bioequivalence of two formulations of orlistat. The comparison of the pharmacokinetics of orlistat might however be useful for the safety assessment.

DRUG NOT ABSORBED OR SPARINGLY ABSORBED AND INTENDED FOR LOCAL ACTIVITY

For this class of drugs, although not or insignificantly absorbed, their bioequivalence would usually be assessed from clinical endpoint studies. On the other hand, although not necessarily acceptable for different agencies, in vitro bioequivalence requirements may also be suggested like for the comparison of drugs intended to bind endogenous compounds as bile salts (i.e., cholestyramine) or phosphate (i.e., sevelamer) (52). Different tests to compare the binding properties of different formulations (i.e., equilibrium and kinetic bindings) would be recommended. Interestingly, even for this class of drugs, alternatives to the clinical endpoint studies are suggested based on the mechanism of action of the drug.

Sucralfate

Sucralfate exerts a generalized gastric cytoprotective effect by enhancing natural mucosal defense mechanisms (53). Studies conducted in humans have demonstrated that sucralfate can protect the gastric mucosa against various irritants such as alcohol, acetylsalicylic acid (ASA), and nonsteroidal anti-inflammatory drugs (NSAIDs) (53).

The action of sucralfate is nonsystemic, as the drug is only minimally absorbed from the gastrointestinal tract. There is no doubt that the assessment of bioequivalence of sucralfate has to be related to a clinical endpoint.

When referring to different clinical studies in combination or not with irritant compounds like ASA or only in presence of duodenal ulcers, most of the clinical endpoints studies assessed the efficacy of sucralfate using endoscopy after 4 to 8 weeks of treatment (54,55).

For the bioequivalence comparison of two sucralfate formulations, a similar design would be recommended: a parallel, randomized, three-treatment (test vs. reference vs. placebo) study in patients with duodenal ulcer. However, the biggest challenge when dealing with this category of drugs is the recruitment and eligibility of the patients, who are being self-treated over acute or irregular periods with an easy access to multiple over-the-counter (OTC) products like famotidine, ranitidine, and other antiulcer agents.

To conclude, although sucralfate is no longer regularly used due to the presence of proton-pump inhibitors and anti-histaminic (H$_2$ receptors) agents, there is no other simple alternative than clinical endpoints to assess the bioequivalence of two sucralfate products.

Vancomycin

Vancomycin is an antibiotic that should be administered orally for the treatment of enterocolitis caused by *Staphylococcus aureus* (including methicillin-resistant strains) and antibiotic-associated pseudomembranous colitis caused by *Clostridium difficile*. Vancomycin can also be administered intravenously, but for different indications (56).

The oral formulation of vancomycin is indicated to act locally in the gastrointestinal tract, as it is poorly absorbed. No blood concentrations are detected and urinary recovery does not exceed 0.76% following oral administration (56).

As vancomycin is indicated for the treatment of infection already resistant to other antibiotics, it should be administered with caution. Vancomycin should be strictly used for the treatment of severe bacterial infections in order to avoid the development of any resistance to this drug. As the oral formulation is specifically indicated for the treatment of enterocolitis and pseudomembranous colitis, all bioequivalence studies with clinical endpoints should be conducted in this group of patients.

On the other hand, according to the U.S. FDA agency, in vitro dissolution tests could be judged acceptable for the comparison of two vancomycin oral products. This decision seems to be based on the guidance entitled "Waiver of In Vivo Bioavailability and Bioequivalence Studies for Immediate-Release Solid Oral Dosage Forms Based on a Biopharmaceutics Classification System (BCS)." The reason to waive an in vivo clinical endpoints study would be related to the bioclassification of vancomycin, belonging to the Class 1 (highly soluble and highly permeable drug substance in immediate-release dosage form) (57). In that case, demonstration of in vivo bioavailability or bioequivalence may not be necessary as long as the inactive ingredients used in the dosage form do not significantly alter the absorption of the active ingredients. However, the final decision remains to be confirmed.

CONCLUSION

It is well established that pharmacodynamic or clinical endpoint studies are required for most drug products acting locally or poorly absorbed. On one hand, clinical study designs recommended by the regulatory agencies are sometimes difficult to understand, as the rationale behind such decisions is not always well described. On the other hand, all agencies are looking for innovative methods that could provide more efficient and accurate comparisons. As indicated by regulatory agencies, although clinical or pharmacodynamic endpoints will ultimately confirm similar efficacy of two drug products, alternatives to clinical endpoints studies are to be explored.

In contrast to regulatory requirements for nonoral, locally-acting drugs, the present chapter illustrates the use of other approaches to demonstrate bioequivalence of oral formulations nonsystemically absorbed or acting within the gastrointestinal tract, for instance, the evaluation of the local distribution of the drug (or metabolite), the use of additional biomarkers, or the use of imaging techniques. Interestingly, the U.S. FDA agency has recommended specific critical path initiatives for the development of generic drugs (58), which are aimed to stimulate exactly this kind of innovation and research by all stakeholders in the generic pharmaceutical industry.

ACKNOWLEDGMENTS
The author would like to thank Mira Francis, Ph.D., Sarah Baker, Ph.D., and Julie Massicotte, all colleagues at Algorithme Pharma Inc., for their support and comments during the preparation of this chapter.

REFERENCES
1. 21 Code of Federal Regulations Part 320. Types of Evidence to Establish Bioavailability or Bioequivalence. US Department of Health and Human Services. Revised April 1, 1999.
2. Guidance for Industry: Bioavailability and Bioequivalence Studies for Orally Administered Drug Products—General Considerations. Center for Drug Evaluation and Research (CDER), FDA, 2003.
3. European Medicines Agency. Note for Guidance on the Investigation of Bioavailability and Bioequivalence. European Medicines Agency, July 2002.
4. Health Products and Food Branch, Health Canada. Guidance for Industry: Conduct and Analysis of Bioavailability and Bioequivalence Studies—Part A: Oral Dosage Formulations Used for Systemic Effects. Health Products and Food Branch, Health Canada, 1992.
5. Issar M, Stark JG, Shargel L. Pharmacodynamic measurements for determination of bioequivalence. In: Kanfer I, Shargel L, eds. Generic Drug Product Development: Bioequivalence Issues, Vol. 176. New York: Informa Healthcare USA, 2007:47–69.
6. Saivin S, Lavit M, Michel F, et al. Pharmacokinetics and bioequivalence of two trimebutine formulations in healthy volunteers using desmethyl-trimebutine levels. Arzneimittelforschung 2000; 50(8):717–721.
7. Mascher HJ, Kikuta C, Millendorfer A, et al. Pharmacokinetics and bioequivalence of the main metabolites of selegiline: Desmethylselegiline, methamphetamine and amphetamine after oral administration of selegiline. Int J Clin Pharmacol Ther 1997; 35(1):9–13.
8. Niopas I, Daftsios AC, Nikolaidis N. Evaluation of the bioequivalence of two tablet formulations of enalapril/hydrochlorothiazide after single oral administration to healthy volunteers. Arzneimittelforschung 2004; 54(3):160–165.
9. Internal Data on Trimebutine. Canada: Algorithme Pharma Inc.
10. Jackson AJ, Robbie G, Marroum P. Metabolites and bioequivalence: Past and present. Clin Pharmacokinet 2004; 43(10):655–672.
11. Jackson A. Role of metabolites in bioequivalence assessment. In: Kanfer I, Shargel L, eds. Generic Drug Product Development; Bioequivalence Issues, Vol. 176. New York: Informa Healthcare USA, 2007:171–183.
12. EMEA letter (valacyclovir) to Algorithme Pharma Inc. Communication on file at Algorithme Pharma Inc., 2007.
13. EMEA confirmation (ramipril) to Algorithme Pharma Inc, 2007.
14. European Medicines Agency. Questions & Answers on the Bioavailability and Bioequivalence Guideline. London: European Medicines Agency, 2006.
15. Health Products and Food Branch, Health Canada. Draft Guidance for Industry: Use of Metabolite Data in Comparative Bioavailability studies. Health Products and Food Branch, Health Canada, 2004.
16. Midha KK, Rawson MJ, Hubbard JW. The role of metabolites in bioequivalence. Pharm Res 2004; 21:1331–1344.

17. Hendy C. Bioequivalence using clinical endpoint studies. In: Kanfer I, Shargel L, eds. Generic Drug Product Development: Bioequivalence Issues, Vol. 176. New York: Informa Healthcare USA, 2007:71–96.
18. Product Labeling. Zetia (ezetimibe), 10 mg Oral Tablet Formulation. Labeling Revision, 2008.
19. Patrick JE, Kosoglou T, Stauber KL, et al. Disposition of the selective cholesterol absorption inhibitor ezetimibe in healthy male subjects. Drug Metab Dispos 2002; 30:430–437.
20. van Heek M, Farley C, Compton DS, et al. Comparison of the activity and disposition of the novel cholesterol absorption inhibitor, SCH58235 and its glucuronide, SCH60663. Br J Pharmacol 2000; 129:1748–1754.
21. Benet LZ, Mitchell JR, Sheiner LB. Pharmacokinetics: The dynamics of drug absorption, distribution, and elimination. In: Goodman Gilman A, Rall TW, Nies AS, Taylor P, eds. The Pharmacological Basis of Therapeutics, 8th ed. New York: Pergamon Press, 1990:3–32.
22. FDA Letter to Algorithme Pharma Inc. Communication on Ezetimibe on File at Algorithme Pharma Inc., 2006.
23. Kosoglou T, Statkevich P, Johnson-Levonas AO, et al. Ezetimibe: A review of its metabolism, pharmacokinetics and drug interactions. Clin Pharmacokinet 2005; 44(5):467–494.
24. Bachrach WH. Sulfasalazine: An historical perspective. Am J Gastroenterol 1988; 83:487–503.
25. Gionchetti P, Campieri M, Belluzzi A, et al. Bioavailability of single and multiple doses of new oral formulation of 5-ASA in patients with inflammatory bowel disease and healthy volunteers. Aliment Pharmacol Ther 1994; 8:535–540.
26. Canadian Pharmacists Association. Electronic Compendium of Pharmaceuticals and Specialties (e-CPS) [Internet]. Ottawa, ON: Canadian Pharmacists Association, Updated November 29, 2006; Accessed. April 30, 2008. Salofalk® (5-ASA). www.e-therapeutics.ca/cps.showMonograph.action.
27. Canadian Pharmacists Association. Electronic Compendium of Pharmaceuticals and Specialties (e-CPS) [Internet]. Ottawa, ON: Canadian Pharmacists Association, Updated November 29, 2006. Accessed April 30, 2008. Pentasa® (5-ASA). www.e-therapeutics.ca/cps.showMonograph.action.
28. Canadian Pharmacists Association. Electronic Compendium of Pharmaceuticals and Specialties (e-CPS) [Internet]. Ottawa, ON: Canadian Pharmacists Association, Updated November 29, 2006. Accessed April 30, 2008. Asacol® (5-ASA). www.e-therapeutics.ca/cps.showMonograph.action.
29. Aumais G, Lefebvre M, Tremblay C, et al. Rectal tissue, plasma and urine concentrations of mesalazine after single and multiple administrations of 500 mg suppositories to healthy volunteers and ulcerative proctitis patients. Aliment Pharmacol Ther 2003; 17:93–97.
30. Aumais G, Lefebvre M, Massicotte J, et al. Pharmacokinetics and pilot efficacy of a mesalamine rectal gel in distal ulcerative colitis. Drugs RD 2005; 6(1):41–46.
31. Aumais G, Lefebvre M, Spenard J. Rectal tissue, plasma and urine concentrations of mesalamine (5-ASA) after single and multiple administration of 4 g enema to ulcerative colitis patients. Gastroenterology 2002; 122(4):A499.
32. Sandborn WJ, Hanauer SB. Systematic review: The pharmacokinetic profiles of oral mesalazine formulations and mesalazine prodrugs used in the management of ulcerative colitis. Aliment Pharmacol Ther 2003; 17:29–42.
33. Forbes A, Cartwright A, Marchant S, et al. Review article: Oral, modified-release mezalamine formulations—Proprietary *versus* generic. Aliment Pharmacol Ther 2003; 17:1207–1214.
34. Internal Data on Mesalamine (5-ASA). Canada: Algorithme Pharma Inc.
35. Wilding I. Bioequivalence testing for locally acting gastrointestinal products: What role for gamma scintigraphy? J Clin Pharmacol 2002; 42:1200–1210.
36. FDA Letter to Algorithme Pharma Inc. Communication on Mesalamine on File at Algorithme Pharma Inc., April 2006.

37. Dew MJ, Ryder REJ, Evans N, et al. Colonic release of 5-aminosalicylic acid from an oral preparation in active ulcerative colitis. Br J Clin Pharmacol 1983; 16:185–187.
38. Physician Desk Reference (PDR), Electronic Library [Internet]. Colazal® (Balsalazide). Updated in December 2006.
39. Internal Data on Balsalazide (Colazal). Canada: Algorithme Pharma Inc.
40. FDA Letter to Algorithme Pharma Inc. Communication on Balsalazide on File at Algorithme Pharma Inc., 2006.
41. Canadian Pharmacists Association. Electronic Compendium of Pharmaceuticals and Specialties (e-CPS) [Internet]. Ottawa, ON: Canadian Pharmacists Association, Updated November 29, 2006. Accessed April 30, 2008. Fosamax® (Alendronate). www.e-therapeutics.ca/cps.showMonograph.action.
42. Cocquyt V, Kline WF, Gertz BJ, et al. Pharmacokinetics of intravenous alendronate. J Clin Pharmacol 1999; 39:385–393.
43. Internal Data on Alendronate. Canada: Algorithme Pharma Inc.
44. Electronic Compendium of Pharmaceuticals and Specialties (e-CPS) [Internet]. Ottawa, ON: Canadian Pharmacists Association; 2008. Updated November 29, 2006; Accessed April 30, 2008. Glucobay™ (Acarbose).
45. Clissold SP, Edward C. Acarbose: A preliminary review of its pharmacodynamic and pharmacokinetic properties, and therapeutic potential. Drugs 1988; 35:314–243.
46. Putter J. Studies on the pharmacokinetics of acarbose. In: Brodbeck U, ed. Enzyme inhibitors. Weinheim, Germany: Verlag Chemie, 1980:139–151.
47. Physician Desk Reference (PDR), Electronic Library [Internet]. Xenical® (Orlistat), 2008.
48. Zhi J, Melia AT, Joly R, et al. Review of limited systemic absorption of orlistat, a lipase inhibitor, in healthy human volunteers. J Clin Pharmacol 1995; 35:1103–1108.
49. Zhi J, Mulligan TE, Hauptman JB. Long-term systemic exposure or orlistat, a lipase inhibitor, and its metabolites in obese patients. J Clin Pharmacol 1999; 39:41–46.
50. Zhi J, Melia AT, Guerciolini R, et al. Retrospective population-based analysis of the dose–response (Fecal Fat Excretion) relationship of orlistat in normal and obese volunteers. Clin Pharmacology Therapeutics 1994; 56:82–85.
51. FDA Letter to Algorithme Pharma Inc. Communication on Orlistat on File at Algorithme Pharma Inc., 2006.
52. FDA Draft Guidance on Sevelamer Hydrochloride, Recommended on July 2008.
53. Canadian Pharmacists Association. Electronic Compendium of Pharmaceuticals and Specialties (e-CPS) [Internet]. Ottawa, ON: Canadian Pharmacists Association. Updated November 29, 2006; Accessed April 30, 2008. Sulcrate® (Sucralfate). www.e-therapeutics.ca/cps.showMonograph.action.
54. Stern AI, Ward F, Hartley G. Protective effect of sucralfate against aspirin-induced damage to the human gastric mucosa. Am J Med 1987; 83(suppl 3B):83–85.
55. Coste T, Rautureau J, Beaugrand M, et al. Comparison of two sucralfate dosages presented in tablet form in duodenal ulcer healing. Am J Med 1987; 83(suppl 3B):86–90.
56. Canadian Pharmacists Association. Electronic Compendium of Pharmaceuticals and Specialties (e-CPS) [Internet]. Ottawa, ON: Canadian Pharmacists Association, Updated November 29, 2006. Accessed April 30, 2008. Vancocin® (Vancomycin). www.e-therapeutics.ca/cps.showMonograph.action.
57. Center for Drug Evaluation and Research (CDER), FDA. Guidance for Industry: Waiver of In Vivo Bioavailability and Bioequivalence Studies for Immediate-Release Solid Oral Dosage Forms Based on a Biopharmaceutics Classification System. Center for Drug Evaluation and Research (CDER), FDA, 2000.
58. Lionberger RA. FDA critical path initiatives: Opportunities for generic drug development. AAPS J 2008; 10(1):103–109.

3 Topical Drug Products—Development, Manufacture and Regulatory Issues

Mary Beth G. Erstad

Paddock Laboratories, Inc., Minneapolis, Minnesota, U.S.A.

Candis Edwards

St. John's University, College of Pharmacy & Allied Health Professions, Queens, New York, U.S.A.

OVERVIEW

The objective of this chapter is to describe the various types of topical drug products, the disease states that they target, and the special considerations that apply specifically to the development of generic topical dosage forms. This chapter also addresses the regulatory requirements to achieve a successful ANDA filing with the Food and Drug Administration (FDA) for generic topical drug products.

TYPES OF TOPICAL DRUG PRODUCTS

Topical drug products are intended to deliver the active ingredient(s) onto or across the skin or mucous membranes so that the desired pharmacological effect can be achieved. These drug products can be categorized by their various characteristics including route of administration, site of action, therapeutic area, disease state, and type of dosage form. Strategies for development, requirements for FDA approval, and marketing issues can vary depending upon these characteristics and the therapeutic objective.

Topical products are applied through several routes of administration and can be categorized as shown in Table 1. The typical sites of action and specific dosage forms are included for each route of administration.

Topical products may be used for local/regional or systemic effects. Products for local effect are applied directly to the site of action, namely, the skin, whereas, topical products for regional or systemic effect require that the drug penetrates deeper into and through the skin, be absorbed into the bloodstream and subsequently to the site of action. Most topical products, however, are formulated for local drug effect and are not intended to be absorbed into the systemic circulation. Hence the application site (skin, eyes, ears, nose, etc.) is usually the intended site of action. Although intended for local action, unintended systemic drug absorption can occur and potential safety concerns may arise. The regulatory agency may require a systemic exposure study to compare the systemic absorption of the drug from the Test product compared to that of the Reference Listed Drug (1) (RLD) (1). Retinoic acid used topically for acne is an example and in such situations the U.S. FDA may require systemic absorption studies to establish safety. The benefit of topical application for systemic absorption is

TABLE 1 Typical Sites of Action and Dosage Forms for Various Topical Routes of Administration

Route of administration	Site of action	Dosage form
Skin	Local or systemic	Aerosols Creams Emulsions Gels Irrigations Lotions Ointments Strips Pastes Powders Solutions Tinctures Suppositories Suspensions Transdermal systems
Eye (ophthalmic)	Local	Emulsions Ointments Solutions Suspensions Strips
Ear (otic)	Local	Emulsions Ointments Solutions Suspensions
Nose (nasal)	Local or systemic	Inhalations Powders Solutions Suspensions
Mouth–(oral respiratory)	Local or systemic	Inhalations
Anus (rectal)	Local or systemic	Aerosols (two-phase, three-phase, and foam) Creams Gels Irrigations Ointments Solutions Suppositories Suspensions
Vagina (vaginal)	Local	Aerosols (two-phase, three-phase, and foam) Creams Gels Ointments Suppositories/pessaries Suspensions Tablets or inserts

TABLE 2 Description and Special Considerations for Ophthalmic Preparations

Ophthalmic preparations	Description	Special considerations
Ointments	Manufactured under rigidly aseptic conditions as formulation ingredients often cannot be sterilized using routine techniques	Sterility Micronization or solubilization of active pharmaceutical ingredient (API) Drug leakage and metal particulates from container
Solutions	Ideally isotonic to lacrimal fluid; buffered to physiological pH; sterilized by membrane filtration and/or autoclaving	Sterility Isotonicity Buffering Preservation Thickening
Suspensions	Sterile liquid preparations containing solid particles which must be micronized to prevent irritation and/or scratching of the cornea	Sterility Micronization of active drug
Strips	Impregnated paper often used for diagnostic purposes	Sterility

usually to minimize side effects of oral delivery and often to minimize first pass metabolism.

Topical products for the skin are usually for non–life-threatening conditions, and the patient may choose which product to use due to personal preference.

Ophthalmic, otic, and nasal products are typically administered to a local site of action and have technological barriers to market entry; for example, these products may need to be sterile, may require unique packaging and/or patented container-closures, and may have unique delivery systems. Ophthalmic products, in addition to the sterility requirement, must be isotonic with respect to the corneum.

Rectal and vaginal products can be used for systemic or local effects and often utilize specialized manufacturing technologies (suppositories).

Dosage Forms

Topical products can be made in a variety of dosage forms, seemingly a much larger variety than oral products. The United States Pharmacopeia (USP) (2) General Chapter <1151> Pharmaceutical Dosage Forms defines the appropriate terminology for dosage forms used in the industry. Definitions in this Chapter are accepted by the FDA and include the following dosage forms intended for topical use.

Lotions and creams are emulsions of oil-in-water or water-in-oil, containing the active pharmaceutical ingredient (API) dissolved in the water or oil phase, or dispersed as a suspension in the emulsion. Creams may also be emulsions but are of higher viscosity, semisolid, and typically do not pour. Lotions are of lower viscosity, typically are pourable, and will flow more easily for better spreadability. Creams and lotions can be used to deliver antibiotics, antiseptics,

TABLE 3 Description and Routes of Administration for USP-Defined Dosage Forms

USP dosage forms	Description	Route of administration
Aerosols	Two-phase (gas and liquid), three-phase (gas, liquid, and solid or liquid), and foam (surfactant included) packaged under pressure using a propellant to produce a fine mist of spray	Skin Nose Mouth Lungs
Concentrate for dip	Animal use	
Creams	Semisolid with drug dissolved or dispersed in water-in-oil, oil-in-water or aqueous microcrystalline dispersions of long-chain fatty acids or alcohols	Skin Rectal Vaginal
Emulsions	Two-phase system, water-in-oil, oil-in-water, in which one liquid is dispersed throughout another liquid	Skin Ophthalmic Otic
Extracts and fluid extracts	Concentrated preparations of vegetable or animal drugs obtained by removal of the active constituents by evaporation, using such processes as percolation and distillation	Skin
Gels	Semisolid suspension of small inorganic particles or large organic molecules interpenetrated by a liquid; two-phase gels have a relatively large dispersed phase and single-phase gels consist of organic macromolecules uniformly distributed throughout a liquid	Skin Rectal Vaginal
Inhalations	Solutions or suspensions nebulized by inert gas (nebulizers) or propellant driven [metered-dose inhalers (MDIs)]; or powders	Nasal Oral respiratory
Irrigations	Sterile solutions for bathing or flushing open wounds or body cavities.	Skin Rectal Vaginal
Lotions	See solutions or suspensions; emulsions	Skin
Ointments	Semisolid preparations falling into four general classes of bases: hydrocarbon, absorption, water-removable, and water soluble	Skin Nasal Rectal Ophthalmic
Strips	Impregnated paper often used for ophthalmic condition diagnostic purposes	Ophthalmic
Pastes	Semisolids such as an aqueous gel or fatty paste that does not flow at body temperature	Skin Mouth
Powders	Dry mixture for dusting	Skin Mouth
Solutions (elixirs)	Liquids containing solubilized active drugs	Ophthalmic Otic Skin Nasal Oral respiratory Mouth
Tinctures	Alcoholic or hydroalcoholic solutions prepared from vegetable materials or chemical substances	Skin Mouth

(*Continued*)

TABLE 3 Description and Routes of Administration for USP-Defined Dosage Forms
(Continued)

USP dosage forms	Description	Route of administration
Suppositories	Solid (or semisolid) that melts at body temperature or dissolves in bodily fluids; bases typically cocoa butter, hydrogenated vegetable oil (fatty acid ester), and hard fat, glycerinated, polyethylene glycol, surfactant	Rectal Vaginal Urethral
Suspensions	Liquid preparations containing dispersed solid particles	Ophthalmic Skin Otic Rectal
Tablets or inserts	Compressed powdered materials or encapsulation in soft gelatin	Vaginal
Transdermal systems	Deliver drug through the skin to the systemic circulation; typically patches containing a reservoir with rate-controlling membrane Ocular system placed in the lower conjunctival fornix from which the drug diffuses Intrauterine system may release drug over an extended period, as long as 1 year	Skin Ophthalmic Vaginal

antifungals, anti-acne agents, and corticosteroids. Lotions are more fluid than creams and more easily applied to large areas of the skin or scalp.

Foams

Foams are formed by trapping gas bubbles in a liquid or solid. In the case of pharmaceutical products this is typically a liquid. Most pharmaceutical foam products are aerosols in nature, in which case the foam is generated by a pressurized liquid propellant—the API typically being dissolved in a liquid in which it is soluble or forms an emulsion with the propellant and is in turn mixed with air upon release from the container to form the gas bubbles of the foam.

Gels

Gels are semisolid systems consisting of either suspensions containing small inorganic particles or large organic molecules interpenetrated by a liquid containing a dissolved or dispersed drug substance. In the case where the gel consists of small inorganic particles, the gel is classified as a two-phased system. This type of gel often needs to be shaken before use to ensure uniformity. Single-phased gels are comprised of organic macromolecules distributed uniformly throughout a liquid phase, typically aqueous, alcoholic, or hydroalcoholic in nature; however, the liquid phase may also be oily in nature.

Ointments

Ointments are viscous semisolid preparations consisting of oleaginous, non-water-soluble-based components (e.g., petrolatum) or water-soluble-based components (e.g., polyethylene glycol) containing a dissolved or dispersed drug substance. The official definition of an ointment was introduced in the USP XV in 1955 and included oil-in-water and water-in-oil emulsions, as well as the oleaginous and water-soluble-type bases mentioned previously. However, use of the term ointment when referring to a pharmaceutical product today is generally understood to mean a viscous semisolid of either the oleaginous or water-soluble type base. The type of formulation used for a given drug is based on optimum delivery, adherence of the drug, and ease or extent of application.

Powders
Powders are typically composed of a drug mixed with a carrier such as talc or corn starch and are used when it is beneficial to keep the skin dry rather than moisturized.

Shampoos
Shampoos are liquid preparations used to deliver drug to the scalp and are systems of mixtures of surfactants designed to cleanse without stripping the hair.

Solutions
Solutions are homogeneous mixtures composed of one or more drug substances [the solute(s)], dissolved in a liquid known as the solvent. From a chemistry perspective, it is possible to have homogeneous mixtures that are solid or gaseous; however, use of the term solution when referring to a pharmaceutical product is generally understood to be a liquid.

Sprays
Sprays are solutions with delivery systems that cause the solution to become a mist of fine droplets of the solution.

Suspensions
Suspensions are dispersions (mixtures) in which a finely divided insoluble API is suspended in a liquid medium. The drug substance may be finely divided and mixed so that it does not rapidly settle out or a suspending agent is added to increase viscosity of the mixture and stabilize the suspension.

Innovator (Brand Name) pharmaceutical companies often market the same drug substance in a variety of topical dosage forms. This approach is a typical marketing technique to maximize product sales by using new dosage forms as line extensions of the original drug. Additional patents may also be granted for new dosage forms and can provide another tactic by which innovator pharmaceutical companies can maintain market share for prolonged periods of time.

Major Therapeutic Areas
A variety of topical products are used to treat patients with conditions in areas such as dermatology, ophthalmology, gastroenterology, gynecology, endocrinology, and trauma (wound-healing).

Disease States
Topical drug products are used to treat many disease states. More than one type of topical drug product may be available for the same therapeutic objective (Table 4). Some examples of common uses are:

Topical antifungals, antibiotics, and products for treating acne comprise a large portion of the topical products market and are available in many dosage forms. Rosacea, or acne rosacea, is a chronic condition characterized by

TABLE 4 Typical Disease States and Common Topical Dosage Forms Used According to Therapeutic Area

Therapeutic area	Disease state	Dosage forms available
Dermatology	Acne	Aerosols
	Ichthyosis and xerosis	Creams
	Psoriasis	Emulsions
	Seborrheic dermatitis	Extracts and fluid
	Keratosis	extracts
	Purpura	Gels
	Dandruff	Irrigations
	Corns and calluses	Lotions
	Warts	Ointments
	Fungal infections	Shampoos
		Strips
		Pastes
		Powders
		Solutions
		Tinctures
		Suspensions
Ophthalmology		Creams
		Emulsions
		Extracts and fluid
		extracts
		Gels
		Irrigations
		Lotions
		Ointments
		Solutions
		Suspensions
Gynecology and Endocrinology	Hormone replacement	Creams
	Infertility	Emulsions
	Amenorrhea	Gels
	Dysmenorrhea	Irrigations
	Menorrhagia	Lotions
	Birth control	Ointments
		Solutions
		Suppositories/pessaries
		Suspensions
		Tablets
Trauma (wound-healing)		Aerosols
		Creams
		Emulsions
		Extracts and fluid
		extracts
		Gels
		Irrigations
		Lotions
		Ointments
		Strips
		Pastes
		Powders
		Solutions
		Tinctures
		Suspensions

inflammation of the central portion of the face with seborrhea, telangiecta-
sia, and acneiform eruptions and is often treated with a topical product in
the form of a cream or gel (i.e., metronidazole, azelaic acid).

Corticosteroids are commonly used for treating a variety of skin conditions
including psoriasis, seborrheic dermatitis, pruritus, and other dermatoses.
Topical products for these uses are often in the form of a cream, lotion, gel,
foam, or solution.

Hormone replacement therapy treatments were historically delivered by oral
products (tablets and capsules) and over the past several years have
changed to topical administration for systemic absorption. The topically
absorbed drugs minimize side effects by avoiding first pass metabolism.
This has been a very successful strategy for hormone replacement therapy
products and has been used for female and male hormone replacement in
gel and cream dosage forms. Several products have patented formulations
consisting of various gelling agents, emulsion particle size and design, and
penetration enhancers.

Pediculicides in the form of shampoos, sprays, or powders are used topically to
rid the body of parasites such as fleas or lice.

Alopecia can be treated with topical products in the form of solutions or
sprays (i.e., minoxidil). There are also topical products indicated for reduction
of facial hair, and keratolytics/moisturizers for severe dry skin (ichthyosis and
xerosis).

Defining the Generic Product and Development Strategy

Determine Marketing Strategy

Product development efforts can begin after definition of the marketing strategy
for the generic product. The Sales and Marketing function of the generic company will typically define the desired product, market approach, and required
product characteristics. For topical products, unless there are restrictive patent
or other considerations, this will most often be a copy of the innovator (Brand
Name) product. This is the product which has been designated as the RLD by
the FDA (1). The generic product can be named with its generic name based
on the active ingredient and the dosage form from the Food and Drug Administration's Approved Drug products with Therapeutic Equivalence Evaluations
(Orange Book) (1), or can be a so-called *Branded Generic*, which has a trade name
associated with the product and/or tied to the generic company name, image,
and company design. Branded generics are more prevalent with topical products, likely because dermatologists, as a group, are a small closed community
of physicians and respond better to promotion. Topical products are also often
applied to the skin and so have cosmetic and aesthetic marketing considerations.
Topical products labeling and presentation are often more attractive and trendy.
Branded products with stylish presentation appeal to patients who have more
impact upon determining what topical products they would use than with other
dosage forms and their respective therapeutic areas. Dermatologists, who pre-
scribe a large percentage of topical products, are also more likely to keep busi-
ness/generate repeat business by prescribing (their preferred) such medications.

Once the Sales and Marketing function has selected a product of interest,
the Product Development function will initiate the activities required to develop

a generic topical product that will ultimately be determined to be bioequivalent to an innovator product or RLD.

DEVELOPMENT OF TOPICAL PRODUCTS

Patent Searches

The first step in the development of generic topical drug products is the patent search. Identification of patents is important for two reasons. The first reason is that all unexpired patents must be identified so that you can develop a time line in relation to the drug development process and final submission to FDA, or a strategy regarding patent challenge to a listed patent. Listed patents (1), and patent challenges (Paragraph IV ANDAs that are one of the four options for patent certification (3)) will not be addressed in this chapter. Identifying patents that have not reached their expiry period will determine the limitations that will be placed on a development program from an intellectual property prospective. The second reason is that review of information provided in the patents may assist a development program, as it may contain information that can be utilized as a basis for formulation and analytical methodology development.

Patent searches are conducted in the following areas:

- Molecule patents
- Method of use patents
- Process patents
- Formulation patents
- Combination with other active ingredients patents
- Pharmacokinetic patents
- Dosage regimen patents

Molecule Patents

Molecule patents describe the discovery of a new chemical entity that will be utilized as the active ingredient in the topical dosage form under development. They often describe a particular crystalline form of the chemical entity (Polymorph patent). The crystalline form may be described in terms of its water of hydration, that is, amorphous, or its salt or ester form. The crystalline form of the API is critical to the development process because it often provides data regarding the phase solubility necessary for multiphase systems such as creams or emulsions. In general, it is preferable to use the same crystalline form provided in the patent unless the strategy is to develop a product that does not infringe on the patent. Changing to a different polymorph than used in the RLD may or may not affect bioequivalence.

Method of Use Patents

A *method of use* patent generally describes the methods for using the drug or more specifically, the approved indications for the topical drug product under development. Considering that the intention is to develop a generic topical product, the key factor in reviewing these types of patents will be to assure that the patent expiry period is not infringed. There can also be multiple indications, only some of which are patented, and so the generic product can be developed using labeling for only the indications which are not patent protected.

Process Patents

A *process patent* most often describes the process utilized to manufacture the API or the finished drug product. As with other patents, determination of the patent expiry period is critical; however, these types of patents may provide technical information regarding specific manufacturing processes. Review of process patents can potentially shorten a development time line, as valuable data may be obtained regarding the process development phase of the topical drug development plan.

Formulation Patents

Formulation patents, often referred to as *composition patents*, may provide details on the qualitative and quantitative formulation of the topical drug product under development. They provide the grade (qualitative) and usually an acceptable quantitative range of each inactive ingredient that is contained in the formulation. Formulation development for a topical drug product that is covered by this type of patent is simplified due to the type of information provided in the patent.

Combination with Other Active Ingredient Patents

Combination with other active ingredient *patents* describe the therapeutic actions of topical drug products containing more than one active ingredient. Valuable clues can sometimes be found in these types of patents regarding the solubility and other physical characteristics of the inactive ingredients and how they can best be incorporated into a multiphase topical dosage form. Information may also be provided regarding the stability of the inactive ingredients in relationship to each other.

Pharmacokinetic Patents

Pharmacokinetic patents are being used by brand companies more recently to extend the life cycle of their products and create additional obstacles for companies to bring generic products to market. Various claims related to AUC, C_{max}, T_{max}, and therapeutic blood levels are included. Topical products intended to penetrate the skin (transdermal) can be subject to these types of patents.

Dosage Regimen Patents

Dosage regimen patents are also being used by brand companies more recently and are based on purportedly new and novel ways to administer a given drug. Many types of topical products could be subject to these types of patents.

Literature Search

Literature searches are important as they help to identify publicly available technical information in specific areas that will be crucial to the development plan for the topical drug product. Considering that the goal is often to develop a product that is identical to the innovator product, review of available information can save time and money in the development process. Various sources for critical data are discussed below.

Approved Drug Products with Therapeutic Equivalence Evaluations
("Orange Book") (1)

The Orange Book is utilized to determine the Reference Listed Drug that will serve as the basis for the topical drug product under development. It is also utilized to determine the various dosage forms, that is, cream, ointment, or gel, which the drug product is commercially available in. Although the various dosage forms may all contain the same inactive ingredient, usually at the same concentration, each dosage form will require a separate development plan and a separate submission to the FDA.

Summary Basis of Approval (SBA)

The Summary Basis of Approval for a topical drug product provides the Agency's summary of their review of the technical data that was used to approve the Reference Listed Drug. This information is available either on CDERs Homepage under the Drugs@FDA link or can be requested under the Freedom of Information Act (4). Although the proprietary information is most often made illegible by strikeouts, valuable information can still be gleaned from this document.

The SBA serves as a valuable tool in the development of the bioequivalence phase of the drug development program. Since most topical products are not absorbed and exhibit their therapeutic action locally, they require clinical end point studies in order to establish bioequivalence to the Reference Listed Drug.

The clinical endpoint bioequivalence studies are conducted in diseased patient populations for which the drug is indicated. The studies are modified Phase III double-blinded placebo-controlled comparative safety and efficacy trials. The study design for a clinical endpoint bioequivalence study is usually a randomized, double-blind, placebo-controlled, parallel designed study comparing *Test* product, *Reference* product, and *Placebo* product. The primary analyses include a bioequivalence evaluation and a superiority evaluation. Bioequivalence is determined by evaluating the difference between the proportion of patients in the *Test* and *Reference* treatment groups who are considered a "therapeutic cure" at the end of study. The superiority of the *Test* and *Reference* products against the *Placebo* is tested using the same dichotomous end point of "therapeutic cure." The study is determined to be successful if both the generic product and the Reference Listed Drug are superior to the placebo and the generic product is equivalent to the Reference Listed Drug from a safety and efficacy perspective. The FDA provides little or no published guidance regarding the conduct of these studies. A protocol may be submitted to the FDA for review; however, FDA response time to such inquiries is generally quite long and may cause significant delays in the development project.

FDA/Office of Generic Drugs

The FDA Office of Generic Drugs Homepage[1] provides various technical and legal documents that serve as the basis for the topical product development plan. Specific information regarding formulation development, analytical methodology development, bioequivalence, and regulatory submission issues can be reviewed in order to optimize the development plan and assure compliance with all Federal Regulations governing ANDA submissions for topical products. Review and comprehension of these documents are essential to assure that

---

[1] http://www.fda.gov/AboutFDA/CentersOffices/cder/ucm119100.htm

a development plan will generate an end product that will be acceptable by current FDA standards.

FDA Guidance Documents

FDA publishes various Guidance Documents[2] that present the Agency's current thinking on a particular subject. These guidances for industry are prepared so that both the FDA review staff and the drug company sponsors who submit drug applications have guidelines regarding the technical information that is included in the ANDA submission. Although most of these Guidances are general in nature, a few guidances deal specifically with topical products.

The FDA currently has two Guidances that present their current thinking on the design of bioequivalence issues related to nonabsorbed locally acting topical products. The first Guidance is a draft Guidance, entitled "Bioavailability and Bioequivalence Studies for Nasal Aerosols and Nasal Sprays for Local Action" (5). This guidance discusses studies that focus on product performance, specifically regarding the release of the drug substance from the drug product. It discusses specific requirements for the formulation and the container closure system that will be utilized in the drug product under development. It provides guidance on in vitro and in vivo bioequivalence study requirements for various topical nasal dosage forms, including solutions and suspensions, where systemic absorption limits are detectable, and also on suspension formulations, where blood or plasma levels are too low for adequate measurement (6). This Guidance is further supported by a Guidance entitled "Nasal Spray and Inhalation Solution, Suspension, and Spray Drug Products—Chemistry, Manufacturing, and Controls Documentation," which provides specific information on the chemistry, manufacturing, and (7) control, documentation required for submission. A Guidance entitled "Bioavailability and Bioequivalence Studies for Nasal Aerosols and Sprays for Local Action Statistical Information form the June 1999 Draft Guidance and Statistical Information for In Vivo Bioequivalence Data Posted on August 18, 1999" provides statistical analysis recommendations associated with these studies (8).

A second Guidance specific to Topical Products is entitled "Topical Dermatologic Corticosteroids: In Vivo Bioequivalence" (9). This document provides specific guidance on the use of a pharmacodynamic effect methodology to assess the bioequivalence of topical corticosteroid products. The approach is based on the property of corticosteroids to produce blanching or vasoconstriction in the microvasculature of the skin. This property presumably relates to the amount of drug entering the skin and thus becomes a possible basis for the comparison of drug delivery from two potentially equivalent topical corticosteroid formulations.

Physicians Desk Reference (PDR)

The Physicians' Desk Reference (PDR) is a commercially published compilation of manufacturers' prescribing information (package insert) on prescription drugs, updated annually.[3] The PDR is a valuable tool to examine the commercial aspects of topical drug products that are currently available on the market. The PDR provides information on how the product is packaged, what sizes are

---

[2] http://www.fda.gov/Drugs/GuidanceComplianceRegulatoryInformation/Guidances/
[3] Thomson Reuters at Montvale, NJ 07645.

available (i.e., 15 and 30 g tubes), and whether or not there are any secondary packaging components required. An example of a secondary packaging component is an applicator for use with a vaginal cream. This information will direct the development of the generic product with regard to how the commercial product is packaged and supplied.

Pharmaceutical Handbooks

Various technical handbooks are available that deal specifically with the development of topical products. Some of these handbooks are followed in the cosmetic industry, because cream or ointment bases used in cosmetic products are very similar to some of the cream and ointment bases that are used in topical drug products.

Examples of handbooks commonly used to develop topical drug products are as follows:

- Handbook of Pharmaceutical Excipients

  Source: Pharmaceutical Press

  Author(s): Rowe, Raymond C; Sheskey, Paul J; Owen, Sian C.

  Handbook of Pharmaceutical Excipients collects, in a systematic and unified manner, essential data on the physical and chemical properties of excipients. Information has been assembled from a variety of sources, including the primary literature and excipients manufacturers.

- Handbook of Pharmaceutical Manufacturing Formulations

  Volume Four

  Semisolid Products

  Second edition, edited by Sarfaraz K. Niazi

  Source: C.H. I. P.S Services

  The fourth volume in the series covers the techniques and technologies involved in the preparation of semisolid products such as ointments, creams, gels, suppositories, and special topical dosage forms. Drug manufacturers need a thorough understanding of the specific requirements that regulatory agencies impose on the formulation and efficacy determination of drugs contained in these formulations.

- Dermatological and Transdermal Formulations (Drugs and the Pharmaceutical Sciences)

  Source: Marcel Dekker Inc.

  by Kenneth A. Walters (Editor)

- Remington: The Science and Practice of Pharmacy

  Source: University of the Sciences in Philadelphia (Editor)

Research Articles

Review of specific literature related to dermatology and cosmetics can serve as a valuable resource in the formulation and process development of topical products.

Reference Listed Drug Characterization

This section describes the process that a development scientist must undergo in order to fully characterize the RLD that will serve as the basis for the ANDA submission. Once the RLD has been characterized, sufficient information will be available to create a drug product development strategy. Characterization of the RLD is accomplished by performing reverse engineering of the RLD. The following tools are utilized, with the goal of creating a skeleton for development of the prototype formulation.

Product Description and Appearance

Fresh and aged (near expiration) samples of the RLD are evaluated in an effort to characterize the targeted drug product. Microscopic analysis under polarized light may be of assistance in evaluating the physical criteria of the drug product intended for development.

The following criteria are examined:

Criteria	Observation
Physical appearance	opaque, transparent, translucent
Color	white to off white, yellowish
Odor	Does the product exhibit a characteristic odor, i.e., sulfurous, citrus?
Consistency	Is the product smooth without lumps, or is it grainy to touch?
Dosage form	Is the product an emulsion, gel, liquid, or ointment?

Evaluation of the physical characteristics will provide the formulator with an endpoint for the design of the formulation.

Identification of the Inactive Ingredients

The list of inactive ingredients in a topical dosage form will appear in the description section of the RLD's Package Insert. Reverse engineering studies are conducted to provide information on the quantitative ranges of the individual inactive ingredients. For example, a Karl Fischer analysis can be conducted to determine the total water present in the formulation. Quantitative ranges for other ingredients can be evaluated utilizing standard analytical tools such as gas chromatography for hydrocarbon content. The reverse engineering studies can also be utilized to confirm the presence of the active ingredients listed in the Package Insert.

Analytical Evaluation

The following table summarizes standard analytical tools that can be of assistance in estimating the quantitative levels of various component of the RLD.

Analytical tool	Quantitative analysis
Drying at 105°C for 3–5 hr	Total solids
Karl Fischer	Water content
GC analysis	Hydrocarbons and fatty acid composition (i.e., emulsifiers, viscosity increasing agents, humectants, solvents, hard fats)
HPLC analysis	Active, preservatives, antioxidants
Atomic absorption	Buffering agents, chelating agents
Rheological studies	Consistency, viscosity

Physiochemical Testing

The RLD product must be evaluated to determine its physiochemical properties. The following properties listed in the table describe the criteria that determine product performance.

Physiochemical property	Product performance criteria
Viscosity	Determines fluidity of product for ease of spreadability.
pH	Important for compatibility of product with skin or mucous membrane and stability of active drug.
Spreadability	The texture of the product determines the ease with which it can be spread on the skin.
Particle size of the active ingredient	Affects the desired degree of solubility of the active ingredient, i.e., dissolved versus suspended.
Number of phases	For multiphase systems, internal phase versus external phase.

If the product is an emulsion, then microscopic evaluation should be conducted to determine how many phases are present, which phase is the internal phase and which phase is the external phase and whether or not the active ingredient is dissolved or suspended. If the product is an oleaginous-based ointment product, then particle size and solubility versus suspension of the active ingredient are critical.

Container/Closure Characterization

Container/closures are critical to the development of topical products. The product will remain in contact with the container/closure system throughout its shelf life, even during use. Additionally, topical products are more fluid in nature and usually contain some degree of water. Therefore, they are more susceptible to container interaction and subsequent degradation from both a physical and chemical perspective.

Container closure components for topical drug products are usually classified as primary and secondary. Primary components are those that make direct product contact. Secondary components are those that are provided with the container closure system to deliver the drug product to the area of application, and only make product contact during the time that they are in use. For example, vaginal or rectal applicators are provided with the product for insertion into the appropriate mucosal area for drug product delivery. Once the drug is applied, disposable (one-time use) applicators are disposed of and reusable applicators are thoroughly cleansed and dried prior to the next use. Some vaginal products that require applicators are actually supplied in prefilled applicators, which serve as the primary packaging component.

The following table describes the most common types of container closure systems used for topical products:

Container	Closure	Material of construction	Special considerations
Tubes	Screw caps	Tubes: aluminum or plastic Caps: plastic	Aluminum tubes may be lined with lacquer to minimize direct exposure of the product to the aluminum. Blind end tubes have a seal (usually a layer of foil) over the opening of the tube. The seal is broken by inserting the inverted end of the cap, which will have a pointed end constructed in the end of the cap for piercing the seal. This mechanism provides tamper evidence. Aluminum tubes are closed by a saddle fold at the end of the tube, which is folded and sometimes glued after filling the product into the tube and are more reactive with the product due to the metal construction. Plastic tubes are closed by heat seal after the tubes are filled and are less protective than aluminum tubes due to the effects of moisture permeation and leaching.
Jars	Screw caps	Wide mouth plastic or glass containers with plastic or metal screw caps containing a liner.	Liners prevent direct contact with the cap and are applied either by pressure (pressure sensitive liners) or induction heat seal, after the product is filled into the container.
Bottles	Screw caps	Same as jars	Same as jars
Canisters	Dosage delivery System, i.e., nasal spray delivery systems, inhalant delivery systems, pumps for foams	Containers are either plastic or aluminum. Closures are usually constructed of plastic.	Closures are specially designed to perform specific dose delivery related functions. Spray systems are constructed of actuator pumps with a nozzle opening that controls the spray droplet size for dose delivery Inhalant systems are designed to deliver very precise doses through a system that vaporizes the drug solution. Foam systems are designed to aerate the drug product to provide a steady stream of foam through the nozzle.
Foil pouches	N/A	Alternating layers of plastic and aluminum	Foil pouches are generally used for unit dose delivery, i.e., suppositories or for one-time product use, i.e., antibiotic preparations for first aid kits.

Although it is not mandatory that the container/closure system of a topical drug product under development be identical to the RLD's container/closure system, it is important that both systems have identical performance characteristics. For example, if the product under development is a tablet for vaginal insertion, then the dosage insertion secondary packaging component must deliver the dose into the vagina similar to the RLD to assure that the drug is delivered to its intended position for comparable drug delivery at the site of action to occur.

Accelerated Stress Testing

Accelerated stress testing of the RLD can provide a plethora of information on both the physical and chemical characteristics of product degradation. The main goal of this testing is to determine the degree of chemical degradation that the product might experience. High-pressure liquid chromatography (HPLC) or other appropriate analytical analysis will often reveal the types of degradants and will give some insight to the stability of the API.

Often, with multiphase topical products, accelerated stress testing may result in phase separation. This information is useful to the formulation developer because it provides an early clue that an alternative to accelerated stability testing for predicting the shelf life of the product may need to be considered.

Sourcing of the API

As with all dosage forms, obtaining API from a manufacturer that already has a Drug Master File (DMF) on file with the FDA, which has already been reviewed for acceptability, is the path with least regulatory risk. Pricing issues are also of consideration. Another important factor is the willingness of the API manufacturer to adjust specifications to meet requirements so that the generic product can match RLD in particle size and/or impurity profile. Additionally, it is key that the manufacturing facility of the active ingredient has passed a recent FDA facility inspection (generally within the last two years). Of particular interest for API manufacturers of topical drug products will be the particle size specifications because many of the APIs incorporated in a large number of topical drug products are in suspension.

The source of API is usually evaluated by performing analytical testing on three batches and assuring adherence to compendial specifications, if applicable, that is, USP/NF. Additionally, the manufacturer may have internal specifications that are not included in the compendia. Examples of these types of specifications include particle size, and impurity profile. It is important to assure that there are no process related impurities or degradants present in the sourced API that are not present in the RLD. If differences do exist, then scientific justification must be provided to assure that these differences do not present a potential safety risk.

Preformulation Development

Once an acceptable API source has been identified and the qualitative and quantitative data on the active and inactive ingredients have been generated from reverse engineering studies, the preformulation development can begin.

Sourcing Inactive Ingredients

This stage of product development involves sourcing to identify CGMP compliant manufacturers/suppliers of the inactive ingredients. Topical products differ from other products in that they are semisolid in nature and the inactive ingredients are often hydrocarbon-based materials or mixtures of materials, which are specifically designed to impart a desired physical characteristic to the final formulation. Therefore, the grade of the inactive ingredient is often critical to the physiochemical and performance characteristics of the final formulation. If the inactive ingredient is commercially available as a mixture, which may be the case with emulsifying agents and or suspending agents, special attention must be paid to select the grade of the inactive ingredient, since these types of inactive ingredients will be critical to emulsion and suspension micellar formations and therefore will be critical to the physical stability of the final drug product.

Active Ingredient Solubility Testing
The next step in the product development process involves API solubility testing. Results from the reverse engineering studies will have provided information on whether or not the API is dissolved or suspended and in the case of emulsion-based products, which phase contains the API. For topical products where the API is dissolved, it is important to determine which inactive ingredient is used to dissolve the API during the manufacturing process. The selected inactive ingredient must provide maximum solubility characteristics with regard to the API. Additionally, the API in combination with the excipients must be stable and contact with the excipients must not result in degradation of the API.

Conversely, if the API is in suspension, then the inactive ingredient that is selected to serve as the vehicle for inclusion of the API into the formulation cannot be soluble with the API. The same stability concerns as with the soluble API ingredient must also be evaluated.

Specifically for multiphase systems, special care must be taken to ensure that the API does not migrate into the other phase. This is critical for bioequivalence evaluation, especially in suspended systems. The API will have to reach its site of local action in the exact same manner as the RLD to have comparable release from the dosage form thereby ensuring therapeutic equivalence.

Inactive Ingredient Compatibility Testing
Once the inactive ingredients have been selected, compatibility studies are conducted by preparing a binary mixture (1:1 ratio) of each individual inactive ingredient with the API and exposing the mixture to elevated temperatures for short period of time (2–3 weeks). At the end of the exposure period, the mixture should be evaluated for the following criteria to determine if any incompatibilities exist.

Criteria	Observation
Physical examination	Is there any discoloration of either of the materials in the mixture?
	Is there any precipitation that might indicate a physical incompatibility of the materials in the mixture or might result in a product that may have a grainy or uneven appearance of feel?
Potency of API	Is the potency of the API maintained after storage?
Impurity profile	Is there a significant increase in impurity levels after storage?
	Are any new impurities present in the mixture that may have resulted from a chemical incompatibility?

Prototype Formulation

The next stage in the drug development process is to create the prototype formulation. This is accomplished by creating various inactive ingredient matrixes in an effort to achieve the desired formulation characteristics, as defined by the reverse engineering studies. The API in most topical formulations is usually included at a very low concentration, i.e., 0.1% to 1.0% Label Claim, therefore the inactive ingredients will define the physical characteristics of the formulation. The critical attributes at this stage are

Physical appearance	i.e., opaque, translucent, clear
Consistency	i.e., smooth, slightly granular to the touch
Color	i.e., white, off white, cream colored, yellowish
Odor	i.e., odorless, characteristic odor (as compared to the RLD), perfumed (when fragrances are included)
Viscosity	
Density	

The prototype formulations that exhibit the desired physical characteristics are then selected for inclusion of the active ingredient.

Formulation Optimization

Formulation optimization studies are undertaken with the selected prototype formulations that have demonstrated favorable compatibility and stability based on short-term evaluation of the following critical quality attributes:

- Active assay
- Impurity profile
- Viscosity
- pH
- Water content
- Preservative assay

Inactive ingredient concentrations are optimized to achieve the desired formulation performance based on comparison to the RLD. As with all generic product development, special care is taken not to exceed the FDA Inactive Ingredient Database limits for the specific topical route of administration. It should be noted that when performing this evaluation, topical routes of administration are not interchangeable, that is, if a specific inactive ingredient is approved at a certain level for a cream intended for administration to the skin, this level may not be acceptable for a suppository that involves a rectal route of administration.

Manufacturing Process

Once the formulation ingredients have been optimized in the selected prototype formulations, the process parameters are defined in preparation for final formulation selection. The lab scale equipment used in the initial development of topical dosage forms is often of simple construction and/or self-assembled since the batches are very small in size (~1 L). As one approaches the final stages of formulation development, the appropriate equipment must be selected to achieve the desired degree of mixing required to maintain the physical stability of the formulation. For example, one must choose between a typical impeller mixer versus a homogenizer. Also of importance are the processing parameters such as mixing times, speed, and temperature required to achieve the desired degree of solubility, suspendability, or aeration. The order of addition of the ingredients is also critical to prevent precipitation and to obtain a robust formulation.

Final Formulation

The following criteria are utilized as a basis for selection of the final formulation(s) that will be subjected to further analysis for formulation screening in an attempt to predict a favorable outcome for bioequivalence evaluation.

- Chemical stability—i.e., active, preservatives, antioxidants, impurity profile
- Physical stability
- Inactive ingredient database evaluation
- Product performance as compared to the RLD—i.e., physical characteristics, physiochemical characteristics

Formulation Screening—Final Formulation Selection

Topical drug products differ from solid oral dosage forms in that there is no single predictor, such as dissolution, that can be used as a reliable tool to establish an in vitro/in vivo correlation. Final formulation selection is generally based primarily on the product performance criteria previously discussed, compared to the RLD. In the case when the topical product is intended for systemic absorption, then traditional multimedia dissolution can be used as a potential formulation screening tool.

In the case when the topical drug product under development is not intended for systemic absorption, the following tools may be utilized as indicators to predict a favorable outcome for bioequivalence testing:

- In-Vitro Release Testing (Franz Cell)
- Cadaver Skin Model
- Skin Stripping (Dermatopharmacokinetics)
- Formulation Screening Clinical Trials, that is, Vasoconstrictor Analysis (for corticosteroids),
- Clinical End-Point Trials

Scale Up of Final Formulation

Scale up of the final formulation occurs in three phases as follows:

Phase	Activity
Pre-Pilot	Small-scale batch to establish finalization of manufacturing equipment and process parameters
Pilot	Submission scale process engineering batch to establish appropriate operating ranges for process parameters
	Establish packaging process
Pivotal	Submission/clinical/stability batch

Analytical Development

Special consideration must be given to ensure that the compendial methodologies are applicable to the specific formulation under test, since complete extraction of the analyte from the formulation matrix can vary depending on the selection of inactive ingredients used in the final formulation. Optimal analytical methodologies for topical drug products should be stability-indicating and capable of quantification of multiple analytes such as the API, preservative (antimicrobial agents), and antioxidants (used to prevent the formulation from oxidative degradation). Also of concern is possible interference from inactive ingredients used in topical formulations.

Strategies for analytical method development of topical dosage forms involve the following in order of preference:

- Adoption of compendial methodologies.
- Development of alternative methodologies based on foreign compendia, that is, British Pharmacopoeia (BP), European Pharmacopoeia (EP), Japanese Pharmacopia (JP), etc.
- Adoption of active drug testing methodologies incorporating special sample preparation techniques to assure compete extraction of the analyte for the formulation matrix.
- Development of in-house methodologies.

Method Development/Validation

Initial method development activities are designed to provide reliable quantification for the estimation of analytes present in the RLD and lab scale formula. Standard method validation criteria are applied to the validation of analytical methods for semisolid drug products. Methods must be demonstrated as specific, precise, accurate, linear, robust, and rugged. Final method validation will include confirmation that the method is stability-indicating by performing stress degradation studies, if applicable.

Microbiological Attributes

Nonsterile products

Topical dosage forms are unique in their ability to support microbiological growth and are almost always formulated using preservative systems. This phenomenon is especially applicable to aqueous systems, which may contain up to 70% water. Keeping in mind that most topical product will be applied to compromised dermal and mucosal membranes, microbiological contamination is of special concern. In addition to the potential risk of contamination by pathogenic organisms, there is a strong possibility for opportunistic nonpathogenic organisms to infect the related areas of treatment in the patient resulting in secondary infections, if the product is not adequately preserved.

USP <61> and <62> address evaluation of the following organisms:

Test	Potential contamination source
Total aerobic plate count	Environment, equipment contact surfaces, air, water, human contact.
Total yeast and mold count	Environment, equipment contact surfaces, air, water, human contact.
Specified organisms	
Bile tolerant Gram negative rods	Untreated waste water, human fecal contaminant
Escherichia coli	Untreated waste water, human fecal contaminant
Salmonella spp.	Untreated waste water, human fecal contaminant, spoilage
Pseudomonas aeruginosa	Water
Staphylococcus aureus	Human contact
Candida albicans	Mucous membranes, skin

These organisms serve as indicator organisms and are not totally representative of the microbial attributes of every product. The selection of limits and indicator organisms is based on the origin of the product (i.e., plant, animal, or synthetic).

Environmental monitoring of the manufacturing areas should be performed routinely so as to establish a base line of the normal flora. The baseline can be used to establish alert and action limits, which will help define when additional sanitation measures are required to maintain the environment from a microbiological perspective. For example, an increase in the observation of non-pathogenic species of *Pseudomonas* in the water system may serve as an alert that the system is approaching the state of being "out of control" and may require sanitation to reduce the microbial load. If the water system is left unsanitized, eventual failure could occur due to the presence of undesired pathogenic species of *Pseudomonas*.

Preservative systems for generic topical products should generally be designed to match the RLD preservative system, especially if the generic drug product has similar inactive ingredients to the RLD. In cases where the generic formulation differs considerably from the RLD, appropriate studies must be undertaken to ensure that the selected preservative system is functional over the shelf life of the product. The USP Preservative Challenge Test <51> must be performed at the target concentrations of the preservative system to confirm adequate microbiological protection at finished product release and at the lower limit of the stability specification to confirm adequate microbiological protection during the shelf life of the product.

Analytical testing of preservative systems may also presents a challenge, since many of the commonly used preservative systems show a high level of instability during the shelf life of the product. Additionally, complete extraction of the preservative analytes from the product matrix often requires vigorous test sample preparation prior to analysis.

Before establishing microbiological specifications for topical products, Suitability of method testing must be performed to evaluate whether or not the product has the potential to support microbial growth. Since many topical products contain active ingredients that may have antimicrobial attributes, that is, topical antibiotics and antifungal compounds, sample preparations may require modification. For example, an inactivating agent such as a neutralizer might be required in the sample preparation in order to neutralize the agent that is exhibiting bacteriostatis to assure that the test method is capable of detecting the presence of a specific organism. If this method is used, this is generally followed by neutralizer efficacy testing. USP <1227> provides methodologies for this type of confirmatory testing.

Sterile Products

Microbiological evaluation of sterile products, such as ophthalmic solutions, is described in the USP <71>. The current requirements include sterilization validation studies and end product testing.

A specimen would be deemed sterile only when it is completely absent of viable microorganisms. However, absolute sterility cannot be practically demonstrated without complete destruction of every finished article. The sterility of a lot purported to be sterile is therefore defined in probabilistic terms, where the likelihood of a contaminated unit or article is acceptably remote. Such a state of sterility assurance can be established only through the use of adequate sterilization cycles and subsequent aseptic processing, under appropriate Current Good Manufacturing Practice, and not by reliance solely on sterility testing.

The principles and implementation of a program to validate an aseptic processing procedure are similar to the validation of a sterilization process. Sterilization validation is described elaborately in USP General Chapter <1211>. In aseptic processing, the components of the final dosage form are sterilized separately and the finished article is assembled in an aseptic manner. In order to comply with currently acceptable and achievable limits in sterilization parameters, it is necessary to employ appropriate instrumentation and equipment to control the critical parameters such as temperature and time, humidity, and sterilizing gas concentration or absorbed radiation. A typical validation program is outlined in the USP for various methods of sterilization: steam, dry heat, gas, ionizing radiations, filtration, and aseptic processing. In addition to the initial validation of the sterilization cycles, in house controls with Biological Indicator checks are also performed as a part of laboratory Quality Control.

The primary means of supporting the claim that a lot of finished articles purporting to be sterile meets the specifications consist of the documentation of the actual production and sterilization record for the lot and additional validation records that demonstrate that the sterilization process has the capability of totally inactivating the established product microbial burden. Further, it should be demonstrated that any processing steps involving exposed product subsequent to the sterilization procedure are performed in an aseptic manner to prevent contamination.

If data derived from the manufacturing process sterility assurance validation studies (e.g., media fills) and from in-process controls are judged to provide greater assurance that the lot meets the required low probability of containing a contaminated unit (compared to sterility testing results from finished units drawn from that lot), any sterility test procedures adopted may be minimal. However, assuming that all the above production criteria have been met, it may still be desirable to perform sterility testing on samples of the lot of finished articles. Such sterility testing is usually carried out directly after the lot is manufactured as a final product Quality Control test. Additionally, environmental monitoring of the facility is required to establish controls in the sterile/aseptic production environment.

Container/Closure Testing

Container/closure testing is performed in four stages.

Stage 1 consists of USP General Chapter <671> Containers-Performance Testing describes the testing required for determination of the amount of moisture that can potentially permeate the container closure system.

Stage 2 consists of USP General Chapter <87> Biological Reactivity Test Biological Reactivity determines the biological reactivity of mammalian cell cultures following contact with elastomeric plastics or other polymeric materials with direct or indirect patient contact. This test addresses the safety of plastic packaging components.

Stage 3 consists of USP General Chapter <87> Biological Reactivity Test In Vitro. This test is used to determine if any of the components of the container have the potential to be extracted or leach into the product or vice versa. The presence of extractables or leachables can compromise either the product or the container and may result in a hazardous situation from a toxicological perspective.

Stage 4 consists of freeze-thaw cycle testing. This testing is performed to simulate the conditions that the product may potentially be exposed to during shipping for commercial distribution. The product is exposed to extreme heat and freezing temperatures for a minimum of three cycles and subsequently analyzed for degradation of the active ingredient and/or physical separation of the components of the product. Multiphase systems, such as suspension and emulsion products are especially sensitive to extreme environmental conditions since they are prone to phase separation. The data generated by freeze-thaw testing will determine if special storage requirements are necessary for shipping the product.

Exhibit Batch Manufacture

Once the final formulation and manufacturing process have been confirmed, a Process Engineering (PE) batch is manufactured in preparation for the final exhibit batch that will be used to conduct all testing required for final FDA product approval. The PE batch will be used to confirm the following criteria for scale up:

- order of addition of ingredients;
- appropriate equipment selection;
- final processing parameters, that is, mixing times and speeds, and processing temperatures; and
- blend uniformity—analysis of final product from predefined areas in the mixing vessel to confirm that the active ingredient(s) and preservative(s) are evenly distributed throughout the batch.

Once all process parameters have been identified and optimized, the master manufacturing document should be created. This document captures all steps that are required to manufacture and package the final product. Using this document, the Exhibit Batch is manufactured and tested based on the preestablished specifications. The final packaged product is then placed under accelerated and control room temperature stability evaluation to establish the shelf life.

Although accelerated stability studies are generally used to predict the shelf life for most dosage forms, multiphase topical products may undergo separation or liquefaction when stored under these conditions. Therefore, this type of study may not be predictive of the actual product shelf life since active ingredient degradation does not usually accompany the physical separation. Very often, the active ingredient will migrate in the container, especially in tubes, and the product consistency may be reestablished once the product is removed from the accelerated stability conditions. This is very often observed in oleaginous-based ointments and wax-based suppository products, which are designed to liquefy at body temperature. In these instances, the only true predictor of shelf life will be real-time stability data resulting from controlled room temperature studies.

Development Report

As with all generic drug product development, the Product Development Report is a summary of all scientific endeavors undertaken to develop the final generic drug product. It summarizes the following information in order to provide the FDA reviewers with the scientific justifications applied during product development.

- RLD characterization
- Preformulation activities
- Formulation development
- Manufacturing process development
- Container/closure system development
- Packaging process development
- Release specification development
- Stability specification development

The Product Development Report also serves as the basis for the Quality Overall Summary document, which is included in the Summary Module 2 of the Common Technical Document format that is submitted to the FDA to support product approval.

Regulatory Requirements

Regulatory strategies to support the development of topical products are somewhat more difficult than strategies for solid oral dosage forms. The path of least resistance for development of a topical product would be to make a product that is "identical" to the RLD. This challenge presents special opportunities in the manufacture of topical drug products and environment since the products under development are usually multiphase and/or have unique delivery systems.

Of special concern is the fact that there are no reliably predictable in vitro tools that can be used to predict the similarity of two topical products from a bioequivalence perspective. Formulation comparisons are often based solely on product performance attributes, as previously described in this chapter. Therefore special care must be taken to assure that the generic product is as close as possible to the RLD to increase the chances of a successful bioequivalence outcome.

The advent of the Common Technical Document (CTD) provides a powerful tool to assure that all FDA requirements for product approval are adequately addressed based on scientific justification. Specifically, the Quality Overall Summary Document of the CTD provides an opportunity to describe the scientific justification that was utilized during product development. As previously stated, this scientific justification serves as the basis for the product performance criteria that will be utilized for product comparison. Ultimate confirmation of bioequivalence will of course be provided by the outcome of the bioequivalence testing, either in vitro or in vivo.

SUMMARY

Topical products are difficult and complicated to develop. They do not lend themselves to the typical development and testing techniques used for more common dosage forms like tablets, capsules, suspensions, and solutions. Many ingredients in topical products, especially emulsions, are mixtures of related molecules and therefore difficult to quantify for purposes of reverse engineering, so matching a formulation of an RLD can be quite a challenge for a generic company. Evaluation of similarity to RLD to anticipate performance of the generic test product in a bioequivalence study is difficult as there is no single predictor, such as dissolution, that can be used as a reliable tool to establish an in vitro/in vivo correlation. Therefore, development of topical dosage forms is considered an "art" and must be undertaken with scientific creativity.

REFERENCES

1. Approved Drug Products with Therapeutic Equivalence Evaluations (Orange Book). www.fda.gov/cder/ob/default.htm.
2. United States Pharmacopeia—National Formulary, USP-NF, The United States Pharmacopeial Convention, Rockville, MD. www.usp.org.
3. 21C. F.R.314.94(a)(2).
4. http://www.fda.gov/RegulatoryInformation/FOI/HowtoMakeaFOIARequest/default.htm.
5. www.fda.gov/downloads/Drugs/GuidanceComplianceRegulatoryInformation/Guidances/ucm070111.pdf.
6. http://www.fda.gov/downloads/Drugs/GuidanceComplianceRegulatory Information/Guidances/ucm070234.pdf.
7. http://www.fda.gov/downloads/Drugs/GuidanceComplianceRegulatory Information/Guidances/ucm070575.pdf
8. http://www.fda.gov/downloads/Drugs/GuidanceComplianceRegulatory Information/Guidances/ucm070118.pdf.
9. http://www.fda.gov/downloads/Drugs/GuidanceComplianceRegulatory Information/Guidances/ucm070234.pdf.

4

Assessment of Topical Dosage Forms Intended for Local or Regional Activity

Isadore Kanfer, Ralph Nii Okai Tettey-Amlalo, and Wai Ling Au
Faculty of Pharmacy, Rhodes University, Grahamstown, South Africa

Betsy Hughes-Formella
bioskin GmbH, Hamburg, Germany

INTRODUCTION

The merits of topical application of medicines have been well documented (1–5). Topical dosage forms intended for local effects have the advantage of facilitating drug delivery to the site of interest to produce a local therapeutic effect while avoiding or minimizing systemic side effects. However, not all drugs are suitable candidates for topical administration due to their physicochemical properties amongst others (6). Ideally, a suitable drug candidate for topical use should have a low molecular mass (<500 Da), high lipophilicity (log P in the range of 1–3), low melting point (<200°C), and high potency (<50 mg/day). Nevertheless, the relative ease of use and increased patient acceptance of topical formulations have made their use widespread (7) and many products for topical administration are currently available. Commercially available topical products include antibacterials, antifungals, antivirals, nonsteroidal anti-inflammatories, analgesics, corticosteroids, and also local anesthetics.

Topical formulations applied to the skin can be subdivided into three categories:

(i) Dermatological formulations (creams, ointments, gel, lotions) intended for the treatment of local (cutaneous) skin disorders—produce a pharmacologic or other effect confined to surface of skin or within the skin and may or may not require percutaneous penetration and deposition.

(ii) Dermatological formulations (creams, ointments, gels, lotions, sprays) intended for regional disorders or symptoms in deeper tissue—the pharmacological action is effected within musculature, vasculature, joints, synovial fluid beneath and around the application site. More selective activity compared to systemic delivery and requires percutaneous absorption and deposition.

(iii) Dermatological formulations and delivery systems (transdermal products) intended for the treatment or prevention of systemic diseases—aimed at achieving systemically active drug concentrations and percutaneous absorption is a prerequisite for activity. Ideally, no local drug accumulation occurs (7,8).

Since penetration through the skin (percutaneous) is a prerequisite for the effectiveness of topical formulations, there is a tendency amongst researchers to misinterpret the intentions of topical delivery systems and a great deal of confusion currently exists with respect to definitions and semantics. It is therefore important to distinguish the actual objectives of topical formulations, that is, whether intended for local, regional, or systemic use, and the associated clinical outcomes.

TOPICAL FORMULATIONS INTENDED FOR LOCAL AND/OR REGIONAL DELIVERY

Objectives
Topical formulations for local activity are intended to treat cutaneous disorders (e.g., acne) or the cutaneous manifestations of a general disease (e.g., psoriasis), whereas topical formulations for regional activity are intended to treat diseases or alleviate disease symptoms in deep tissues (e.g., inflammation) beneath the site of application (9).

Principles
The intention of local delivery is to confine the pharmacological effect of the drug to the surface of the skin (i.e., at the site of application) or within the skin, whereas the intention of regional delivery is to effect the pharmacological action of the drug within musculature, vasculature, joints, and tissues beneath and around the site of application (9). Since the skin is the site of delivery for drugs intended for either local or regional activity, high dermal drug concentrations with low exposure to systemic circulation are desirable (10,11).

Types of Formulations
Semi-solid formulations dominate systems for local delivery but foams, sprays, medicated powders, solutions, and even adhesive systems are also used. Regional delivery has been traditionally accomplished by administration of ointments and creams onto the skin as well as using large adhesive patches, plasters, poultices, and cataplasms (9).

Formulation Characteristics
Formulations are generally applied (rubbed) over diseased or inflamed skin with no visible mass left on the skin surface after repeated application. In most cases, the formulation is left unoccluded, which permits the components of the formulation to be absorbed through the skin, evaporated from the skin, or sloughed off the skin. This class of topical formulation usually contains several excipients that may partition into the skin in accordance with their physicochemical properties. Excipients present in the formulation may change the integrity of the *stratum corneum* (SC), which in turn can alter the solubility of the active ingredient within the horny layer and/or facilitates the ease with which they diffuse through the affected tissue. The act of physically rubbing the formulation on the skin usually results in changes in physicochemical and thermodynamic conditions, which may enhance the permeation of the drug through the skin following application (10,11).

BIOAVAILABILITY AND BIOEQUIVALENCE

Definitions

- *Bioavailability* is defined by regulatory bodies as the *rate* and *extent* to which the active ingredient or active moiety is absorbed from the drug product and becomes available at the site of action (12–14).
- *Relative bioavailability* is defined as the *rate* and *extent* to which the active ingredient or therapeutic moiety becomes available in the organism from a dosage form, compared with a reference standard administered by the same route (9).
- *Absolute bioavailability* is the *extent* to which the active ingredient or therapeutic moiety becomes available in the organism from a dosage form in comparison with an intravenously administered reference standard, which is taken to be 100% bioavailable (9).
- *Pharmaceutical alternatives* are dosage forms that have the same chemical moiety but differ in chemical form, dosage form type, or strength of the therapeutic moiety (15).
- *Bioequivalence* is the absence of a significant difference in bioavailability between pharmaceutical equivalents or pharmaceutical alternatives when administered at the same molar dose under similar conditions in an appropriately designed study (9,14,16).
- *Formulation performance* is defined as the release of the drug substance from the drug product leading to bioavailability of the drug substance and eventually leading to one or more pharmacologic effects, both desirable and undesirable (14,17).
- *Pharmaceutical equivalents* are defined as dosage forms that contain the same active ingredient(s), are of the same dosage form and route of administration, and are identical in strength or concentration when compared with a reference product (15).
- *Multisource drug products* are pharmaceutically equivalent or pharmaceutically alternative products that may or may not be therapeutically equivalent. Multisource pharmaceutical products that are therapeutically equivalent are interchangeable (16,18).
- *Therapeutic equivalent products* are pharmaceutical products which when administered in the same molar dose, their effect with respect to both efficacy and safety are essentially the same when administered to patients by the same route under the conditions specified in the labeling (16,18).

In the early 1960s, reports of physiological availability of vitamins were found to be erratic, which provided initial indications of the potential for bioavailability and bioequivalence problems with multisource drug products (9). The variations in absorption profiles of active ingredients from generic drug products when compared to the innovator or "Brand" product were soon recognized as a potential health hazard when episodes of drug toxicity were reported. Such variations were noted when excipients in pharmaceutical delivery systems or in manufacturing processes were changed or altered. Advances in bioanalytical technology led to further investigations that permitted the determination of bioavailability of marketed drug products through the measurement of their concentration in biological fluids. Reports of bioinequivalence of a number of drugs such as digoxin, phenytoin, and chloramphenicol, among others (9) were published in the scientific literature. The bioavailability of drugs and/or bioequivalence of drug products have therefore emerged as important national and international regulatory and scientific issues (19) and assessment of bioavailability has therefore being employed as a tool by regulatory authorities to monitor the quality, safety, and efficacy of multisource drug products (generic), labeled as "test" versus the innovator or "Brand" product, labeled as "reference." For generic drug products to be interchangeable with the innovator drug product, a generic drug product must not only be pharmaceutically equivalent or a pharmaceutical alternative, but must be bioequivalent to an acceptable relevant reference product.

Innovator/"Brand" companies utilize bioavailability and/or bioequivalence in formulation development to assess formulation performance between two or more pharmaceutically equivalent drug products (20). For example, based on the outcome of a clinical study that made use of the initial formulation, manufacturers effected minor changes to the formulation or altered the manufacturing protocol to improve the formulation. The revised formulation (test) is compared with the initial clinical trial product (reference). Therefore, bioequivalence not only plays an important role in assuring the therapeutic quality of multisource drug products but is also useful during formulation development.

Topical Bioavailability and Bioequivalence

Undoubtedly, the major focus of attention for the assessment of drug products has been in the area of drugs administered extravascularly and intended to be absorbed into the systemic circulation. More and more attention has currently been focused on the assessment of nonabsorbed drugs, that is, topical dosage forms intended for local action.

Bioavailability relates to the "release/availability" of drug into (usually) the systemic circulation following extravascular administration or, when not intended for the systemic circulation, "availability" at the site of action following topical application. For orally administered drugs, it indicates "absorption," whereas for topically applied products intended for local and/or regional action, no "absorption" per se is intended. On the other hand, the use of bioavailability as a surrogate measure of safety or efficacy is premised on the assumption that there is a relationship between safety/efficacy and the concentration of drug in the systemic circulation. However, such surrogate measures cannot be justified on same basis as for drugs intended to be absorbed—no such relationship expected for topical products not intended to be absorbed. Hence, the assessment of bioavailability of topical formulations intended for local and/or regional activity should therefore be performed by measuring drug concentrations within the skin (i.e., at the site of action) (21,22) and not in blood/plasma/serum or urine (23). In other words, bioavailability of topically applied drug may be assessed by (surrogate) measurements intended to reflect the *rate* and *extent* to which the active ingredient or moiety becomes available at the site of action. This can be accomplished by measuring the release of the active ingredient or its "availability in vivo."

Bioavailability Determination of Dermatological Formulations

The determination of the bioavailability of drugs from dermatological formulations requires a consideration of the anatomy and physiology of human skin.

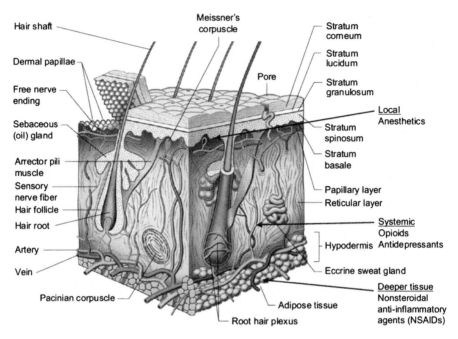

FIGURE 1   Anatomy and physiology of the skin with the potential target or sites of action of selected analgesics. *Source*: Adapted from Ref. 24.

The structure of the human skin and its barrier functions have been described (24). Skin is the largest organ of the human body and is composed of three layers, namely, the SC, which is the outermost part of the epidermis, the living epidermis, and the dermis penetrated by a highly complex network of capillaries, which facilitate the transport and distribution of drugs from the skin into the systemic circulation. In addition, several pilosebaceous and sweat glands are dispersed throughout the skin, in various numbers and size, depending on body site (25). A detailed illustration of the structure of the skin is shown in Figure 1.

The most important function of human skin is its role as a barrier, which reduces water loss while barring the percutaneous absorption of harmful or unwanted molecules from the external environment. The SC contributes to the rate-limiting step in the sequence of processes involved in percutaneous absorption, although the viable tissue can hinder the penetration of very hydrophobic compounds (26). Biophysical, morphological, and biochemical data indicate that the SC forms a continuous sheet of protein-enriched corneocytes embedded in an intercellular matrix enriched in nonpolar lipids and organized as lamellar lipid layers (27). It has a thickness of about 10 to 20 μm and comprises about 20 cell layers (25,28). Penetration of drug substances through the skin involves complex processes but the major barrier to permeation resides within the SC. The reason for the impermeability of the skin is that a diffusing molecule has first to pass through a tortuous route around the dead corneocyte cells and through the intercellular lipids, which comprise mainly ceramides, cholesterol, and cholesterol derivatives (29). Generally, small nonpolar, lipophilic molecules are most readily absorbed, while high water solubility confers poor percutaneous absorptive capacity through normal skin. The octanol/water partition coefficient (log P) of a drug is an indication of its lipophilicity and substances with a log P value of two are absorbed best across the skin (30).

However, despite its barrier properties, the skin is an important route of entry into the body for many topically applied drugs. Topical formulations for local use are assumed to exert their effect through the penetration of the active ingredient directly into deeper tissues, although controversy exists over whether the dermal blood supply removes many drugs before they reach their site of action (31). The sites of action of selected analgesics when applied topically from formulations intended for local and/or regional activity are illustrated in Figure 1. The occurrence of a response, its time of onset, duration, and magnitude when a topical formulation is intended for local and/or regional activity, depends upon the relative efficiency of three sequential processes:

  (*i*)   release of the drug from the vehicle,
 (*ii*)   penetration of the drug through the skin barriers, and
(*iii*)   activation by the drug of the desired pharmacological effect (8,32).

Bioequivalence Assessment of Topical Formulations

Bioequivalence assessment between two drug products as stipulated by the FDA (33) can be evaluated by four methods in order of preference: pharmacokinetic, pharmacodynamic, comparative clinical trials, and in vitro studies.

Currently, in the United States, with the exception of topical corticosteroids, the only means by which a generic company can demonstrate bioequivalence of a topical formulation intended for local and/or regional activity to a formulation manufactured by an innovator/"Brand" company is through comparative clinical trials with a bioequivalence endpoint (34,35).

An innovator/"Brand" company that wishes to replace an approved post-1962 topical dermatological product with a new formulation that involved appreciable compositional changes must demonstrate bioequivalence using data from clinical studies (36,37). For topical corticosteroid formulations, the demonstration of bioequivalence of two pharmaceutically equivalent (i.e., cream vs. cream) products may be accomplished by performing a human skin–blanching assay (HSBA) following the appropriate protocol outlined in the FDA Guidance (33).

Clinical endpoint trials, generally considered as the "*gold standard*" in establishing bioequivalence of drug products usually involve the use of large population sizes, which makes these studies time-consuming and expensive. A clinical endpoint study that involved the topical application of tretinoin gel formulations in the management of acne reported that, in order for bioequivalence to be established with significant statistical power, a study population size of between 275 and 300 is required (22). Therefore, there is increasing interest in developing and validating appropriate methodologies including a method permitting the measurement of drug concentrations within the skin to assess bioavailability. However, there is still some uncertainty as to which is the most appropriate layer of skin to measure drug concentrations for a given topical product (38).

Statistical Procedure and Criteria for Bioequivalence Assessment

A two one-sided test procedure is the currently recommended statistical method for use in bioequivalence assessment (39,40). The area under the plasma concentration versus time curve (AUC) is used as an index of the *extent* of drug penetration (topical application) and maximum concentration (C_{max}) is used as an index of the *rate* of drug penetration (topical application). By convention, bioequivalence data are expressed as a ratio of the average response (AUC and C_{max}) for Test/Reference. The statistical criteria for acceptance of bioequivalence are that both the AUC and C_{max} confidence intervals (CIs) for the generic product (test) must fall within 80% to 125% of the innovator/"Brand" product (reference) based on log-transformed data. These statistical tests are carried out using an analysis of variance (ANOVA) procedure and calculating the 90% CI for each pharmacokinetic parameter (AUC and C_{max}) (14).

Wide sample collection intervals and an insufficient number of enrolled human subjects are two main reasons why a clinical study may fail to establish bioequivalence between two pharmaceutically equivalent products unless the products are indeed inequivalent. The use of wide sampling intervals affects the C_{max} values due to the possibility that the true C_{max} value may be missed. Inherent intrasubject variability requires that a sufficient number of subjects are included. This is necessary to provide the requisite power to be able to detect a 20% difference in bioavailability. A pilot bioequivalence study using relatively few subjects (i.e., 8–12) can provide useful information relating to the choice of the appropriate sampling times and also provide data on intrasubject variability which can then be used to estimate the number of subjects to be enrolled (14).

In Vitro Determination of Dermal Drug Concentration

Considerable effort has been directed towards the development of in vitro methods for the determination of bioavailability of drugs or bioequivalence of topical formulations. However, these methodologies have been associated with high variability when determining drug release from topical dosage forms and therefore provided neither adequate nor consistent information. The high variability associated with these studies usually resulted from the source and treatment of skin tissues used as membranes to study drug diffusion in vitro. With excised human skin, the problem encountered was that the skin tissue was not only obtained from various body parts of cadavers, but the absence of a standard protocol for the pretreatment of such skin tissues had also been identified as the major cause of variability. This problem was also observed with animal tissues (32). Moreover, the use of excised skin (i.e., either human or animal) tissue does not take into account the possible effects of skin flora and skin metabolism on drug release from topical formulations. Animal skins have also been reported to be more permeable to drugs than human skin (7,41), thus making animal skin an inaccurate predictor for percutaneous penetration in man.

The use of commercially available synthetic membranes has been demonstrated to be useful in the development of topical formulations (5). Synthetic membranes function as an inert support that separates the formulation from the receptor phase (19). Several reports using various diffusion cells for the determination of drug release from topical formulations have been published (42–44). Diffusion cells frequently used are the Franz diffusion cell (FDA approved) and the European Pharmacopeia diffusion cell (5,45,46).

While these in vitro experiments are generally conducted under controlled laboratory conditions, in vitro data bear little relation to delivery kinetics in vivo because vehicles also have an effect on the barrier properties of the skin and such vehicle effects cannot be assessed in vitro (47).

While in vitro drug diffusion experiments are useful to evaluate quality, batch-to-batch uniformity, evaluation of changes in manufacturing process, and drug release from the dosage form, the use of in vitro methods for the assessment of bioavailability/bioequivalence has not found acceptance by most regulatory agencies around the world.

In Vivo Determination of Dermal Drug Concentration

Although various in vivo techniques such as suction blister sampling or biopsies have been developed to study kinetic and dynamic parameters of topical drug applications, they are either too nonspecific or too traumatic to be used readily and repeatedly in both normal and diseased skin in human subjects because of their invasive nature (9). Additionally, the cost, technical demands, and ethical considerations limit the applicability of these techniques in human volunteers (27,48). None of these techniques permit the continuous monitoring of drug penetration and metabolism in the same individual on the same test area. Moreover, the determination of drug concentrations in the skin requires that the skin be excised, homogenized, extracted, and the extract analyzed. This methodology defeats the purpose of monitoring drug concentration at the site of action, that is, either the SC (local) or tissues (regional). Furthermore, these in vivo techniques are quite restrictive with respect to the amount of information generated and also require a large number of subjects or a large number of sampling sites per subject, which increases invasiveness.

Currently, there are no adequate sampling techniques that can demonstrate dermal bioavailability and bioequivalence of topical formulations intended for local and/or regional activity, which are both minimally invasive and provide an indication of tissue concentration at the target site. Hence, recourse to clinical endpoint studies remains the only route to assess bioequivalence of topical drug products. It is thus apparent that the development of a method to assess topical bioavailability of drugs or bioequivalence of topical drug products not intended for the systemic circulation without the need to conduct expensive and time-consuming clinical studies in humans should make a significant contribution to this particular area of endeavor.

MICRODIALYSIS

Microdialysis (MD) is an in vivo sampling technique used to measure endogenous and/or exogenous compounds in extracellular spaces (49–51). The technique involves the implantation of a semipermeable membrane into a specific region of a tissue or fluid-filled space (52). The technique was originally developed for use in neuroscience research, which monitored rodent behavior, subsequently the procedure was adapted for use in humans from which studies in many tissues have been reported (49,53). Although the MD technique was introduced for pharmacokinetic studies in animals in 1972, it was not until 1987 that the first pharmacokinetic study in humans was published (54). Preliminary clinical studies that employed MD demonstrated the potential of this technique for the determination of drug concentrations in the *interstitium* of target tissues. This allowed relative changes of concentration–time profiles of drugs to be described. While MD is used in neuroscience and metabolic studies to detect metabolic disorders, its application in clinical pharmacokinetics provides information on concentration–time profiles of drugs in the interstitial fluid (ISF) (54). The development and refinement of this technique over the past two decades has led to its increased acceptance in studies of drug distribution, metabolism, and pharmacodynamics (52,55). This technique has been successfully used for the continuous sampling of low-molecular-weight compounds including glucose, lactose, pyruvate, glycerol, glutamate, and urea as well as pharmacologically active agents in extracellular fluid (ECF) (56). MD has also been used to investigate basic physiology and endogenous substances, as well as the pathophysiology of inflammation and allergic responses, pharmacokinetics and pharmacodynamics of topical and systemic drugs, skin barrier function, and drug penetration into the skin (27,49,57,58).

Theoretical Principles
The principle of the MD technique is based on passive diffusion of compounds down a concentration gradient across the semipermeable membrane of a dialysis fiber (59). This technique mimics the functions of a capillary blood vessel that permits the exchange of solutes in and out of the ECF (41,49,60,61). Movement of drug across the membrane is based on Fick's law of diffusion (62), although facilitated diffusion has also been reported with the use of certain types of perfusates (63).

Dermal Microdialysis
Dermal microdialysis (DMD) is a relatively new application of MD, which allows continuous monitoring of endogenous and/or exogenous solutes in the interstitial fluid (ISF) of dermal tissue with minimal tissue trauma (52). The technique, as illustrated in Figure 2, involves the placement of small perfused membrane systems at given depths within the dermis.

When a topical formulation is applied onto the skin and perfusate is pumped through the implanted membrane system, drug molecules from the topical formulation present in the dermal ISF diffuse (driven by the concentration gradient) into the lumen of the membrane, resulting in the presence of net gain of drug in the perfusion medium collected as dialysate. The dialysate is sampled

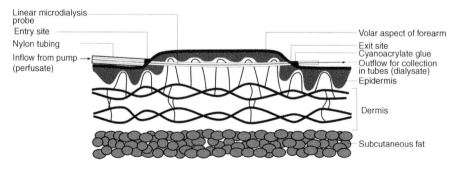

FIGURE 2 Membrane system implanted into the dermis (41).

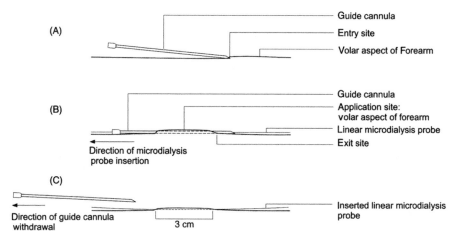

FIGURE 3 Implantation of linear DMD probes in the skin. (**A**) Guide cannula insertion at the entry point marked on the skin. (**B**) Guide cannula pierced through the exit point and MD probes inserted into the guide cannula. (**C**) Guide cannula withdrawal leaving the MD probe within the dermis.

at timed intervals and the drug concentration in the dialysate can be determined quantitatively (52).

Since the first report of DMD, this technique has been used successfully in human volunteers to study the endogenous cutaneous release of histamine in response to various topical *stimuli* and the penetration of a number of topically applied organic solvents (e.g., ethanol, isopropanol) (31), to measure inflammatory mediators in the dermis, to study skin metabolism, and to determine the absorption of drugs or other agents in the skin. DMD has also been utilized as an alternate route of drug administration (64).

Dermal tissue is an attractive sampling site since the tissue is relatively uniform with the ECF in constant equilibrium with the systemic circulation. Moreover, the implantation of the membrane system in the dermis involves a relatively simple procedure, although training is imperative (59). The implantation procedure of DMD probes within the volar aspect of the forearm of human volunteers is illustrated in Figure 3 (49).

DMD has been considered as a promising technique for the assessment of bioavailability and bioequivalence of topical formulations and has garnered a lot of interest among research scientists, dermatologists, and the pharmaceutical industry (36,37,49,65). The technique is minimally invasive and capable of producing concentration–time profiles sampled directly in the dermis, the target tissue, and is therefore suited to study the local and/or regional delivery of drugs following topical administration. Table 1 depicts the various applications of MD to study drug diffusion through human and animal skin.

Membrane System Design

Membrane systems used for MD differ extensively in shape and type of material, depending on the tissue being sampled (51). Two types of membrane systems commonly used in skin studies are either linear or concentric. Linear systems are

TABLE 1 Dermal, Cutaneous, and Subcutaneous MD Studies

Drug	Probe specification	Perfusate	Analytical method	Species	References
17-β-Oestradiol	Concentric, 20 kDa	Ringer's solution	Radioimmuno assay	Healthy human volunteers	(66)
5-Fluorouracil	Linear, 9 kDa		HPLC-UV	Fuzzy rats	(67)
Acyclovir	Linear, 2 kDa	Ringer's solution	HPLC-UV	Healthy human volunteers	(30)
Betamethasone-17-valerate	Linear, 2 kDa	Ringer's solution or sterile PBS	HPLC-UV	Healthy human volunteers and Hairless rats	(68)
Diclofenac	Concentric, 20 kDa	Ringer's solution	HPLC-UV	Healthy human volunteers	(69)
Ethanol	Concentric, 20 kDa	Ringer's solution	GC-Flame ionization	Healthy human volunteers	(70)
Fluconazole	Linear, 5 kDa	0.01 M Na2HPO4/NaCl 8.288 g/l	HPLC-UV	Hairless rats	(20,71)
Fluconazole	Concentric, 20 kDa	Normal saline	HPLC-UV	Healthy human volunteers	(48)
Flurbiprofen	Linear, 5 kDa	Isotonic PBS (pH 7.4)	HPLC-Fluorescence	Hairless rats	(72)
Fucidic acid	Linear, 2 kDa	Ringer's solution	HPLC-UV	Healthy human volunteers and Hairless rats	(68)
Ibuprofen	Concentric, 20 kDa	Normal saline	HPLC-MS/MS	Healthy human volunteers	(73)
Ketoprofen	Concentric, 20 kDa	Normal saline	HPLC-MS/MS	Healthy human volunteers	(74)
Lidocaine	Linear, 2 kDa		HPLC-MS/MS	Healthy human volunteers	(65,75)
Nicotine	Concentric, 20 kDa	Ringer's solution	HPLC-UV	Healthy human volunteers	(66,76)
Pencyclovir	Linear, 2 kDa	Ringer's solution	HPLC-UV	Healthy human volunteers	(30)
Salicylic acid	Linear, 2 kDa	Sterile PBS (pH 7.4)	HPLC-UV	Healthy human volunteers (perturbed skin)	(77)
Salicylic compounds	Linear, 5 kDa	0.05 M PBS (pH 7.4)	HPLC-UV, Liquid scintillation	Hairless rats	(78,79)
Toluene	Linear, 3000 kDa	Albumin solution (5%)	GC-Electrochemical detector	Hairless rats	(80)

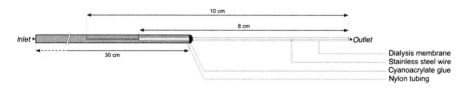

FIGURE 4 Linear DMD probe for in vivo applications.

normally employed in DMD studies (41,65,68,77,81), whereas concentric systems are normally employed in deeper tissue studies involving the cutaneous and subcutaneous regions (69,73,74). The application of the different types of membrane systems employed in dermal, cutaneous, and subcutaneous MD studies are reported in Table 1.

Linear membrane systems are fabricated from hollow fibers (hemodialysis cylinders) often taken from artificial kidneys. These types of membranes are not generally commercially available but are simple to manufacture in the laboratory. Their use requires entry and exit punctures by means of a guide cannula through the skin when placed in the tissue. In contrast, commercially available concentric membrane systems require one entrance puncture (49).

MD membrane systems are commonly referred to as "probes," which comprise the permeable membranes being attached at either one or both ends to impermeable nylon or other inert materials such as Peek® tubing. Henceforth MD membrane systems will be referred to as MD probes. Figure 4 illustrates a schematic diagram of an in-house fabricated linear DMD probe.

The membrane materials used in MD probes are available in different types and pore sizes (27). Common substances used as membrane materials are cellulose acetate, polyacrylonitrile, polycarbonate, AN-69 copolymer (which consists of polyacrylonitrile and methylsulphonate), polyethersulphone, and cuprophan (regenerated cellulose membrane) (27,51,61,70,82–85). MD membranes are porous in nature, which permits diffusion to occur within the pores of the membrane (84).

Different membrane materials have different chemistries that can affect the recovery and/or delivery of drugs. The molecular weight cut-off (MWCO), a physical property commonly used by manufacturers of these membranes, describes the ability of a membrane to reject 90% of molecules with a specific molecular weight. The MWCO value, however, gives little insight into whether or not a particular membrane will result in a higher recovery for a particular drug except for larger drugs such as neuropeptides and hormones where their size is the major constraining factor. It is important to note that the MWCO value does not describe the porosity of the membrane, that is, how many pores per unit area exist for drug diffusion to occur, nor does it predict possible drug interactions with the material of the membrane (84). However, even if the drug molar weight falls below the MWCO value, acceptable extraction efficiency (*EE*) will only be attained with substances having a molar weight lower than approximately one-fourth of the value.

Although most membrane pore sizes range from 6 to 3000 kDa, the majority of MD experiments have been conducted using membranes with MWCO of 20 kDa (27). Researchers make do with this MWCO size because it is small enough to permit diffusion of a large number of drugs, while restricting the entry of large endogenous compounds such as proteins and other macromolecules (86). A study of the effect of diffusion of acetaminophen, 3,4-dihydroxyphenylacetic acid, 5-hydroxyindoleacetic acid, and homovanillic acid by three different semipermeable membranes indicated that although in vitro differences were observed between membranes, no significant difference was observed in vivo (87).

The choice of membrane type is an essential element in searching for the optimal probe for a particular application. It is important that the membrane as well as any other component of the MD system does not interact with the drug since this would reduce the drug concentration in the dialysate (85). The membranes incorporated in linear probes are usually reinforced with a stainless steel guide wire during manufacture for mechanical strength. Klimowicz et al. (88) reported no significant effect on the presence of an intraluminal guide wire in linear MD probes during sampling. When manufacturing MD probes, the length and inner diameter of the outlet tubing should be considered in order to minimize mixing of the dialysate and to prevent hydrostatic pressure build-up across the probe membrane (85).

Probe Calibration

MD probes should be checked to ensure reproducible recovery of the study drug to minimize probe-to-probe differences. This is especially important for in-house fabricated MD probes (84,89). Calibration of probes may be performed in vitro and in vivo. For in vitro studies, the surrounding medium is referred to as the periprobe, whereas the surrounding medium for in vivo studies is the tissue ISF.

Since MD is a dynamic technique with the perfusate continuously being pumped through the probe, equilibrium is not established and dialysate concentrations represent only a fraction of actual concentrations in the tissue ISF or in the periprobe (90,91). The fraction obtained is referred to as extraction efficiency (EE), which has to be determined in order to quantitatively relate drug dialysate concentrations in either the tissue or in the periprobe. However, if the desired information from a MD experiment is the relative change in drug concentration, knowing the in vivo EE is not absolutely necessary. Knowledge of the in vitro EE, however, provides information on the reproducibility and patency of the MD probe being used (91). The general working definition of EE is $(C_p - C_d)/(C_p - C_s)$, where C_d is the drug dialysate concentration, C_s is the known concentration in the external medium, and C_p is the drug concentration in the perfusate. If C_p equals zero, EE is also referred to as the relative recovery (RR). The EE describes the overall mass transport of drugs to and from the MD probe and is commonly used as a means to calibrate the device (83,92).

Several approaches have been reported to determine EE. The most frequently used calibration methods are

- the low-flow rate method,
- the no-net-flux (or zero-net-flux) method,
- the dynamic (or extended) no-net-flux method, and
- retrodialysis by drug or by calibrator methods (52).

The assumption employed with the use of most calibration techniques, such as no-net-flux, extrapolation to zero flow rate, and the use of very slow perfusion rates, is that at very slow flow rates equilibrium is established across the dialysis membrane (91,93). These approaches are, however, time-consuming and require long steady-state periods before any kinetic information is obtainable (82).

The simplest approach to calibrate a MD probe is by using a standard solution. For in vitro calibration studies, since the drug concentration in the periprobe is known and the perfusate contains no drug (i.e., $C_p = 0$), diffusion of the drug occurs from the periprobe into and through the membrane and is collected as dialysate. The *EE* obtained from a measure of drug concentration recovered in the dialysate (*EE_r*) is described as C_d/C_s. *EE_r* is also referred to as recovery by gain.

For in vivo and/or in vitro calibration studies, *EE* may be determined by using a standard solution as perfusate with no drug in the tissue or periprobe (i.e., $C_s = 0$). Diffusion occurs from the perfusate into the tissue or periprobe. The *EE* in this instance, referred to as retrodialysis, recovery by delivery or recovery by loss (*EE_d*), is defined as the ratio of loss of drug from the perfusate relative to the perfusate drug concentration as $(C_p - C_d)/C_p$.

Retrodialysis experiments can either be performed using the same drug as the calibrant (i.e., retrodialysis by drug; RD_D) or using a calibrant with physicochemical properties identical to the drug and which does not interfere with the experiment (i.e., retrodialysis by calibrant; RD_C). Although the advantage of performing RD_D involves the use of the actual study drug for calibration, the RD_D cannot be performed during the MD experiment due to possible tissue contamination. It must be performed before and/or after the actual experiment. It is, however, assumed that *EE* does not change (91).

Initially, the *EE_r* determined in vitro was used as *EE* for experiments performed in vivo. This was mostly because the *EE_r* was easily determined in vitro, while calibration in vivo was cumbersome. However, considerable evidence has shown that it is not reliable to use *EE* determined in vitro for the *EE* in vivo. For most in vivo systems, transport through the tissue is the rate-limiting step determining *EE* rather than transport through the dialysis membrane as is the case for most systems in vitro. It is advisable to perform in vitro experiments before human use, to check for in vitro adsorption to tubing, time delays in drug movement, and to compare drug gain and loss (49,52).

The diffusion process is directly proportional to temperature and therefore the MD study should be conducted at a constant, preferably body temperature environment. The relationship with temperature is depicted as $D = (k_b T/6\pi\eta\sigma)$, which describes the Stokes–Einstein equation, where k_b is the Boltzmann constant (1.38×10^{-23} J/K), T is absolute temperature, η is the viscosity of the suspending fluid, and σ is the particle radius (51).

An influential factor on *EE* is the flow rate. In general, low flow rates result in higher recoveries and vice versa, as described according to $EE = (1 - e^{-rA/F}) \times 100$, where r is the mass transport coefficient, A is the surface area of the MD membrane, and F is the flow rate. Low perfusion rates are often limited by the small sample volumes and the quantitation limit of the analytical method. Therefore, it is not advisable to choose minimal flow rates, as this would increase the sample collection interval and consequently result in worse temporal resolution (51,94). On the other hand, it is also not advisable to choose high flow rates (>10 µL/min), as this would significantly result in reduced *EE*. Increasing the flow rate might also be conducive to the process of ultrafiltration due to the build-up of pressure in the dialysis tubing, resulting in a net flow out of the probe (51).

EE is also influenced by several solute and tissue-related factors. Among these factors are the physicochemical properties of the solute of interest and its diffusion coefficient in the tissue, the ECF volume fraction, and the processes for elimination from the tissue, including active transport mechanisms (27,52).

Assessment of DMD (In Vivo) Probe Depth

The effect of probe depth on drug concentration has been the subject of debate. Whereas Benfeldt et al. (77) found increased drug concentrations in the dialysate with superficial probe insertions, in contrast, Benfeldt and Serup (81), Hegemann et al. (76), Müller et al. (69), and Simonsen et al. (79) reported no such correlation. A probe depth of 0.6 to 1.0 mm is considered acceptable for DMD studies (49). The depth of the probe insertion, that is, the distance of the dialysis membrane within the skin to the skin surface can be measured by ultrasound imaging using a frequency of 20 MHz (49,77,81).

Composition of Perfusates

Perfusates used in MD experiments vary widely in composition and pH. Ideally, the composition, ion strength, osmotic value, and pH of the perfusate should be identical to those of the ECF of the dialyzed tissue (85), that is, the perfusate chosen should be physiologically compatible with the dermis environment (86). This prevents the excessive migration of molecules into or out of the periprobe fluid due to osmotic differences. The perfusate is normally perfused at low flow rates of 1 to 10 μL/min (90). Although isotonic perfusates have been employed during MD experiments, fluid losses in the dialysates have been reported (84). These losses have been attributed to sample evaporation during the experiment at ambient conditions. Table 1 shows a list of commonly used perfusates [Ringer's solution, isotonic phosphate buffers (PBS), and normal saline] in skin MD, and a comprehensive list of different perfusates has been reported (85). Perfusates should be sterile when used in human and animal experiments.

The choice of perfusate used in MD studies affects drug recovery. Studies have demonstrated that the inclusion of β-cyclodextrin in the perfusate as a complexing agent enhanced the *RR* of ibuprofen in an in vitro experiment by a factor of 1.5 to 2.0 (95). Cyclodextrins have also been used to prevent adsorption of hydrophobic materials onto plastics (83). Trickler and Miller (96) studied the inclusion of an osmotic agent, bovine serum albumin (BSA), to the perfusate, which increased the recovery of macromolecules such as tumor necrosis factor and interleukin-1. In vivo studies have also demonstrated the use of Intralipid®, a fat emulsion used for nutritional disorders, and 2-hydroxypropyl-β-cyclodextrin as perfusates for recovery enhancement of lipophilic, highly protein-bound compounds (63).

Invasiveness and Trauma

Although DMD has been described by authors as a minimally invasive technique (27,48,62,90,97,98), human subjects under study do experience some degree of reversible trauma caused by the insertion of probes (51). The trauma thus experienced is a result of inflammatory reactions that occur due to the implantation of the probes but the inflammation is usually reversible with little bleeding and edema (86).

Insertion of MD probes has been reported to increase local blood flow, which has been confirmed with laser Doppler perfusion imaging (70). The skin blood flow has been observed to return to normal by 60 minutes after insertion. Probe implantation also causes histamine release into the skin, which returns to baseline after 40 minutes (70).

Histological examinations have shown no cellular infiltration or tissue disruption around dialysis probes in the skin within the first 6 to 10 hours (67). However, to our knowledge, no DMD studies in human volunteers have been published that extended this time interval (27). Apart from the inflammatory response, the probe might also introduce bacteria therefore causing infection and purulent response. It is important to sterilize the probes before they are implanted in the dermis. Ethanol has been reported (70) not to damage the dialysis membrane, and ethanol solution (70%; v/v) has been used as sterilization medium before implantation in human tissue.

Analytical Challenges

A major limitation with the use of MD is the production of extremely small volumes (~1–30 μL) of dialysate. Because of the low perfusion flow rates (~1–10 μL/min) normally employed in MD, long sampling collection times may be necessary in order to collect sufficient volumes of dialysate for reliable quantitation (99). Volumes deemed as sufficient will be dictated by the minimum volume required by the analytical instrumentation required for reproducible analysis. Generation of more sample volume by increasing the perfusion rate, thereby decreasing the sample collection interval, results in sample dilution. Therefore, an analytical method that can either make use of small sample volumes collected during the MD process or be of sufficient sensitivity to measure the drug concentration in the dialysate, that is, the lower quantification limit (LLOQ) of the analytical method, is essential (51). Moreover, the temporal resolution is determined by a combination of perfusion rate through the MD probe and sample volume requirement of the analytical technique (71,99). Commonly used analytical techniques employed in DMD studies are reported in Table 1.

Advantages and Limitations

DMD has several advantages for in vivo sampling of drugs. No endogenous fluid is removed, so continuous sampling can be performed with minimal disruption of the physiological system (86). This technique is also valuable in patients with minimal blood supply such as children or neonates (51). Although minimally invasive, the procedure is well tolerated by subjects and dermal implantation of the probes may be achieved without the use of drug anesthesia (100). The probe can be implanted directly into the tissue of interest, and since the probe collects only free fraction of the drug, the therapeutically active portion of the dose can be monitored (94).

Provided that the perfusate used is simple and the type of the analytical method employed for the analysis of samples is rapid, the MD technique eliminates the need for complex, elaborate, and time-consuming sample preparation, which is normally the case with plasma samples (51,85,90). The membrane can also act as an effective enzyme inhibitor because it excludes enzymes that could cause degradation of the drug especially during sample storage (86).

DMD may also address the issue of bioequivalence of topical formulations intended for local and/or regional delivery as acknowledged at FDA workshops (36,37,52). Since more than one probe can be inserted in a subject, this reduces the number of subjects needed for pharmacokinetic investigations (55,85). Drug metabolism in tissues can also be studied locally (91) and the method can be used to evaluate drug permeation across both normal and diseased skin. MD may also have the potential to quantify a biomarker, or concentration of another surrogate measure, for therapeutic activity. Concomitant pharmacokinetic and pharmacodynamic evaluations are therefore possible (25). Finally, MD can be performed in almost any organ or tissue of the body (i.e., brain, blood, liver, muscle, heart, subcutaneous tissue, etc.) (96).

The MD technique has a few main limitations. As previously noted, the small sample volumes coupled with associated extremely low drug concentrations, necessitate for the need for a very sensitive analytical method (63). Lipophilic compounds may adsorb onto the polymeric materials that are used in the manufacture of MD probes as well as the inlet and outlet tubing (83,93). Müller et al. (66) observed that 17-β-oestradiol, even with the addition of albumin to the perfusion medium, was shown to be dialyzable only in small amounts in vitro, whereas no detectable 17-β-oestradiol concentration was obtained in vivo. Benfeldt and Groth (68) also reported a similar observation when an attempt to measure dermal concentrations of betametasone-17-valerate after topical drug administration failed since no in vivo concentrations of the corticosteroid were found.

Many pharmaceutically active compounds demonstrate substantial in vivo dermal protein binding. Therefore, the actual recovery of total drug concentration at the site of interest within the skin in an in vivo MD study can be well below 1%. This limitation may be overcome with high doses or the use of a particularly sensitive analytical technique. However, this overall poor recovery at therapeutic concentrations has limited the routine application of MD in some areas of research (63).

Another limitation seems to be the high inter- and intraindividual variability associated with the recovery of exogenous substances. The variability has been reported (27) not to be caused by the MD technique itself, but because of variations in the dermal concentration after the penetration of an exogenous compound. However, some aspects of the MD clinical methodology may contribute towards the variability. One of such aspects is the difficulty in the standardization of a dose application procedure.

Finally, the MD technique requires training clinical investigators particularly with the fabrication of MD probes and the probe implantation procedure. A summary of the advantages and limitations (52) of DMD sampling is presented in Table 2.

Dermatological Application

The application of topical formulations intended for local and/or regional activity is an attractive way of attaining high dermal concentrations of drug without significant systemic activity but as previously mentioned it is often not clear whether adequate drug concentrations are reached at the site of action within the

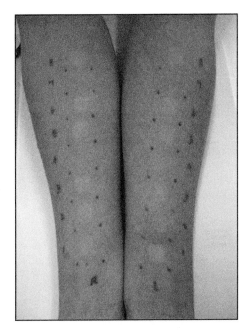

FIGURE 4.6 A typical blanching response after application of Dermovate® cream (0.05% clobetasol propionate).

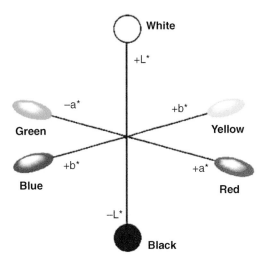

FIGURE 4.7 The L*a*b* color space (192).

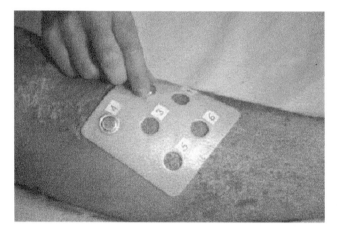

FIGURE 4.17 Treatment of a psoriatic plaque using a hydrocolloid bandage (Psoriasis Plaque Test).

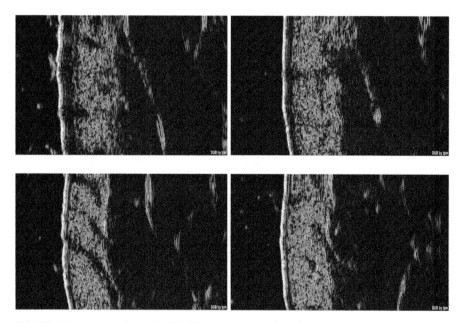

FIGURE 4.18 Image taken with 20 MHz sonography at baseline and after one, two and three weeks of treatment (left to right) (Psoriasis Plaque Test).

TABLE 2 Advantages and Limitations of DMD Sampling

Advantages	Limitations
Highly dynamic continuous sampling	Training of skills required
High-resolution real-time sampling	Insertion of probes
Both drug and metabolites in one sample	Probe manufacture
Minimally invasive	Sensitive analysis needed
Multiple sites in one subject	Drug-specific problems
Sampling and/or delivery via the probe	Lipophilic drugs
Purified samples (protein free)	Highly protein-bound drugs
Highly reproducible	Absolute tissue levels more difficult to estimate
Simultaneous use of auxiliary techniques	Recovery dependent on tissue and time

skin. Moreover, there are currently no in vivo methods available for the direct characterization of drugs applied topically for a local and/or regional effect. DMD provides the opportunity to address this problem due to its minimally invasive nature and its ability to generate concentration–time profiles at a target site with good time resolution provided a sufficiently sensitive analytical method is available (101). DMD may also be used to establish in vivo formulation performance with the possibility of optimizing doses of topical formulations to produce minimum effective concentrations at the site of interest within the skin (52,89).

In a clinical study (102) involving 18 human subjects, with four probes inserted on the left volar aspect of the forearms of each subject, DMD had been successfully employed for the assessment of bioavailability of a ketoprofen topical gel formulation. The same formulation was placed on all four sites on each subject and the dialysate concentrations determined with a validated analytical UPLC-MS/MS method (103). The mean dialysate concentration–time profiles are illustrated in Figure 5. The authors report intra- and intersubject variability of 10% and 68%, respectively. Bioequivalence was subsequently confirmed with a power of greater than 90% (Table 3), thus validating DMD for the determination of topical formulations intended for local and/or regional activity.

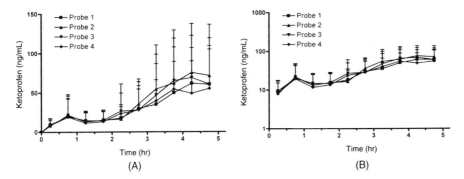

FIGURE 5 Mean dialysate concentration (**A**) and semi-log (**B**) time profiles ($\pm$SD) HS II and III ($n = 18$). Experimental: 4 probe insertions, 4 application sites, 1 probe per site, probes were 1.5 cm apart, probes covered approximately 2 quarters of the volar aspect of the forearm of each volunteer, 18 subjects, Formulation: Fastum® gel.

TABLE 3 Bioavailability Comparison of Sequences ($n = 18$)

Sequence	PK parameter	Units	Arithmetic means (mean ± SD)		% Ratio (S1/S2)	90% CI (lower limit, upper limit)	Power of ANOVA (%)	ANOVA CV (%)
			Test	Reference				
A	AUC_{0-5}	ngh/mL	155.51 ± 98.89	149.98 ± 107.27	106.16	(97.39, 115.72)	92.88	14.88
B	AUC_{0-5}	ngh/mL	152.04 ± 99.23	153.45 ± 103.93	99.01	(89.86, 109.09)	95.95	16.72
C	AUC_{0-5}	ngh/mL	139.89 ± 87.28	165.60 ± 116.67	86.69	(80.37, 93.50)	53.99	13.04
D	AUC_{0-5}	ngh/mL	149.39 ± 112.14	150.68 ± 97.03	95.74	(82.20, 111.52)	51.04	26.30

TAPE STRIPPING

Tape stripping (TS) is a method used to measure drug concentrations within the SC (32) and is based on the reservoir principle of the SC (104). The SC has the property to store drugs applied to the skin depending on the drug, the formulation, the application procedure, and the state of the skin (21). If a formulation is applied to the skin for a limited period of time and removed, the amount of drug in the upper layers of the SC will be predictive of the drug bioavailability in the skin (105–107) and hence this technique may also be employed for local bioavailability and/or bioequivalence of topical formulations intended for local and/or systemic effect.

TS involves sequentially removing microscopic layers (~0.5–1.0 μm thick) of SC by placing an adhesive tape strip onto the skin surface, followed by gentle pressure to ensure good contact and subsequent removal by a sharp upward movement (25), which may be repeated 10 to more than 100 times (9,108). This technique, although painless and noninvasive (25), disrupts the integrity of the water barrier properties of the SC (109), which is rapidly repaired by a homeostatic response in the dermis.

TS has been employed in the dermatological and pharmaceutical fields to evaluate the percutaneous penetration of drugs (21,25,65,110–118) barrier function (119–122), use of animal skin as a surrogate for human skin, chemical toxicity (123), SC mass and thickness (124,125), wound healing processes, and pharmacodynamic and clinical parameters (108). TS may also be useful for selecting or comparing vehicles for drugs to be applied topically.

The initial TS methodology outlining the bioavailability/bioequivalence protocol for topical formulations intended for local and/or regional activity, published in a draft guideline (126) was subject to criticism that resulted in its withdrawal. A number of limitations associated with the draft guidance are listed below but many of them still remain unresolved.

(i) Dosing details such as size and duration of application as well as frequency of sampling were not clearly delineated.
(ii) The number of tape strips needed is considerable, which renders a dermatopharmacokinetic (DPK) evaluation time-consuming. For example, 10 applications per arm require 10 strips per site.
(iii) Discarding of the first two tape strips was not substantiated.
(iv) The same number of tape strips does not remove the SC from all subjects or even potentially within the same subject at different sites (127).

Moreover, there are several sources of variability (104) associated with this technique, namely:

(i) drug application procedure
(ii) type of tape
(iii) size of tape
(iv) pressure applied by investigator
(v) duration of application of pressure
(vi) drug removal procedure
(vii) drug extraction procedure
(viii) analytical methods
(ix) temperature

(x)   relative humidity
(xi)   skin type
(xii)   skin surface uniformity

Although most of these variables (i, ii, iii, vii, viii, ix, and x) can be con-trolled, the degree of variability contributed by the others (iv, v, vi, xi, and xii) is still significant (128–131). Löffler et al. (109) demonstrated significant influences of anatomical site, application pressure, application duration, and TS removal from the application site during the skin stripping technique, whereas the prop-erties of adhesive tapes have been studied by Tsai et al. (119).

Among the list of variables, the number of tape strips needed to remove the SC has received much attention lately. As mentioned previously, the same number of strips does not remove the SC in all subjects due to age, gender, and possibly ethnicity (132), and therefore determination of the SC removed from each strip would provide objective data, which is employed in normalizing the data obtained thus making comparison between different subjects possible.

Sequential skin stripping permits fractions of the SC to be obtained after which the tape strips are extracted to measure the amount of drug absorbed, obtained from either combined or individual tape strips. Various analytical techniques have been employed for the quantitative analysis of drugs in tape stripped skin ranging from HPLC-UV, HPLC-florescence, HPLC-MS/MS, GC, and infrared (IR) spectrophotometry among others (127). Besides tape strip weighing before and after SC harvesting, which is often time-consuming and prone to error because of the SC moisture content (133), spectroscopic measure-ments determining the protein absorption in the UV range, measurements in the visible spectral range after staining of the corneocytes (134), spectral signal con-nected with the SC, or the microscopic determination of the density of the cor-neocytes can also be employed (109,121,128,133,135).

Transepidermal Water Loss (TEWL)

TEWL is a noninvasive bioengineering technique that describes the outward diffusion of water through the skin (136). TEWL monitors the integrity of the SC water barrier function and is an indicator of skin water barrier alteration (109,132), with increased readings often indicating impairment of skin barrier function (136). Healthy SC typically has water content of 10% to 20% and TEWL can be dramatically altered if barrier function is perturbed by physical, chemical, therapeutic, and/or pathological factors (132). Typical basal values of TEWL in adults with healthy skin are between 5 and 10 $g^{-2} \cdot h^{-1}$ (28).

Kalia et al. (28) reported the use of TEWL in TS experiments to determine the thickness of SC in an attempt to normalize the data by incorporating the SC thickness from each subject. This involves weighing each tape strip before and after SC harvesting to determine the amount (m) of tissue removed. Each stripped amount of SC can be converted to a distance (x), which reflects thickness of the removed skin strip, ($x = m/A\rho$), where (A) is the area of the application site, (ρ) is the density of the SC, reported as ~0.88–1.42 g/cm^3 (70).

The additional standardization to determine SC thickness by incorporating TEWL measurements was based on the SC functioning as a homogenous barrier to water transport in vivo, with the diffusional resistance equally distributed and not restricted to a particular tissue layer (28). Baseline TEWL ($TEWL_0$) across nonstripped SC of thickness L (μm) is given by Fick's first law of diffusion expressed as $TEWL_0 = DK\Delta C/L$, where D and K are the diffusion coefficient of water in the SC and the SC viable tissue partition coefficient of water, respectively, and ΔC is the water concentration difference ($55\ M \equiv 1\ g/cm^3$) across the membrane (28). The procedure removes a depth x of SC, which subsequently increases TEWL to a new value by $TEWL_x = DK\Delta C/(L - x)$ of which an inversion yields linear relationship between TEWL and x, where the intercept on the x-axis equals the SC thickness, L (μm) (28). This equation allows the TS data to be expressed as an amount per normalized fraction of SC removed (x/L), a strategy that allows results from disparate subjects of different SC thickness to be normalized and compared.

The TS procedure coupled to TEWL measurements is time-consuming and may pose a problem for fast diffusing drugs. Static electricity on tapes and misleading weights due to the presence of formulation excipients complicates gravimetric measurements.

The use of TEWL with TS, however, demonstrates that the number of tape strips (i.e., 10), as suggested in the FDA Draft Guidance (126) is a poor indicator of the actual amount of SC tissue removed, since no information on the relative position within the SC is known and moreover 10 tape strips fail to permit meaningful comparisons between individuals (28). TS is simple, inexpensive, relatively painless, and noninvasive, given that only dead cells (corneocytes) are removed (109,131) and is commonly employed in routine DPK for the assessment of drug amounts in SC. TS has been reported as being applicable to all drugs that are topically applied for local action (137). Although the TS technique is a single-point determination, it is possible to derive pharmacokinetic parameters such as AUC, C_{max}, and T_{max} by sampling different sites progressively with application time, thereby providing a means to assess topical bioavailability of dermatological formulations (22).

Since the TS technique is accessible only to the SC but not the deeper tissues, for example, the viable epidermis and dermis, this technique may not be applicable to drugs that have their activity in deeper tissues. Vehicle components of products influence both the adhesive properties of the tape as well as the cohesion of the corneocytes and even though TS is considered to be essentially noninvasive, stripped sites in certain dark-skinned individuals may remain pigmented for several months after healing (138).

THE HUMAN SKIN BLANCHING ASSAY (HSBA) FOR TOPICAL CORTICOSTEROIDS

The determination of the bioavailability/bioequivalence of oral dosage forms involves the comparison of drug concentrations found in biological fluids (blood plasma/serum/urine) following administration of the dosage from. However, this approach cannot be used for medicinal products not intended for absorption into the systemic circulation, such as topical dosage forms used for local action.

Topical corticosteroid products have been extremely effective for the treatment of various skin disorders such as eczema, psoriasis, and keloids, among others (139–143). The advent of generic topical corticosteroid products necessitated the development of a suitable technique for the assessment of bioequivalence of such products that resulted in the publication of a Guidance by the U.S. Food and Drugs Administration (FDA) in 1995 (33).

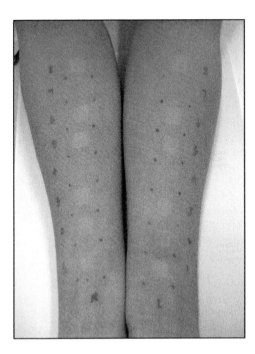

FIGURE 6 (*See color insert*) A typical
blanching response after application of
Dermovate® cream (0.05% clobetasol
propionate).

Various techniques have been studied to assess the effectiveness of topical
corticosteroids. These include the vasoconstrictor assay (36,116,144,145), ultravi-
olet (146–148) and croton oil inflammation suppression studies (149–151), psoriasis assay (152,153), and adhesive tape stripping methods (21,22,127,154–158).

The vasoconstrictor assay or human skin-blanching assay (HSBA) is a reliable and convenient assay for the comparison of the bioavailability and bioequivalence of topical corticosteroids products (159). This assay, initially introduced
by McKenzie and Stoughton in 1962 (144), relies on the unique ability of topical
corticosteroids to produce a blanching response (a skin whitening effect) on the
skin following application and is illustrated in Figure 6. This blanching response
relates to the amount of corticosteroid that has penetrated into the skin (33).

This method was initially utilized for the potency ranking of corticosteroids. However, it was also found that the method has limitations and does
not necessarily rank the corticosteroids in the same order when compared to
other assessment methods (160). In practice, however, topical corticosteroids are
usually applied to diseased skin, thus penetration of the drug will differ due to
less resistance by the *stratum corneum* barrier. As a result, information on topical bioavailability (160) obtained from studies on healthy skin may not reflect
the penetration of the drug into diseased skin. Notwithstanding, the HSBA has
been found to be very effective for the determination of the effect of formulation
on the activity and efficacy of topical corticosteroid products and to examine the
comparative bioavailabilities of such topical preparations (161) as indicators of
the efficacy of those products.

Initially, skin blanching was investigated using a simple Yes/No assessment to establish whether a blanching response was present or absent after

application (162–164). However, this type of assessment does not provide useful information on the degree of bioavailability of the product and it is thus not suitable for the assessment of bioequivalence by comparing bioavailability between different topical corticosteroids and/or their formulations.

The HSBA was subsequently optimized to facilitate comparisons by evaluating time–response profiles, potencies, and formulations of topical corticosteroids. The optimizations included the application of appropriate statistics, establishment of requisite duration of application of the drug or drug product (dose duration) and intervals of time following application at which the response should be assessed and also a scoring system to facilitate visual assessment (33,145,165–167).

Mechanism of Skin Blanching

The exact mechanism of blanching produced by topical corticosteroids remains unresolved. Several investigators have suggested that a relationship exists between noradrenalin and corticosteroids (168–170). However, some investigators have found that although noradrenalin may be involved, there are also other existing factors that may account for the blanching response. Solomon et al. (170) postulated that there are several explanations for the cause of skin blanching, which are as follows:

(i)   Corticosteroids, similar to sympathomimetic agents, directly affect sodium transfer systems across the smooth muscle cellular membranes causing contractions of the smooth muscles of vessels (171).

(ii) Locally bound stores of norepinephrine are released by corticosteroids thereby causing smooth muscle contraction.

(iii) Various enzyme systems including those involved in the release of bradykinin, norepinephrine, and serotonin are affected by corticosteroids, which upsets normal vascular tone. It has thus been suggested that steroids mediate the release of norepinephrine in normotensive subjects only (170).

It has also been proposed that glucocorticoids may act by opposing various natural vasodilators such as histamine, bradykinin, and prostaglandins (172,173). Haynes (174) stated that the slow development of vasoconstriction provides sufficient time for major alterations to occur in the tissue under the influence of glucocorticoids and suggested that since they act to regulate synthesis of proteins, the altered vascular tone may be the result of a basic change in the proteins in the blood vessels or some components of the tissue that affect the blood vessels.

Another mechanism that could cause a skin blanching response may be the effect of melanocytes. Edwards et al. (175) studied the effect of corticosteroids on melanosome aggregation in frog skin. They found that cortisol reversed skin darkening by isoproterenol, caffeine as well as by melanocyte-stimulating hormone (MSH). The results suggested that cortisol binds specifically to a novel type of receptor that has not yet been characterized on the melanophore and which may influence certain steps between the adenylate cyclase and melanosome movement thereby causing the skin whitening effect. It has been postulated that the cause of hypopigmentation may be due to interference by steroids with the synthesis of melanin by smaller melanocytes, leading to patchy areas of decreased pigmentation (176,177).

Methods for the Evaluation of Skin Blanching

Visual Assessment

The initial methodology for the assessment of topical corticosteroids formulations involved the visual assessment of the degree of skin blanching following application to the skin. Visual assessment remained for some period of time as the most commonly used tool when applying the HSBA to compare skin blanching activities between different topical corticosteroids and also formulations (145,178–180) as well as for the assessment of bioavailability/bioequivalence. However, a Guidance (33) document was issued by the U.S. FDA in 1995, wherein an instrumental method using a chromameter was recommended, although the visual assessment method was also retained.

Color is a matter of perception of electromagnetic radiations in the wavelength range of 400 to 700 nm by the eye, followed by subjective interpretation by the brain. Color vision is trichromatic combining the blue, red, and green registrations. Although, the human eye is sensitive enough for the discrimination of small color changes in skin blanching, different people usually draw upon different references and may express the exact same color in different words. As a consequence, this is perceived as an apparent weakness when using the human eye to evaluate skin whitening owing to the observer's subjectivity. A further criticism on visual assessment relates to the inability to validate the eye as one is able to do when using an instrument (181–183). However, if observers undergo sufficient training and gain experience in visual evaluation of skin blanching, reproducibility, and reliability of visual assessment can be established (184–186).

Visual assessment of skin blanching involves the use of a scoring system to measure skin blanching intensity. As mentioned previously, the initial scoring system only used a Yes/No assessment, which simply indicated whether or not a skin blanching response occurred following application of the topical corticosteroid, that is, using a score of 1 or 0, respectively. The scoring system subsequently evolved to a graded response based upon the following criteria: absent, faint, faint--moderate, moderate–strong, and strong–intense blanching using the scores of 0, 1, 2, 3, and 4, respectively. This remains the commonly used visual assessment scoring system based on the 0–4 scale introduced by Barry and Woodford (187,188). The data are reported in terms of the percentage of the total possible score calculated as follows (145):

The maximum score per site $= 4$
The number of independent observers $= n$
The number of sites per preparation per arm $= S$
The number of volunteers $= V$
Total possible score (TPS) $= 4 \times n \times S \times V$
Percent total possible score (%TPS) $=$ (Actual score/TPS) $\times 100$

Chromameter Assessment

An instrumental method involving a tristimulus colorimeter was subsequently introduced as an objective and thus "preferred" method. The Minolta® chromameter, which is a portable instrument that uses tristimulus colorimetry involving reflectance spectroscopy, was adapted to measure skin blanching. This approach had subsequently been used for the objective measurement of skin

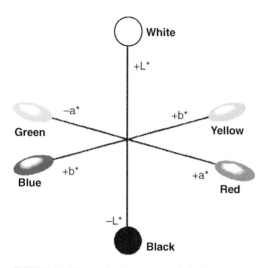

FIGURE 7 (*See color insert*) The L*a*b* color space (192).

color (189,190). The chromameter functions by emitting a white light (using a pulsed xenon arc lamp) onto the chosen area of assessment and measuring the intensity of reflected light through three particular wavelength filters (analyzed at wavelengths of 450, 560, and 600 nm) or using a photodiode array in more recent instruments. The detected signal is converted into three coordinates: L* (luminosity), a* (the amount of green or red), and b* (the amount of yellow or blue). These three coordinates record color in a three-dimensional color system (Fig. 7) recommended by CIE (Commission International de l'Eclairage) (182,183,191).

The skin blanching response is measured relative to the color change in the skin. As the skin blanching response develops, the skin becomes lighter and its redness fades. As the skin becomes more pale the L* scale increases, a* scale decreases, and b* scale increases very slightly. It (193) has been shown that the L* and a* coordinates are more discriminative than the b* coordinate in determining skin blanching responses, thus the latter coordinate is omitted from data analysis. However, following release of the FDA guidance, only the a-scale data has been recommended for use in the statistical analysis (33). This is possibly due to better correlation with visual skin blanching data found by Pershing et al. (116).

The chromameter can offer reliable and repeatable results provided that certain drawbacks are avoided such as manipulation of the measuring head of the instrument that can affect the quality of the data produced. Skin compression by the measuring head and the angle alignment of the chromameter play a role in obtaining repeatable data (184,191,192). To obtain optimal results, each subject's assessment site as well as ambient temperature should ideally be controlled. It is also important for the operator to hold the chromameter head in such a way that variation in pressure is avoided (184). The presence of hair and variations in skin glossiness related to the amount of water and lipid on the skin surface, scarring, uneven skin tone, etc., can influence the data obtained (194). As a result, it is important to avoid these areas of the skin to achieve reliable and reproducible data.

Study Designs

Types of Studies—Pilot and Pivotal
The FDA Guidance recommends that two in vivo studies, that is, a pilot and pivotal study, be conducted in order to determine bioequivalence between topical corticosteroid products. A pilot study provides information on an appropriate dose duration required for the subsequent bioequivalence testing in a pivotal study. The pilot study utilizes a dose duration–response approach, which controls the dose of topical corticosteroid being delivered by comparing different times of exposure of the product on the skin (dose duration is the period of time that the formulation/product is left in contact with the skin). This study is usually conducted only using the reference product.

Dose durations required for the pivotal study as recommended by the Topical Corticosteroid FDA guidance (33) are ED_{50}, D_1 (i.e., $^1/_2ED_{50}$) and D_2 (i.e., $2ED_{50}$), where ED_{50} is the dose duration at which 50% of the maximum blanching response is achieved. The ED_{50} is chosen since it represents the portion of a dose–response relationship plot where the optimum discrimination of relevant differences can be detected. Using longer dose durations may dampen the assessment of relatively small but significant differences in blanching between a test and a reference product. Furthermore, using shorter dose durations will influence the reliability and repeatability of the assessments.

The development and validation of a dose–response curve is therefore essential to determine ED_{50}, D_1, and D_2. These values are determined from an E_{max} model, in accordance to the relevant FDA guidance (33), shown as follows:

$$E = E_0 + \frac{E_{max} \times D}{ED_{50} + D}$$

where E is the effect elicited, E_0 the baseline effect in the absence of ligand, E_{max} the maximum effect elicited, and ED_{50} the dose duration (D) at which effect is half-maximal.

The pivotal study is then conducted where a comparison between the responses of a test and reference product is investigated for bioequivalence using the ED_{50}. Furthermore, the Guidance recommends that a subject must be a "detector" in order for inclusion of their data for statistical analyses supporting in vivo bioequivalence assessment. Hence, subjects' responses are expected to meet the specified minimum D_2/D_1 ratio of AUEC values in the pivotal study as shown in the equation below.

$$\frac{AUEC \text{ at } D_2}{AUEC \text{ at } D_1} \geq 1.25$$

Comparison Between Visual and Chromameter Assessment
A pilot study (195) was conducted where Dermovate® cream (containing 0.05% clobetasol propionate) was used as the reference product. The study was performed using the volar aspect of the forearms of 11 healthy human subjects. The subjects were previously screened for skin blanching response to be included in the study. Approximately 10 mg of the cream formulation was applied onto the relevant demarcated sites using dose durations of 0.25, 0.5, 1.0, 2.0, 4.0, and 6.0 hours. The blanching responses were assessed by three trained observers and

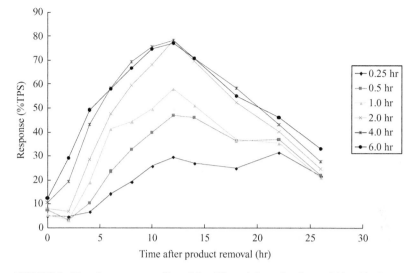

FIGURE 8 Visual response profiles of the different dose durations of 11 subjects.

also with a chromameter at various time intervals over a period of 26 hours after removal of the cream.

Figures 8 and 9 represent skin blanching profiles from the visual and chromameter assessment methods, respectively. The two graphs show very similar blanching profiles at the different dose durations, demonstrating the increase in skin blanching response with increasing dose duration. Skin blanching reached a maximum at dose durations longer than 4 hours, the maximum blanching occurring at 12 hours after removal of the applied formulations at all dose durations for visual (Fig. 8) and chromameter (Fig. 9), respectively.

Figures 10 and 11 depict the E_{max} model fitting using AUEC data from the visual and chromameter assessments, respectively. As seen in the E_{max} profiles (Figs. 10 and 11), the AUEC of the blanching/vasoconstriction response approaches a maximum at ~1 hour dose duration. Thereafter, it is seen that there is a small increase in AUEC and the choice of any of these dose durations if used for the pivotal study, will result in minimal discriminatory power. The dose duration for optimum discrimination between formulation differences will therefore be up to about 1 hour dose duration, which is the steepest portion of the curve. Hence, an ED_{50} corresponding to 50% of the maximum blanching response is chosen.

The ED_{50} values found for both visual and chromameter assessments are shown in Table 4. Based on these values, 0.6 hour was utilized as the ED_{50} in the pivotal study. D_1 and D_2 were determined to be 0.3 and 1.2 hours, respectively.

A pivotal study was subsequently conducted using the above-mentioned dose durations (Table 4) and 34 healthy human subjects were enrolled into the study. The HSBA pivotal study was implemented similarly to that of the pilot study as described previously with the exception that Dermovate® cream was utilized as both reference and test product for the determination of

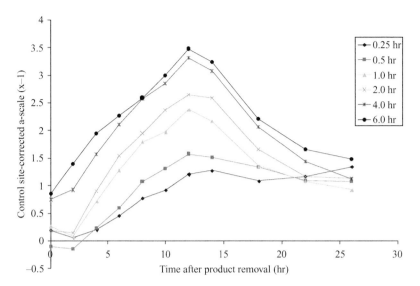

FIGURE 9 Chromameter blanching response profiles of the different dose durations of 11 subjects.

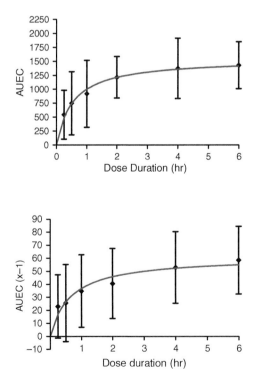

FIGURE 10 E_{max} model of the visual AUEC data where the curve represents the best fit to the experimental data.

FIGURE 11 E_{max} model of the chromameter AUEC data where the curve represents the best fit to the experimental data.

TABLE 4 Parameters Obtained from the E_{max} Model Pertaining to Visual and Chromameter Data

Parameters	Visual	Chromameter
ED_{50} (hr)	0.5457	0.6582
Standard deviation	0.2101	0.3529
D_1 (hr)	0.2729	0.3291
D_2 (hr)	1.0914	1.3164
Dose durations for pivotal study		
ED_{50} (hr)	0.6025	
D_1 (hr)	0.3012	
D_2 (hr)	1.2049	

bioequivalence. Skin blanching was evaluated over a period of 30 hours after the removal of the applied products.

The results revealed that 23 subjects were found to be "detectors" in the pivotal study, but the data for all 34 subjects were included for comparison purposes. The skin blanching profiles shown in Figures 12 to 15 were very similar when comparing the data between the two different assessment methods, or between "detectors" and "nondetectors." This indicates that the visual and chromameter assessment methods are comparable to each other and both are equally applicable for HSBA. The inclusion of "nondetectors" data did not seem to have a significant effect on the skin blanching profiles nor on the outcomes of the comparisons for the assessment of bioequivalence.

Assessment of Bioequivalence
Generally, the acceptance criteria for the declaration of bioequivalence specify that the 90% confidence intervals (CIs) for the ratios of the log-transformed C_{max} and AUC for orally administered test and reference products must fall within the range of 80% to 125% (196). However, for the HSBA only the ratios of the AUEC are recommended for use as assessment criteria. As shown in Table 5, both assessment methods comply with the bioequivalence criteria except for the data for "detectors" using the chromameter. This usually suggests that the power

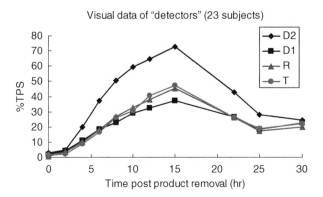

FIGURE 12 Mean visual blanching response profiles for the "detectors."

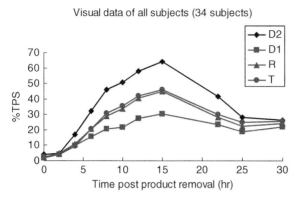

FIGURE 13　Mean visual blanching response profiles for 34 subjects.

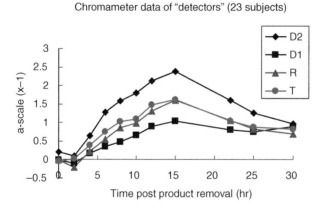

FIGURE 14　Mean chromameter blanching response profiles for the "detectors."

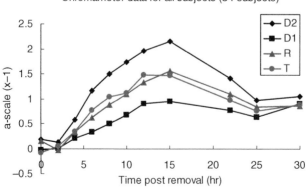

FIGURE 15　Mean chromameter blanching response profiles for 34 subjects.

TABLE 5 90% Confidence Intervals Calculated Using Locke's Method Using Visual and Chromameter Data

	Visual		Chromameter	
	Mean ratio % (T/R)	90% CI	Mean ratio % (T/R)	90% CI
Detectors (n = 23)	104.6	99.3–111.6	104.6	86.5–129.3
All Subjects (n = 34)	102.9	97.9–109.2	104.3	90.2–120.7

Abbreviations: T, test; R, reference; CI, confidence interval.

of the study is too low and indicates that more subjects should be included in order to increase the power of the study when using the chromameter as the assessment tool. On the other hand, the visual assessment data clearly indicate that the study population of 23 subjects is sufficient for this HSBA bioequivalence study.

It is thus seen that although the use of the chromameter is generally recommended as the "preferred" assessment method in favor of visual assessment due to the subjectivity, the above data has shown that visual assessment is a reliable and appropriate assessment technique using the HSBA (195).

Evaluation of the Tape Stripping Method Versus the HSBA for the Assessment of BE

In order to validate the use of the TS method for the bioequivalence assessment of topical formulations, data from a TS study were compared with data generated on the same product using the HSBA.

An initial pilot TS study was undertaken to determine the ED_{50} value to determine bioequivalence using TS where dose durations of 0.5, 1, 2, 4, 6, 8, 10, and 12 hours were evaluated on the volar aspect of the forearm of six healthy human subjects. A blank site was reserved to estimate *stratum corneum* thickness for each individual using transepidermal water loss (TEWL) as previously mentioned (*vide infra* 5.1), and the AUC_{corr} value was determined by correcting for skin thickness (28). The normalized skin thickness was used to compare intra- and interindividual data.

Approximately 5 mg/cm^2 of Dermovate® cream (reference product) was applied to the relevant sites. At the end of the dose duration, the product was removed and tape stripping was commenced. The ED_{50} was found to be 2.4 hours. Two hours was chosen as the dose duration to assess bioequivalence by using TS in the pivotal studies. Scotch® tape (no. 810, 3M) was used as the adhesive tape strips for *stratum corneum* removal.

A further pilot TS study was carried out to determine the number of subjects required for the pivotal TS study in order to attain an acceptable power. This study utilized Dermovate® cream where it was compared against itself as the test and reference product.

The results showed that interindividual variability (CV%) was ~14%, which indicated that approximately 30 subjects would be required to achieve a power of at least 80% (197).

TABLE 6 Validation Study—Bioequivalence Assessment of Identical Products (Test: Dermovate® Cream, Reference: Dermovate® Cream)

	Mean T/R ratio (%)		90% CI (%)	
	Untransformed (Locke's)	Log-transformed	Untransformed (Locke's)	Log-transformed
HSBA ($n = 34$) (195)				
Chromameter	104.3	–	90.2–120.7	–
Visual	102.9	–	97.9–109.2	–
Tape stripping ($n = 7$)				
Pilot study (AUC$_{corr}$)	101.8	101.4	88.0–118.3	87.4–117.7

Abbreviations: T, test R; reference; CI, confidence interval; HSBA, human skin blanching assay; AUC$_{corr}$, area under the curve of corrected tape stripping data.

Upon comparing the product to itself, the results of AUC$_{corr}$ values showed similar confidence intervals and AUC$_{test}$/AUC$_{ref}$ ratios using untransformed (Locke's) and also log-transformed data as shown in Table 6. It is interesting to note that the results are similar to those of a previously conducted pivotal HSBA study (195). Hence, the TS method is seen to be comparable to the HSBA method, as it produced the same bioequivalence outcome.

The pivotal TS studies assessed bioequivalence of a cream and an ointment formulation against a reference cream formulation (Dovate® cream vs. Dermovate® cream; Dermovate® ointment versus Dermovate® cream). It is important to note that creams and ointments are considered not to be pharmaceutically bioequivalent and bioequivalence assessment are normally not done between these two different types of formulations. However, a bioequivalence assessment study was conducted to determine whether the TS method had the necessary sensitivity to determine differences between these two types of formulations, if differences do indeed exist.

The AUC$_{corr}$ data (Table 7) established that Dovate® cream was bioequivalent to Dermovate® cream whereas the opposite was found when comparing Dermovate® ointment against Dermovate® cream. These studies indicated that the TS method was able to determine similarities and differences between the various dosage forms studied.

TABLE 7 Pivotal TS Studies of Clobetasol Propionate Creams and Ointment Products Using AUC$_{corr}$ Data

	Mean T/R ratio (%)		90% CI (%)	
Pivotal TS studies ($n = 30$)	Untransformed (Locke's)	Log-transformed	Untransformed (Locke's)	Log-transformed
Dovate® cream (T) vs. Dermovate® cream (R)	93.8	92.8	84.7–103.6	82.9–103.9
Dermovate® ointment (T) vs. Dermovate® cream (R)	66.3	55.2	48.8–82.2	46.1–66.1

Abbreviations: TS, tape stripping; T, test; R, reference; CI, confidence interval.

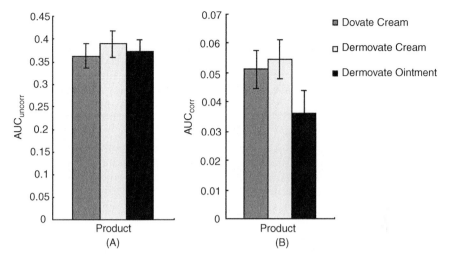

FIGURE 16 Mean AUC_{uncorr} or AUC_{corr} ± SEM ($n = 30$) values of the different formulations.

An aspect found to be quite interesting is the use of different data-processing methods that may affect the bioequivalence results. The data-processing methods, which has commonly been used to calculate AUC_{uncorr} values (22,65) without correcting for skin thickness, were compared with data using AUC_{corr} values where skin thickness had been normalized. When the bioequivalence results using AUC_{corr} and AUC_{uncorr} data were compared (Tables 7 and 8); surprisingly, the use of AUC_{uncorr} data yielded results that indicated that Dermovate® ointment is bioequivalent to Dermovate® cream. When both AUC_{corr} and AUC_{uncorr} data were plotted (Fig. 16), it is seen that the use of AUC_{uncorr} is less discriminatory [Fig. 16(A)]. Hence, the use of AUC_{corr} values is recommended since it enhances the discriminatory power of the TS method for the bioequivalence testing of topical products. Additionally, it was seen that the use of the different statistical methods did not significantly alter the results (Tables 7 and 8).

TABLE 8 Pivotal TS Studies of Clobetasol Propionate Creams and Ointment Products Using AUC_{uncorr} Data

Pivotal TS studies ($n = 30$)	Mean T/R ratio (%)		90% CI (%)	
	Untransformed (Locke's)	Log-transformed	Untransformed (Locke's)	Log-transformed
Dovate® cream (T) vs. Dermovate® cream (R)	93.4	93.6	86.3–101.2	86.22–101.5
Dermovate® ointment (T) vs. Dermovate® cream (R)	95.9	96.3	86.8–106.1	86.6–107.1

Abbreviations: TS, tape stripping; T, test; R, reference; CI, confidence interval.

The above indicates that the TS method can be used as an alternative approach for the bioequivalence assessment of topical clobetasol propionate formulations. This method should therefore be equally applicable to determine bioequivalence of any topical corticosteroid formulations and also applicable with other topical dosage forms intended for local use. In addition, the application of either of the statistical methods described above for a bioequivalence assessment of topical dosage forms using the TS method may be used.

INNOVATIVE APPROACHES

With the exception of the vasoconstrictor assay for corticosteroids, there are no validated methodologies or guidelines currently available for establishing bioequivalence of topical therapies. Equivalence for drug classes other than corticosteroids must generally be established in therapeutic trials.

Models using bioengineering methods offer innovative approaches to determine bioequivalence of topical therapies. With adequate validation these methods may be used as surrogate endpoints for clinical effects. At present, the use of these methods is mainly limited to formulation screening or pilot studies. Before these methods gain acceptance for use in pivotal bioequivalence trials, reliability and reproducibility of the measurements must be established as well as unequivocal correlation of the endpoint with results of therapeutic trials. In the following, alternative approaches for measuring bioequivalence in several of the major dermatological indications are presented.

Psoriasis

Equivalence of antipsoriatic treatments other than corticosteroids is generally established in therapeutic trials. The psoriasis plaque test, also known as the microplaque assay, is a potential alternative methodology for bioequivalence studies for antipsoriatics in drug classes other than corticosteroids. In particular this design has proved useful for formulation screening before performance of a large therapeutic equivalence study. The psoriasis plaque test has been used to investigate efficacy of diverse actives, in particular corticosteroids and vitamin D analogs (198–201).

In the psoriasis plaque test, multiple test fields located on one or two plaques are treated topically in a single panel. In the original test design described by Dumas and Scholtz in 1972 (152), the efficacy of corticosteroids for treatment of plaque-type psoriasis was evaluated clinically following standardized occlusive application over several days. Test fields were graded as follows: unchanged or less than full involution (0) or complete return of epidermis to normal (+). Only a slight residual erythema without further signs of inflammation such as infiltrate or scaling was scored as "normal" epidermis. No attempt was made to estimate partial improvement. The plaque test results for corticosteroids were shown to reflect known clinical efficacy.

Whereas this design with simple grading of clearing is well suited for evaluation of corticosteroids of varying potencies, it is not sensitive enough to differentiate between formulations with identical active ingredients.

Meanwhile bioengineering methods have been used in plaque tests to objectively measure the inflammatory alterations accompanying psoriasis. Measurement of infiltrate thickness with sonography, intensity of erythema by colorimetry, skin blood flow by laser-Doppler flowmetry, skin surface

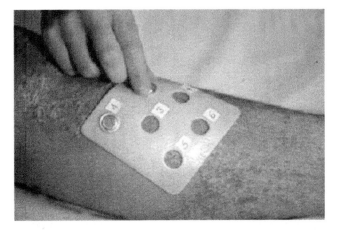

FIGURE 17 (*See color insert*) Treatment of a psoriatic plaque using a hydrocolloid bandage (Psoriasis Plaque Test).

characteristics by profilometry, and skin temperature have all been used in plaque test designs (199–202). One of the most important clinical endpoints in psoriasis is the extent of the psoriatic infiltrate. Therefore, objective measurement of the infiltrate depth with 20 MHz sonography is probably the most relevant outcome that can be used as a biomarker or surrogate outcome for treatment effects.

Patients suffering from psoriasis vulgaris with stable plaques already existing for several months or years are suitable for inclusion in the psoriasis plaque test. Plaques exhibiting spontaneous regression as well as exacerbation of psoriasis are excluded. In the standard design used routinely at *bioskin*®,[1] test fields measuring 12 mm ø are treated over a 12 to 28 day period. Ideally, all test fields are located on a single psoriatic plaque. The plaque thickness and anatomical location of the plaque should be similar in all test fields. Prior to baseline measurements and the first treatment, scales are removed from plaques. This is necessary since it is not possible to obtain sonographic images with clearly demarcated infiltrate if images are taken through layers of scales. A hydrocolloid bandage is attached to the plaque in which the test fields have been punched out (Fig. 17). Treatments are performed in an occluded or open manner. The hydrocolloid bandage itself does not have an influence on the extent of the psoriatic infiltrate over the course of the treatment period (unpublished observations, bioskin®).[1] The primary variable is the depth of the psoriatic infiltrate measured in sonographic images taken at defined intervals during the treatment period. As a secondary variable, clinical improvement is assessed. Because of the small size of the test fields and difficulties encountered in clinical assessment, separate clinical grading of infiltrate depth and erythema are not performed, rather an assessment of general improvement or worsening compared with the untreated plaque beneath the hydrocolloid dressing. Scaling cannot be evaluated in the test.

---

[1] bioskin GmbH, Institute for Dermatological Research and Development, Poppenbuetteler Bogen 25, Hamburg 22399, Germany.

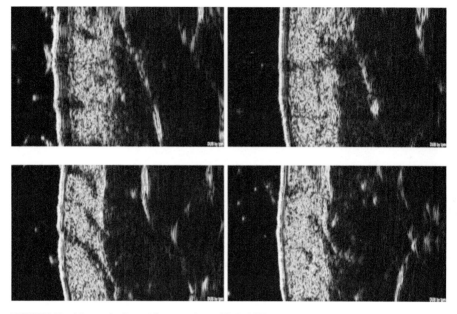

FIGURE 18 (*See color insert*) Image taken with 20 MHz sonography at baseline and after one, two and three weeks of treatment (left to right) (Psoriasis Plaque Test).

The intensity of erythema can also be measured in the test fields by colorimetry. This has proven to be particularly useful in combination with sonography (198,199) and for the evaluation of corticosteroids where vasoconstriction plays a prominent role in the measurable effect. Using this method, residual erythema can be included in the assessment that is often the only remaining finding after all other skin alterations have cleared. Typical sonographic images obtained in the plaque test following treatment with corticosteroids or calcipotriene are shown in Figure 18.

For formulation screening, 15 subjects with chronic stabilized psoriatic plaques are sufficient for obtaining relevant data for comparing the study preparation with a reference listed drug.

The change in infiltrate thickness from baseline is the primary endpoint. Confidence limits calculated for the change in infiltrate thickness and for the difference between the study preparation and a reference listed drug can already give information for assessment of bioequivalence.

Atopic Dermatitis

Clinical severity of AD is usually evaluated in therapeutic trials with a general severity score, for example, Severity Scoring of Atopic Dermatitis (SCORAD) (203) or Eczema Area and Severity Index (EASI) (204), which includes grading of extent (involved surface area), intensity (erythema, induration/population, lichenification, xerosis) and subjective symptoms (e.g., insomnia, itching). An alternative method for the establishment of bioequivalence is within-patient comparison of lesional areas of similar severity and anatomical location.

Considering the large number of endogenous and exogenous factors which contribute to the clinical course of AD and which cannot be standardized between different panels, the use of intraindividual comparison is particularly advantageous for this skin disease.

Bioengineering methods can be used to more objectively assess skin status. In particular, a close correlation has been shown between clinical severity of AD and skin barrier function measured by transepidermal water loss (TEWL) (205–209). In fact, it has been suggested that TEWL measurements may be more sensitive than SCORAD assessments in detecting subtle gradations in clinical activity and in relapse prediction (209). TEWL values are increased in lesional skin of AD and disease improvement under treatment is accompanied by a decrease in TEWL (207). A significant correlation has also been shown between lichenification/xerosis scores and skin capacitance measurements as an indicator of skin hydration (209). Erythema intensity can be quantified using colorimetry.

In a typical design with intraindividual comparison, inclusion criteria include patients with mild-to-moderate atopic dermatitis diagnosed by Hanifin and Rajka (210) with two target lesions in comparable body locations (e.g., antecubital fossae) that exhibit a diminished water permeability barrier, for example, TEWL ≥ 12 g/m^2 · hr. One lesion is treated with the test formulation, the other with the reference listed drug. In order to establish superiority of the test formulation to vehicle, a second patient group must be included comparing test formulation to the vehicle. A sample size of at least 30 per group is usually required. Lesions are treated once or twice daily over a two- to four-week test period by the patients at home. Bioengineering measurements and clinical assessments are done at baseline and once every week or biweekly.

Inflammation

Anti-inflammatory efficacy of steroidal and nonsteroidal anti-inflammatories (NSAIDs) can be determined in the UV-Erythema Test (211–216). In this model, experimental inflammation is induced by simultaneous irradiation of multiple test fields with UVB. Intraindividual comparison of the test formulation, reference listed drug, and vehicle can be performed in parallel. This model has the advantage that it is performed in healthy volunteers as opposed to patients with inflammatory skin disease.

In any model of inflammation, success of the test is dependent on standardizing the extent of inflammation in individuals with different inherent sensitivities. In the UV-erythema test, this is achieved by irradiating with a defined UVB dose, usually within the range of 1 to 2.5 minimal erythemal doses (MED). The MED is the smallest amount of UVB light producing distinct erythema and differs from individual to individual. The MED for each subject must be determined beforehand by parallel exposition of small test fields to graduated UV dosages. Ideally six to eight small fields are exposed to UVB dosages increasing in small increments, for example, 15% increase in dosage from one field to the next. If the increments are too large it is not possible to accurately determine the MED. After determination of the individual MED, test fields on the back are irradiated with the desired UVB doses. In most test designs, inflammation is induced by uniform exposure of the test fields to a single dose of UVB light. However, the optimal dose for determination of efficacy of a test formulation cannot always be predicted beforehand. This difficulty can be circumvented by varying the strength of inflammation by irradiating separate test fields with differing UVB dosages, for example, 1.25, 1.6, and 2 MED, in a single panel (215).

Immediately after irradiation, the test fields are treated with the test formulations, reference listed drug, and vehicle. In order to establish test sensitivity, it is advisable to additionally include comparators of higher and lower strength than the reference listed drug in each panel. Since treatments are performed after irradiation a sun protective effect can be excluded. An irradiated, untreated test field is included as a negative control. This is essential to provide a reading of the course of the inflammatory process in the absence of treatment. Test products can be applied in an occlusive or open manner. In the case of weak anti-inflammatory drugs, for example, hydrocortisone, it may only be possible to measure an anti-inflammatory effect under occlusive conditions. The frequency of application (single or repeated applications) and the length of the treatment period (e.g., 8–48 hours) can be varied, depending on the study aim. In a typical bioequivalence design, the test formulations would be applied once immediately after irradiation. Alternatively, repeated applications can be performed once or twice daily.

The endpoint is the degree of erythema in the irradiated test fields. Diminished erythema compared to the irradiated, untreated fields is indicative of anti-inflammatory efficacy. Objective measurements of erythema are done using colorimetry (a^*-value in the CIELAB color space). Visual assessment of erythema suppression compared to an untreated, irradiated test field may additionally be done. Since the erythema only begins to develop 2 to 3 hours after irradiation and reaches a maximum after approximately 16 to 48 hours, first measurements are usually performed 6 to 8 hours after irradiation and can be repeated for up to 48 to 72 hours after irradiation.

For noninferiority or equivalence analyses, the area under the curve (AUC) is calculated for skin redness (a^*-values) changes from baseline. Test results for a UV-erythema test with repeated product application can be seen in Figure 19. The test formulation was found to be equivalent to the reference listed drug.

Peripheral Pain

UV-induced erythema can also be used as a model to test antihyperalgesic drugs by thermal sensory analysis (217). Thermal hyperalgesia is induced by irradiation with a standardized UVB dose, for example, 3 MED. Heat pain threshold is assessed prior to irradiation and at intervals beginning 6 to 12 hours following irradiation by linearly increasing the skin temperature with a thermode and advising subjects to stop the heating by pressing a button as soon as the heat becomes painful. The threshold temperature is recorded as endpoint. The irradiated fields are usually treated with the test products immediately after irradiation; however, it is also possible to mimic the clinical condition sunburn by application of the products after development of symptoms, for example, 6 to 10 hours after irradiation. The heat pain threshold is lowered by anti-hyperalgesic drugs.

Acne

There are few alternatives for determination of bioequivalence of acne medications except for clinical assessment (lesion counts and investigator's global assessment). However, for some products, for example, retinoids, it is discussed

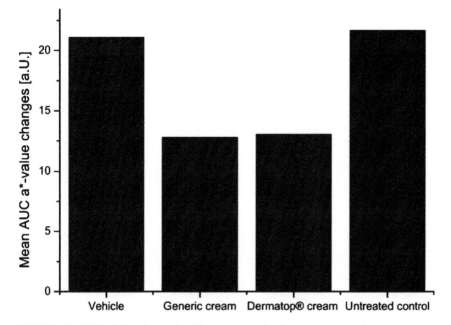

FIGURE 19 AUC of the change in skin redness (a*-values) to baseline for a prednicarbate (0.004%) generic cream, the vehicle, the reference listed drug (Dermatop® cream 0.004 %) and the untreated control field (once daily application immediately after irradiation with 2.0 MED and again at 24 and 48 hour; measurements at 24, 48 and 72 hour after irradiation.

that the irritant potential is correlated with clinical efficacy (218). For these products, it is conceivable that measurement of irritant potential could be used as a surrogate marker of efficacy. Objective assessment of irritant potential can be easily done by measurement of TEWL. It remains to be seen to what extent this can be applied to future bioequivalence studies.

There are a number of bioengineering methods that can be used in innovative designs to test bioequivalence. In any design, it is critical to identify the correlation of the chosen methods with the clinical endpoint. While at present there are clear benefits to be gained from the use of these methods as screening tools, acceptance of innovative designs as pivotal studies depends on future standardization and validation of the methodologies.

REFERENCES

1. Finnin BC, Morgan TM. Transdermal penetration enhancers: Limitations and potential. J Pharm Sci 1999; 88:955–958.
2. Hadgraft J. Dermal and transdermal delivery. In: Rathbone MJ, Hadgraft J, Roberts MS, eds. Modified-Released Drug Delivery Technology. New York, USA: Marcel Dekker Inc., 2003:471–480.
3. Guy RH. Current status and future prospects of transdermal drug delivery. Pharm Res 1996; 13:1765–1769.
4. Naik A, Kalia YN, Guy RH. Transdermal drug delivery: Overcoming the skin's barrier function. Pharm Sci Tech Today 2000; 3:318–326.

5. Tettey-Amlalo RNO. In Vitro Release of Ketoprofen from Proprietary and Extemporaneously Manufactured Gels. Grahamstown, South Africa: Rhodes University, 2005.
6. Walters KA, Brain KR. Dermatological formulations and transdermal systems. In: Walters KA, ed. Dermatological and Transdermal Formulations. New York, USA: Marcel Dekker, Inc., 2002:319–399.
7. Shah VP, Flynn GL, Guy RH, et al. In vivo percutaneous penetration/absorption. Int J Pharm 1991; 74:1–8.
8. Guy RH, Guy AH, Maibach HI, et al. The bioavailability of dermatological and other topically administered drugs. Pharm Res 1986; 3:253–262.
9. Surber C, Davies AF. Bioavailability and bioequivalence of dermatological formulations. In: Walters KA, ed. Dermatological and Transdermal Formulations. New York, USA: Marcel Dekker Inc. 2002:401–497.
10. Shah VP, Midha KK, Dighe S, et al. Analytical methods validation: Bioavailability, bioequivalence and pharmacokinetic studies. Pharm Res 1992; 9:588–592.
11. Shah VP, Behl CR, Flynn GL, et al. Principles and criteria in the development and optimization of topical therapeutic products. Int J Pharm 1992; 82:21–28.
12. Saudi Arabia, Saudi Food and Drug Authorities. Bioequivalence Requirements Guidelines. Riyadh, Saudi Arabia: SFDA, 2005.
13. The European Agency for the Evaluation of Medicinal Products. Note for Guidance on the Investigation of Bioavailability and Bioequivalence. London, United Kingdom: The European Agency for the Evaluation of Medicinal Products, 2000.
14. Kanfer I, Shargel L. Introduction-bioequivalence issues. In: Kanfer I, Shargel L, eds. Generic Drug Product Development Bioequivalence Issues. New York, USA: Informa Healthcare, 2008:1–14.
15. Walker RB, Verbeeck R, Kanfer I. Pharmaceutical alternatives: Considerations for generic substituition. In: Kanfer I, Shargel L, eds. Generic Drug Product Development: Bioequivalence Issues. New York, USA: Informa Healthcare Inc., 2008: 31–45.
16. Walker BR, Kanfer I, Skinner MF. Bioequivalence assessment of generic products: An innovative South African approach. Clin Res Regul Aff 2006; 23:11–20.
17. Conner DP, Davit BM. Bioequivalence and drug product assessment, in vivo. In: Shargel L, Kanfer I, eds. Generic Drug Product Development Solid Oral Dosage Forms. New York, USA: Marcel Dekker, Inc., 2005:227–255.
18. World Health Organisation (WHO). Multisource (generic) pharmaceutical products: Guidelines on registration requirements to establish interchangeability. Geneva, Switzerland: World Health Organisation, 2008.
19. Shah VP. IV-IVC for topically applied preparations—A critical evaluation. Eur J Pharm Biopharm 2005; 60:309–314.
20. Mathy F, Ntivunwa D, Verbeeck R, et al. Fluconazole distribution in rat dermis following intravenous and topical application: A microdialysis study. J Pharm Sci 2005; 94:770–780.
21. Pellanda C, Ottiker E, Strub C, et al. Topical bioavailability of triamcinolone acetonide: Effect of dose and application frequency. Arch Dermatol Res 2006; 298: 221–230.
22. Pershing LK, Nelson J, Corlett JL, et al. Assessment of dermatopharmacokinetic approach in the bioequivalence determination of topical tretinoin gel products. J Am Acad Dermatol 2003; 48:740–751.
23. Wester RC, Maibach HI. In vivo methods for percutaneous absorption measurements. In: Bronauugh RL, Maibach HI, eds. Percutaneous Absoprtion: Drugs-Cosmetics-Mechanisms-Methodology, 3rd ed. New York, USA: Marcel Dekker Inc., 1999:215–227.
24. Stanos SP. Topical agents for the management of musculoskeletal pain. J Pain Symptom Manage 2007; 33:342–355.
25. Herkenne C, Naik A, Kalia YN, et al. Ibuprofen transport into and through skin from topical formulations: In vitro-in vivo comparison. J Invest Dermatol 2007; 127: 135–142.

26. Morganti P, Ruocco E, Wolf R, et al. Percutaneous absorption and delivery systems. Clin Dermatol 2001; 19:489–501.
27. Schnetz E, Fartasch M. Microdialysis for the evaluation of penetration through the human skin barrier—A promising tool for future research? Eur J Pharm Sci 2001; 12:165–174.
28. Kalia YN, Alberti I, Sekkat N, et al. Normalisation of stratum corneum barrier function and transepidermal water loss in vivo. Eur J Pharm Sci 2000; 12:165–174.
29. Hadgraft J, Whitefield M, Rosher PH. Skin penetration of topical formulations of ibuprofen 5%: An in vitro comparative study. Skin Pharmacol Appl Skin Physiol 2003; 16:137–142.
30. Morgan CJ, Renwick AG, Friedmann PS. The role of stratum corneum and dermal microvascular perfusion in penetration and tissue levels of water-soluble drugs investigated by microdialysis. Brit J Dermatol 2003; 148:434–443.
31. Cross SE, Anderson C, Roberts MS. Topical penetration of commercial salicylate esters and salts using human isolated skin and clinical microdialysis studies. Br J Clin Pharmacol 1998; 46:29–35.
32. Shah VP. Progress in methodologies for evaluating bioequivalence of topical formulations. Am J Clin Dermatol 2001; 2:275–280.
33. FDA. Guidance for industry: Topical dermatological corticosteroids: In vivo bioequivalence. Maryland, USA: Center for Drug Evaluation, Food and Drug Administration, 1995.
34. Issar M, Stark JG, Shargel L. Pharmacodynamic measurements for determination of bioequivalence. In: Kanfer I, Shargel L, eds. Generic Drug Product Development: Bioequivalence Issues. New York, USA: Informa Healthcare Inc., 2008:47–70.
35. Hendy C. Bioequivalence using clinical endpoint studies. In: Kanfer I, Shargel L, eds. Generic Drug Product Development: Bioequivalence Issues. New York, USA: Informa Healthcare Inc., 2008:71–96.
36. Shah VP, Flynn GL, Yacobi A, et al. Bioequivalence of topical dermatological dosage forms—Methods of evaluation of bioequivalence. Pharm Res 1998; 15:167–171.
37. Shah VP, Flynn GL, Yacobi A, et al. Bioequivalence of topical dermatological dosage forms—Methods of evaluation of bioequivalence. Skin Pharmacol Appl Skin Physiol 1998; 11:117–124.
38. Jackson AJ. Determination of in vivo bioequivalence. Pharm Res 2002; 19:227–228.
39. Schuirmann DJ. A comparison of the two one-sided tests procedure and the power approach for assessing the equivalence of average bioavailability. J Pharmacokinet Biopharm 1987; 15:657–680.
40. U.S. Food and Drug Administration (FDA). Guidance for Industry, Bioanalytical Method Validation. Rockville, MD, USA: US Food and Drug Administration, 2001.
41. Benfeldt E. In Vivo Microdialysis for the Investigation of Drug Levels in the Dermis and the Effect of Barrier Perturbation on Cutaneous Drug Penetration [Ph.D. Thesis]. Stockholm, Sweden: Scandinavian University Press, 1999.
42. Chattaraj SC, Kanfer I. 'The insertion cell': A novel approach to monitor drug release from semi-solid dosage forms. Int J Pharm 1996; 133:59–63.
43. Chattaraj SC, Kanfer I. Release of acyclovir from semi-solid dosage forms: A semi-automated procedure using a simple plexiglass flow-through cell. Int J Pharm 1995; 125:215–222.
44. Chattaraj SC, Kanfer I. A simple difuusion cell to monitor drug release from semi-solid dosage forms. Int J Pharm 1995; 120:119–124.
45. Liebenberg W, Engelbrecht E, Wessels A, et al. A comparative study of the release of active ingredients from semisolid consmeceuticals measured with Franz, Enhancer or Flow-through cell diffusion apparatus. J Food Drug Anal 2004; 12:19–28.
46. Fares HM, Zatz JL. Measurements of drug release from topical gels using two types of apparatus. Pharm Technol 1995; 19:52–58.
47. Hadgraft J, Guy RH. Feasibility assessment in topical and transdermal delivery: Mathematical models and in vitro studies. In: Guy RH, Hadgraft J, eds. Transdermal Drug Delivery, 2nd ed. New York, USA: Marcel Dekker, Inc., 2003:1–23.

48. Sasongko L, Williams KM, Day RO, et al. Human subcutaneous tissue distribution of fluconazole: Comparison of microdialysis and suction blister techniques. Br J Clin Pharmacol 2003; 56:551–561.
49. Groth L, Ortiz PG, Benfeldt E. Microdialysis methodology for sampling in the skin. In: Serup J, Jemec GBE, Grove GL, eds. Handbook of Non-invasive Methods and the Skin, 2nd ed. Florida, USA: CRC Press Taylor and Francis Group, 2006: 443–454.
50. Groth L. Cutaneous Microdialysis: Methodology and Validation [Ph.D. Thesis]. Stockholm, Sweden: Scandinavian University Press, 1996.
51. Plock N, Kloft C. Microdialysis: Theoretical background and recent implementation in applied life-sciences. Eur J Pharm Sci 2005; 25:1–24.
52. Chaurasia CS, Müller M, Bashaw ED, et al. AAPS-FDA Workshop White Paper: Microdialysis principles, applications and regulatory perspectives report from the joint AAPS-FDA workshop, November 4–5, 2005, Nashville, TN. Pharm Res 2007; 24:1014–1025.
53. Müller M. Microdialysis in clinical drug delivery studies. Adv Drug Deliver Rev 2000; 45:255–269.
54. Joukhadar C, Müller M. Microdialysis: Current applications in clinical pharmacokinetic studies and its potential role in the future. Clin Pharmacokinet 2005; 44: 895–913.
55. Li Y, Peris J, Zhong L, et al. Microdialysis as a tool in local pharmacodynamics. AAPS J 2006; 8:E222–E235.
56. Abrahamsson P, Winsö O. An assessment of calibration and performance of the microdialysis system. J Pharm Biomed Anal 2005; 39:730–734.
57. Leis S, Drenkhahn S, Schick C, et al. Catecholamine release in human skin—A microdialysis study. Exp Neurol 2004; 188:86–93.
58. Schwalbe O, Buerger C, Plock N, et al. Urea as an endogenous surrogate in human microdialysis to determine relative recovery of drugs: Analytics and applications. J Pharm Biomed Anal 2006; 41:233–239.
59. Mathy F, Denet AR, Vroman B, et al. In vivo tolerance assessment of skin after insertion of subcutaneous and cutaneous microdialysis probes in the rat. Skin Pharmacol Appl Skin Physiol 2003; 16:18–27.
60. Peña A, Liu P, Derendorf H. Microdialysis in peripheral tissues. Adv Drug Deliver Rev 2000; 45:189–216.
61. Zhao Y, Liang X, Lunte CE. Comparison of recovery and delivery in vitro for calibration of microdialysis probes. Anal Chim Acta 1995; 316:403–410.
62. Brunner M, Derendorf H. Clinical microdialysis: Current applications and potential use in drug development. Trends Anal Chem 2006; 25:674–680.
63. Ward KW, Medina SJ, Portelli ST, et al. Enhancement of in vitro and in vivo microdialysis recovery of SB-265123 using Intralipid® and Encapsin® as perfusates. Biopharm Drug Dispos 2003; 24:17–25.
64. Fulzele SV, Babu RJ, Ahaghotu E, et al. Estimation of proinflammatory biomarkers of skin irritation by dermal microdialysis following exposure with irritant chemicals. Toxicol 2007; 237:77–88.
65. Benfeldt E, Hansen SH, Vølund A, et al. Bioequivalence of topical formulations in humans: Evaluation by dermal microdialysis sampling and the dermatopharmacokinetic method. J Invest Dermatol 2007; 127:170–178.
66. Müller M, Schmid R, Wagner O, et al. In vivo characterization of transdermal drug transport. J Control Release 1995; 37:49–57.
67. Ault JM, Riley CM, Meltzer NM, et al. Dermal microdialysis sampling in vivo. Pharm Res 1994; 11:1631–1639.
68. Benfeldt E, Groth L. Feasibility of measuring lipophilic or protein-bound drugs in the dermis by in vivo microdialysis after topical or systemic drug administration. Acta Derm Venereol 1998; 78:274–278.
69. Müller M, Mascher H, Kikuta C, et al. Diclofenac concentrations in defined tissue layers after topical administration. Clin Pharmacol Ther 1997; 62:293–299.

70. Anderson C, Andersson T, Molander M. Ethanol absorption across human skin measured by in vivo microdialysis technique. Acta Derm Venereol 1991; 71: 389–393.
71. Mathy F, Vroman B, Ntivunwa D, et al. On-line determination of fluconazole in blood and dermal rat microdialysates by microbore high-performance liquid chromatography. J Chromatogr B Analyt Technol Biomed Life Sci 2003; 787:323–331.
72. Mathy F, Lombry C, Verbeeck R, et al. Study of the percutaneous penetration of flurbiprofen by cutaneous and subcutaneous microdialysis after iontophoretic delivery in rat. J Pharm Sci 2005; 94:144–152.
73. Tegeder I, Muth-Selbach U, Lotsch J, et al. Application of microdialysis for the determination of muscle and subcutaneous tissue concentrations after oral and topical ibuprofen administration. Clin Pharmacol Ther 1999; 65:357–368.
74. Tegeder I, Lötsch J, Kinzig-Schippers M, et al. Comparison of tissue concentrations after intramuscular and topical administration of ketoprofen. Pharm Res 2001; 18:980–986.
75. Kreilgaard M. Assessment of cutaneous drug delivery using microdialysis. Adv Drug Deliver Rev 2002; 54:S99–S121.
76. Hegemann L, Forstinger C, Partsch B, et al. Microdialysis in cutaneous pharmacology: Kinetic analysis of transdermally delivered nicotine. J Invest Dermatol 1995; 104:839–843.
77. Benfeldt E, Serup J, Menné T. Effect of barrier perturbation on cutaneous salicylic acid penetration in human: In vivo pharmacokinetics using microdialysis and non-invasive quantification of barrier function. Brit J Dermatol 1999; 140:739–748.
78. Simonsen L, Petersen MB, Groth L. In vivo penetration of salicylic compounds in hairless rats. Eur J Pharm Sci 2002; 17:95–104.
79. Simonsen L, Jørgensen A, Benfeldt E, et al. Differentiated in vivo skin penetration of salicylic compounds in hairless rats measured by cutaneous microdialysis. Eur J Pharm Sci 2004; 21:379–388.
80. Klede M, Schmitz H, Göen T, et al. Transcutaneous penetration of toluene in rat skin a microdialysis study. Exp Dermatol 2005; 14:103–108.
81. Benfeldt E, Serup J. Effect of barrier perturbation on cutaneous penetration of salicylic acid in hairless rats: In vivo pharmacokinetics using microdialysis and non-invasive quantification of barrier function. Arch Dermatol Res 1999; 291:517–526.
82. Lönnroth P, Strindberg L. Validation of the 'internal reference technique' for calibrating microdialysis catheters *in situ*. Acta Physiol Scand 1995; 153:375–380.
83. Stenken JA, Chen R, Yuan X. Influence of geometry and equilibrium chemistry on relative recovery during enhanced microdialysis. Anal Chim Acta 2001; 436: 21–29.
84. Stenken JA. Methods and issues in microdialysis calibration. Anal Chim Acta 1999; 379:337–358.
85. de Lange ECM, de Boer AG, Breimer DD. Methodological issues in microdialysis sampling for pharmacokinetic studies. Adv Drug Deliver Rev 2000; 45:125–148.
86. Davies MI, Cooper JD, Desmond SS, et al. Analytical considerations for microdialysis sampling. Adv Drug Deliver Rev 2000; 45:169–188.
87. Hsiao JK, Ball BA, Morrison PF, et al. Effects of different semipermeable membranes on in vitro and in vivo performance of microdialysis probes. J Neurochem 1990; 54:1449–1452.
88. Klimowicz A, Bielecka-Grzela S, Groth L, et al. Use of an intraluminal guide wire in linear microdialysis probes: Effect on recovery? Skin Res Technol 2004; 10:104–108.
89. Chamboko BV. Evaluation of the safety and efficacy of topical mometasone furoate formulations [MSc. Thesis]. Grahamstown, South Africa: Rhodes University, 2007.
90. Brunner M, Langer O. Microdialysis versus other techniques for the clinical assessment of in vivo tissue drug distribution. AAPS J 2006; 8:E263–E271.
91. Song Y, Lunte CE. Comparison of calibration by delivery versus no net flux for quantitative in vivo microdialysis sampling. Anal Chim Acta 1999; 379:251–262.

92. Steuerwald AJ, Villeneuve JD, Sun L, et al. In vitro characterization of an *in situ* microdialysis sampling assay for elastase activity detection. J Pharm Biomed Anal 2006; 40:1041–1047.
93. Ståhle L. On mathematical models of microdialysis: Geometry, steady-state models, recovery and probe radius. Adv Drug Deliver Rev 2000; 45:149–167.
94. Weiss DJ, Lunte CE, Lunte SM. In vivo microdialysis as a tool for monitoring pharmacokinetics. Trends Anal Chem 2000; 19:606–616.
95. Khramov AN, Stenken JA. Enhanced microdialysis extraction efficiency of ibuprofen in vitro by facilitated transport with β-cyclodextrin. Anal Chem 1999; 71:1257–1264.
96. Trickler WJ, Miller DW. Use of osmotic agents in microdialysis studies to improve the recovery of macromolecules. J Pharm Sci 2003; 92:1419–1427.
97. Lorentzen H, Kallehave F, Kolmos HJ, et al. Gentamicin concentrations in human subcutaneous tissue. Antimicrob Agents Chemother 1996; 40:1785–1789.
98. Tettey-Amlalo RNO. Application of Dermal Microdialysis and Tape Stripping Methods to determine the Bioavailability and/or Bioequivalence of Topical Ketoprofen Formulations. Grahamstown, South Africa: Rhodes University, 2008.
99. Verbeeck R. Blood microdialysis in pharmacokinetic and drug metabolism studies. Adv Drug Deliver Rev 2000; 45:217–228.
100. Tong S, Yuan F. An equivalent length model of microdialysis sampling. J Pharmceut Biomed Anal 2002; 28:269–278.
101. McCleverty D, Lyons R, Henry B. Microdialysis sampling and the clinical determination of topical dermal bioequivalence. Int J Pharm 2006; 308:1–7.
102. Tettey-Amlalo RNO, Kanfer I, Skinner MF, et al. Application of dermal microdialysis for the evaluation of bioequivalence of a ketoprofen topical gel. Eur J Pharm Sci 2009; 36:219–225.
103. Tettey-Amlalo RNO, Kanfer I. Rapid UPLC-MS/MS method for the determination of ketoprofen in human dermal microdialysis samples. J Pharm Biomed Anal 2009; 50:580–586.
104. Brain KR, Walters KA, Watkinson AC. Methods for Studying Percutaneous Absorption. In: Walters KA, ed. New York, USA: Marcel Dekker, Inc., 2002:197–269.
105. Rougier A, Lotte C, Dupuis D. An original predictive method for in vivo percutaneous absorption studies. J Soc Cos Chem 1987; 38:397–417.
106. Rougier A, Lotte C, Maibach HI. The measurement of the stratum corneum reservoir. A predictive method for in vivo percutaneous absorption studies: Influence of application time. J Invest Dermatol 1984; 83:460–462.
107. Rougier A, Dupuis D, Lotte C, et al. In vivo correlation between stratum corneum reservoir function and percutaneous absorption. J Invest Dermatol 1983; 81: 275–278.
108. Surber C, Schwarb FP, Smith EW. Tape-stripping technique. J Toxicol Cutaneous Ocul 2001; 20:461–474.
109. Löffler H, Dreher F, Maibach HI. Stratum corneum adhesive tape stripping: Influence of anatomical site, application pressure, duration and removal. Brit J Dermatol 2004; 151:746–752.
110. Surber C, Wilhelm KP, Hori M, et al. Optimization of topical therapy: Partitioning of drugs into stratum corneum. Pharm Res 1990; 7:1320–1324.
111. Alberti I, Kalia YN, Naik A, et al. In vivo assessment of enhanced topical delivery of terbinafine to human stratum corneum. J Control Release 2001; 71:319–327.
112. Cal K, Krzyzaniak M. Stratum corneum absorption and retention of linalool and terpinen-4-ol applied as gel or oily solution in humans. J Dermatol Sci 2006; 42:265–267.
113. Alberti I, Kalia YN, Naik A, et al. Effect of ethanol and isopropyl myristate on the availability of topical terbinafine in human stratum corneum, in vivo. Int J Pharm 2001; 219:11–19.
114. Lodén M, Åkerström U, Lindahl K, et al. Bioequivalence determination of topical ketoprofen using a dermatopharmacokinetic approach and excised skin penetration. Int J Pharm 2004; 284:23–30.

115. Pershing LK, Silver BS, Krueger GG, et al. Feasibility of measuring the bioavailability of topical betamethasone dipropionate in commercial formulations using drug content in skin and a blanching bioassay. Pharm Res 1992; 9:45–51.

116. Pershing LK, Lambert LD, Shah VP, et al. Variability and correlation of chromameter and tape-stripping methods with the visual skin blanching assay in the quantitative assessment of topical 0.05% betamethasone dipropionate bioavailability in humans. Int J Pharm 1992; 86:201–210.

117. Pershing LK, Bakhtian S, Poncelet CE, et al. Comparison of skin stripping, in vitro release and skin blanching response methods to measure dose response and similarity of triamcinolone acetonide cream strengths from two manufactured sources. J Pharm Sci 2002; 91:1312–1323.

118. Pershing LK, Reilly CA, Corlett JL, et al. Effects of vehicle on the uptake and elimination kinetics of capsaicinoids in human skin in vivo. Toxicol Appl Pharmacol 2004; 200:73–81.

119. Tsai J, Weiner ND, Flynn GL, et al. Properties of adhesive tapes used for stratum corneum stripping. Int J Pharm 1991; 72:227–231.

120. Zhai H, Poblete N, Maibach HI. Stripped skin model to predict irritation potential of topical agents in vivo humans. Int J Dermatol 1998; 37:386–389.

121. Bashir SJ, Chew A, Anigbogu A, et al. Physical and physiological effects of stratum corneum tape stripping. Skin Res Technol 2001; 7:40–48.

122. Curdy C, Naik A, Kalia YN, et al. Non-invasive assessment of the effect of formulation excipients on stratum corneum barrier function in vivo. Int J Pharm 2004; 271: 251–256.

123. Søeborg T, Basse LH, Halling-Sørensen B. Risk assessment of topically applied products. Toxicology 2007; 236:140–148.

124. Berardesca E, Pirot F, Singh M, et al. Differences in stratum corneum pH gradient when comparing white Caucasian and black African-American skin. Brit J Dermatol 1998; 139:855–857.

125. Wagner H, Kostka K, Lehr C, et al. pH profiles in human skin: Influence of two in vitro test systems for drug delivery testing. Eur J Pharm Biopharm 2003; 55:57–65.

126. U.S. Food and Drug Administration (FDA). Guidance for Industry, Topical Dermatological Drug Product NDAs and ANDAs—In Vivo Bioavailability, Bioequivalence, In Vitro Release and Associated Studies, 1998.

127. Herkenne C, Alberti I, Naik A, et al. In vivo methods for the assessment of topical drug bioavailability. Pharm Res 2008; 25:87–103.

128. Weigmann HJ, Ulrich J, Schanzer S, et al. Comparison of transepidermal water loss and spectroscopic absorbance to quantify changes of the stratum corneum after tape stripping. Skin Pharmacol Physiol 2005; 18:180–185.

129. Tokumura F, Ohyama K, Fujisawa H, et al. Seasonal variation in adhesive tape stripping of the skin. Skin Res Technol 1999; 5:208–212.

130. Tokumura F, Yoshiura Y, Homma T, et al. Regional differences in adhesive tape stripping of human skin. Skin Res Technol 2006; 12:178–182.

131. Choi MJ, Zhai H, Löffler H, et al. Effect of tape stripping on percutaneous penetration and topical vaccination. Exog Dermatol 2003; 2:262–269.

132. Zhai H, Dika E, Goldovsky M, et al. Tape-stripping method in man: Comparison of evaporimetric methods. Skin Res Technol 2007; 13:207–210.

133. Dreher F, Modjtahedi BS, Modjtahedi SP, et al. Quantification of stratum corneum removal by adhesive tape stripping by total protein assay in 96-well microplates. Skin Res Technol 2005; 11:97–101.

134. Teichmann A, Jacobi U, Waibler E, et al. An in vivo model o evaluate the efficacy of barrier creams on the level of skin penetration of chemicals. Contact Dermat 2006; 54:3–13.

135. Jacobi U, Weigmann HJ, Ulrich J, et al. Estimation of the relative stratum corneum amount removed by tape stripping. Skin Res Technol 2005; 11:91–96.

136. Levin J, Maibach HI. The correlation between transepidermal water loss and percutaneous absorption: An overview. J Control Release 2004; 103:291–299.

137. Wiedersberg S, Leopold CS, Guy RH. Bioavailability and bioequivalence of topical glucocorticoids. Eur J Pharm Biopharm 2008; 68:453–466.

138. Surber C, Smith EW, Schwarb FP, et al. Drug concentration in the skin. In: Bronaugh RL, Maibach HI, eds. Percutaneous Absorption: Drugs-Cosmetics-Mechanism-Methodology, 3rd ed. New York, USA: Marcel Dekker Inc., 1999:347–374.

139. Afifi T, de Gannes G, Huang C, et al. Topical therapies for psoriasis: Evidence-based review. Can Fam Physician 2005; 51:519–525.

140. Ahluwalia A. Topical glucocorticosteroids and the skin-mechanisms of action: An update. Mediat Inflamm 1998; 7:183–193.

141. Horii KA, Simon SD, Liu DY, et al. Atopic dermatitis in children in the United States, 1997–2004: Visit trends, patient and provider characteristics, and prescribing patterns. Pediatrics 2007; 120:e527–e534.

142. Bos JD, Spuls PI. Topical treatments in psoriasis: Today and tomorrow. Clin Dermatol 2008; 26:432–437.

143. Krakowski AC, Dohil MA. Topical therapy in pediatric atopic dermatitis. Semin Cutan Med Surg 2008; 27:161–167.

144. McKenzie AW, Stoughton RB. Method for comparing percutaneous absorption of steroids. Arch Dermatol 1962; 86:608–610.

145. Haigh JM, Kanfer I. Assessment of topical corticosteroids preparations: The human skin blanching assay. Int J Pharm 1984; 19:254–262.

146. Scott A, Kalz F. The effect of the topical application of corticotrophin, hydrocortisone, and flurocortisone on the process of cutaneous Inflammation. J Invest Dermatol 1956; 26:361–378.

147. Jacobi H, Kadner H, Pinzer B. The suppressive effect of topical corticosteroids in UV erythema. Dermatol Monatsschr 1977; 163:970–974.

148. Woodbury RA, Kligman LH, Woodbury MJ, et al. Rapid assay of the anti-inflammatory activity of topical corticosteroids by inhibition of a UVA-induced neutrophil infiltration in hairless mouse skin. I. The assay and its sensitivity. Acta Derm Venereol 1994; 74:15–17.

149. Witkowski JA, Kligman AM. A screening test for anti-inflammatory activity using human skin. J Invest Dermatol 1959; 32:481–483.

150. Ortega E, Rodriguez C, Burdick K, et al. The croton oil inflammation suppression assay as a measure of topical corticosteroid potency. Acta Derm Venereol Suppl (Stockh) 1971; 52:95–97.

151. Kaidbey KH, Kligman AM. Assay of topical corticosteroids by suppression of experimental inflammation in humans. J Invest Dermatol 1974; 63:292–297.

152. Dumas KJ, Scholtz JR. The psoriasis bio-assay for topical corticosteroid activity. Acta Derm Venereol 1972; 52:43–48.

153. Cornell RC. Clinical trials of topical corticosteroids in psoriasis: Correlations with the vasoconstrictor assay Int J Dermatol 1992; 31:38–40.

154. Wells GC. The effect of hydrocortisone on standardized skin-surface trauma. Br J Dermatol 1956; 69:11–18.

155. Carr RD, Wieland RG. Corticosteroid reservoir in the stratum corneum. Arch Dermatol 1966; 94:81–84.

156. Polano MK, Hagenouw JRB, Richter JR. Advances in topical corticosteroid therapy. Dermatologica 1976; 152:1–276.

157. Wendt H, Frosch PJ. Clinico-Pharmacological Models for the Assay of Topical Corticoids. Switzerland: S. Karger AG, 1982.

158. Weigmann HJ, Lademann J, von Pelchrzim R, et al. Bioavailability of clobetasol propionate—Quantification of drug concentrations in the stratum corneum by dermatopharmacokinetics using tape stripping. Skin Pharmacol Appl 1999; 12: 46–53.

159. Barry BW, Woodford R. Activity and bioavailability of topical corticosteroids: In vivo/in vitro correlations for the vasoconstrictor test. J Clin Pharm 1978; 3:43—65.

160. Baker H, Sattar HA. The assessment of four new fluocortolone analogues by a modified vasoconstriction assay. Brit J Dermatol 1968; 80:46–53.

161. Coldman MF, Lockerbie L, Laws EA. The evaluation of several topical corticosteroid preparations in the blanching test. Br J Dermatol 1971; 85:381–387.
162. McKenzie AW. Comparison of steroid induced vasoconstriction. Br J Dermatol 1966; 78:182–183.
163. Poulsen BJ, Burdick K, Bessler S. Paired comparison vasoconstrictor assays. Arch Dermatol 1974; 109:367–371.
164. Heseltine WW, McGilchrist JM, Gartside R. Comparative vasocontrictor activities of corticosteroids applied topically. Br J Dermatol 1964; 76:71–73.
165. Clanachan I, Devitt HG, Foreman MI, et al. The human vasoconstrictor assay for topical steroids. J Pharmacol Method 1980; 4:209–220.
166. Pepler AF, Woodford R, Morrison JC. The influence of vehicle compostion on the vasoconstrictor activity of betamethasone-17-valerate. Br J Dermatol 1971; 85:171–176.
167. Woodford R, Barry BS. The placebo response to white soft paraffin and propylene glycol in the skin blanching test. Br J Dermatol 1973; 89:53–59.
168. Fritz I, Levine R. Action of adrenal cortical steroids and nor-epinephrine on vascular responses of stress in adrenalectomized rats. Am J Physiol 1951; 165:456–465.
169. Frank L, Rapp Y, Biro L, et al. Inflammation mediators and the inflammatory reaction. Arch Dermatol 1964; 89:55–67.
170. Solomon LM, Wentzel E, Greenberg MS. Studies in the mechanism of steroid vasoconstriction. J Invest Dermatol 1965; 44:129–131.
171. Friedman SM, Friedman CL. Ionic basis of vascular response to vasoactive substances Can Med Assoc J 1964; 90:167–173.
172. Altura BM. Role of glucocorticoids in local regulation of blood flow. Am J Physiol 1966; 211:1393–1397.
173. Juhlin L, Michaëlsson G. Cutaneous vascular reactions to prostaglandins in healthy subjects and in patients with urticaria and atopic dermatitis. Acta Derm Venereol 1969; 49:251–261.
174. Haynes RC. Biochemical mechanisms of steroid effects. In: Azarnoff DL, editor. Steroid Therapy. Philadelphia, PA: WB Saunders Company, 1975:15–26.
175. Edwards PM, Jacquemyns CR, Rousseau GG. Melanosome aggregation of corticosteroids: Evidence for a novel type steroid action. J Steroid Biochem 1981; 15: 17–23.
176. Hengge UR, Ruzicka T, Schwartz RA, et al. Adverse effects of topical glucocorticosteroids. J Am Acad Dermatol 2006; 54:1–15.
177. Christophers E, Schöpf E, Kligman AM, et al. Topical corticosteroid therapy: A novel approach to safer drugs. New York, USA: Raven Press Ltd., 1988.
178. Coleman GL, Kanfer I, Haigh JM. Comparative blanching activities of proprietary diflucortolone valerate topical preparations. Dermatologica 1978; 156:224–230.
179. Coleman GL, Magnus AD, Haigh JM, et al. Comparative Blanching activities of locally manufactured proprietary fluocinolone acetonide topical preparations. SA Med J 1979; 56:447–449.
180. Magnus AD, Haigh JM, Kanfer I. Assessment of some variables affecting the blanching activity of betamethasone 17-valerate cream. Dermatologica 1980; 160: 321–327.
181. Clarys P, Wets L, Barel A, et al. The skin blanching assay with halcinonide: Influence of halcinonide concentration and application time. J Eur Acad Dermatol Venereol 1995; 5:250–257.
182. Montenegro L, Ademola JI, Bonina FP, et al. Effect of application time of betamethasone-17 valerate 0.1% cream on skin blanching and *stratum corneum* drug concentration. Int J Pharm 1996; 140:51–60.
183. Piérard GE. EEMCO guidance for the assessment of skin colour. J Env Eng Div Asce 1998; 10:1–11.
184. Schwarb FP, Smith EW, Haigh JM, et al. Analysis of chromameter results obtained from corticosteroid-induced skin blanching assay: Comparison of visual chromameter data. Eur J Pharm 1999; 47:261–267.

185. Haigh JM, Smith EW. Topical corticosteroid-induced skin blanching measurement: Eye or instrument? Letter to the editor. Arch Dermatol 1991; 127:1065–1065.
186. Skelly JP, Shah VP, Peck CC. Topical corticosteroid-induced skin blanching measurement: Eye or instrument? Response to the editor. Arch Dermatol 1991; 127: 1065–1065.
187. Barry BW, Woodford R. Comparative bio-availability of proprietary topical corticosteroid preparations; vasoconstrictor assays on thirty creams and gels. Br J Dermatol 1974; 91:323–338.
188. Barry BW, Woodford R. Bioavailability and activity of betamethasone 17-benzoate in gel and cream formulations: Comparison with proprietary topical corticosteroid preparation in the vasoconstrictor assay. Curr Ther Res 1974; 16:338–345.
189. Király K, Soós GY. Objective measurement of topically applied corticosteroids. Dermatologica 1976; 152:133–137.
190. Gras PW, Bason ML, Esteves LA. Evaluation of a portable colour meter for assessment of the colour of milled rice. J Stored Prod Res 1990; 26:71–75.
191. Taylor S, Westerhof W, Im S, et al. Noninvasive techniques for the evaluation of skin color. J Am Acad Dermatol 2006; 54:s282–s290.
192. Waring MJ, Monger L, Hollingsbee DA, et al. Assessment of corticosteroid-induced skin blanching: Evaluation of the Minolta Chromameter CR200. Int J Pharm 1993; 94:211–212.
193. Chan SY, Po ALW. Quantitative skin blanching assay of corticosteroid creams using tristimulus colour analysis. J Pharm Pharmacol 1992; 44:371–378.
194. Fullerton A, Fischer T, Lahti A, et al. Guidelines for measurement of skin colour and erythema: A report from the Standardization Group of the European Society of Contact Dermatitis. Contact Dermatitis 1996; 35:1–10.
195. Au WL, Skinner M, Kanfer I. Bioequivalence assessment of topical clobetasol propionate products using visual and chromametric assessment of skin blanching. J Pharm Pharm Sci 2008; 11:147–153.
196. U.S. Food and Drug Administration (FDA). FDA, Guidance of industry: Bioavailability and bioequivalence studies for orally administered drug products—General considerations. MD, USA: Centre for Drug Evaluation, Food and Drug Administration, 2000.
197. Diletti E, Hauschke D, Steinijans VW. Sample size determination for bioequivalence assessment by means of confidence intervals. Int J Clin Pharmacol Ther Toxicol 1991; 29:1–8.
198. Gaßmüller J, Klinger B, Levy J. The ultrasound-erythema index (USE-Index) for monitoring of therapeutic effects on the psoriatic plaque by 20 MHz ultrasound and colorimetry. Regional meeting of the International Society for Bioengineering and the Skin, 1993.
199. Willers CP, Frase T, Schmidt A. The USE- (Ultrasound-Erythema-) Index in antipsoriatic testing. 20th World Congress of Dermatology, Paris, France, 2002.
200. Bangha E, Elsner P. Evaluation of topical antipsoriatic treatment by chromametry, visiometry and 20-Mhz ultrasound in the psoriasis plaque test. Skin Pharmacol 1996; 9:298–306.
201. Remitz A, Reitamo S, Erkko P. Tacrolimus ointment improves psoriasis in a microplaque assay. Br J Dermatol 1999; 141:103–107.
202. Wolff HH, Kreusch JF, Wilhelm KP. The psoriasis plaque test and topical corticosteroids: Evaluation by computerized laser profilometry. Curr Probl Dermatol 1993; 21:107–113.
203. Consensus Report of the European Task Force on Atopic Dermatitis. Severity scoring of atopic dermatitis: The SCORAD index. Dermatology 1993; 186:23–31.
204. Hanifin JM, Thurston M, Omoto M. The eczema area and severity index (EASI): Assessment of reliability in atopic dermatitis. EASI Evaluator Group. Exp Dermatol 2001; 10:11–18.
205. Shahidullah M, Raffle EJ, Rimmer AR. Transepidermal water loss in patients with dermatitis. Br J Dermatol 1969; 81:722–730.

206. Seidnari S, Giusti G. Objective assessment of the skin of children affected by atopic dermatitis: A study of pH, capacitance and TEWL in eczematous and clinically normal skin. Acta Derm Venereol 1995; 75:429–433.
207. Kristiina A. Improvement of skin barrier function during treatment of atopic dermatitis. J Am Acad Dermatol 1995; 33:969–972.
208. Chamlain SL, Kao J, Frieden IJ. Ceramide-dominant barrier repair lipids alleviate childhood atopic dermatitis: Changes in barrier function provide a sensitive indicator of disease activity. J Am Acad Dermatol 2002; 47:198–208.
209. Kim D, Park J, Na G. Correlation of clinical features and skin barrier function in adolescent and adult patients with atopic dermatitis. Int J Dermatol 2006; 45:698–701.
210. Hanifin JM, Rajka G. Diagnostic feature of atopic dermatitis. Acta Derm Venereol Suppl (Stockh) 1980; 92:44–47.
211. Farr PM, Diffey BL. A quantitative study of the effect of topical indomethacin on cutaneous erythema induced by UVB and UVC radiation. Br J Dermatol 1986; 115:453–466.
212. Väänänen A, Hannuksela M. UVB erythema inhibited by topically applied substances. Acta Derm Venereol 1989; 69:12–17.
213. Juhlin L, Shroot B. Effect of drugs on the early and late phase UV erythema. Acta Derm Venereol 1992; 72:222–223.
214. Takiwaki H, Shirai S, Kohno H. The degrees of UVB-induced erythema and pigmentation correlate linearly and are reduced in a parallel manner by topical anti-inflammatory agents. J Invest Dermatol 1994; 103:642–646.
215. Hughes-Formella BJ, Bohnsack K, Rippke F. Anti-inflammatory effect of hamamelis lotion in a UVB erythema test. Dermatology 1998; 196:316–322.
216. Jocher A, Kessler S, Hornstein S. The UV erythema test as a model to investigate the anti-inflammatory potency of topical preparations—Reevaluation and optimization of the method. Skin Pharmacol Physiol 2005; 18:234–240.
217. Bickel A, Dorfs S, Schmelz M. Effects of antihyperalgesic drugs on experimentally induced hyperalgesia in man. Pain 1998; 76:317–325.
218. MacGregor JL, Maibach HI. The specificity of retinoid-induced irritation and its role in clinical efficacy. Exog Dermatol 2002; 2:68–73.

5

Rectal Dosage Forms and Suppositories*

K. Rosh Vora and Mohammed N. AliChisty

G & W Laboratories, Inc., South Plainfield, New Jersey, U.S.A.

INTRODUCTION TO RECTAL DOSAGE FORMS AND SUPPOSITORIES

In this era, various medicated rectal dosage forms are available in the market such as, creams, ointments, gels, douches, aerosol foams, enemas, and suppositories. Historically, the use of suppositories was referred to in the ancient times of Egyptians, Greeks, and Romans. Suppository was one of the several choices of drug delivery systems that were suitable, especially for very young children and elderly people, a notion first recorded by Hippocrates, when oral drug delivery system was found to be difficult for these particular groups of people (1).

The suppositories are a type of solid dosage form and in general they contain medications that are intended for rectal, vaginal, and to a much lesser extent, urethral use. Although three suppository types are available, the systemic absorption in general is limited to the rectal absorption, and the other two forms are mostly intended for local action. Fat-based suppositories melt at the site of application and release the drug for local action or systemic absorption. In contrast, the polyethylene glycol (PEG)–based suppositories dissolve in the rectal environment and release the drug for systemic absorption. The suppositories may be used as a protectant or a palliative to the local rectal tissue or as a carrier of therapeutic agents intended for local action or systemic absorption. Although rectal, vaginal, or urethral routes are not the first choice of route of administration, the rectal route has some definite advantages over the other routes of administration, especially when the drug is destroyed in the gastrointestinal tract, or administration of the oral dosage form is not possible because of nausea, vomiting, incapability of swallowing, or if the patient is unconscious. Then the rectal administration is the alternate route of choice.

TYPES OF SUPPOSITORIES AND DOSE CHARACTERISTICS

Various types and shapes of suppositories are shown in Figure 1.

Rectal Suppositories: Rectal suppositories are found in various weights and shapes; tapered at one end and usually weighing about 2 g each for the adult dose, but in some instances a suppository may weigh about 3 g. In general, pediatric suppository is about half the weight of the adult suppository. Usual length of the suppositories ranges from 40 to 75 mm.

Vaginal Suppositories: Usually vaginal suppositories are oviform or globular and weigh about 3 to 5 g depending on the size and shape. Vaginal tablets are

---

* This manuscript and the writing are neither presented on behalf of nor endorsed by G & W Laboratories, Inc. The writing represents the views of the authors only.

FIGURE 1 Various types and shapes of suppositories. *Source*: From Ref. 4.

also called pessaries; however, they do meet the definition of suppository because of their form, manufacture, and route of administration.

Urethral Suppositories: Urethral suppositories are an uncommon and rarely encountered dosage form, and therefore, not specifically described in the United States Pharmacopeia (USP). The suppository is a cylindrical-shaped dosage form 3 to 5 mm in diameter; 25 to 70 mm in length for female and up to 125 mm for male use.

Suppository Use and Its Advantages and Disadvantages

In general, suppositories are more acceptable in southern Europe and in Latin American countries than in northern Europe and United States. In Asian countries the use of suppositories is also limited.

USP Monograph Suppositories

Several drugs are listed as the official suppositories in the United States Pharmacopeia and are indicated for variety of treatments. The official USP suppository drugs from USP 29 are listed in Table 1.

General Use

Acetaminophen is used to reduce fever (antipyretic), and minor aches and pain (analgesic). Aminophylline is indicated in the treatment of asthma, bronchial pneumonia, bronchitis, and pulmonary, or renal edema. Aspirin is used to reduce

TABLE 1 Official Suppositories in the USP

Acetaminophen	Miconazole nitrate (vaginal)
Aminophylline	Morphine sulfate
Aspirin	Nystatin (vaginal)
Bisacodyl	Oxymorphone HCl
Chlorpromazine	Prochlorperazine
Ergotamine tartrate and caffeine	Progesterone (vaginal)
Glycerin	Promethazine HCl
Indomethacin	Thiethylperazine maleate

Source: From Ref. 13.

fever and aches and pain, but it is also indicated as a maintenance therapy to prevent heart attack or stroke by virtue of its blood thinning effect. Chlorpromazine has a wide range of activity that includes antiemetic, antipruritic, sedative, and some antihistaminic properties. Ergotamine tartrate and caffeine are used to treat migraine headache and pain. Indomethacin has anti-inflammatory and analgesic properties and it is used in the treatment of rheumatoid arthritis, ankylosing spondylitis, osteoarthritis, and acute gout.

Morphine is a narcotic analgesic and used for the relief of moderate-to-severe pain. It can be used as a sedative where sleeplessness is due to pain. Oxymorphone is an analgesic with the similar uses as morphine.

Prochlorperazine possesses activities that include antiemetic, antipruritic, and sedative actions and some antihistaminic properties. It can also be used to treat vertigo. Thiethylperazine Maleate is a phenothiazine derivative and is a sedating antihistamine. It is indicated for preventing nausea and vomiting. Promethazine HCl is a phenothiazine derivative and is a sedating antihistamine. Promethazine HCl is used as for the symptomatic relief of allergic conditions including urticaria, angioedema, rhinitis, and pruritic skin disorders. It is also used as an antiemetic to prevent nausea and vomiting.

Miconazole nitrate is a broad-spectrum antifungal agent and is indicated in the suppository form for the treatment of vaginal candidiasis. Nystatin suppository is also an antifungal antibiotic product, which is used for the prophylaxis and treatment of vaginal candidiasis.

Progesterone is a hormone that may be derived from either natural or synthetic origin. Progesterone suppository is indicated in the treatment of menstrual disorder or in the hormone replacement therapy for postmenopausal women.

Bisacodyl suppositories are used in the treatment of bowel irregularity or constipation as a stimulant laxative. It also can be used for bowel evacuation before having an invasive or surgical procedure in the rectum. Glycerin suppositories are used to improve bowel movement, and thereby reduce constipation.

As is known with the other conventional dosage forms, use of suppositories also has some advantages and disadvantages. For better understanding of this rectal dosage form and its use, the following advantages and disadvantages are briefly described:

Advantages:

- It allows the application of the drug directly to the affected area, for example, internal hemorrhoids, anal fissures, and localized infections, etc.
- It allows the administration of a drug to an unconscious patient when oral route is not possible and the injectable form may not be desirable.
- It allows administering a drug to an infant or child who is unwilling or unable to take oral medicine and has fear of injection.
- It minimizes the *first pass effect* of those drugs that go under extensive first pass metabolism upon oral administration.
- It improves the drug absorption by eliminating the instability or decomposition aspects of a drug in the gastric juice.
- It allows administration of a relatively large dose (e.g., 500–1000 mg) of a drug with better patient compliance.

- It allows administration of drug to a patient who is suffering from nausea and vomiting.
- It can achieve rapid drug absorption and can also show therapeutic effect comparable to the injection dosage form.

Disadvantages:

- In general, patients may feel uncomfortable in the use of the suppository that requires special physical positioning to facilitate the drug activity or absorption process.
- If user directions are not followed properly, melted suppository base may come out from the anorectal site without exerting its full therapeutic activity.
- Early defecation may interrupt the drug absorption process.
- The absorbing surface area of the rectum is much smaller compared to the small intestine and for some drugs that may lead to the lower amount of total drug absorption.
- Some drugs may degrade in the presence of microbial flora in the rectum.
- Rectal fluid content is very little and some drugs may need large amount of fluid to dissolve.

To ensure better patient compliance, the prescribing and dispensing health care professionals should take extra care in educating patents in the use of suppositories or similar type of dosage form.

ANORECTAL PHYSIOLOGY

The rectum is the terminal portion of the large intestine beginning at the juncture of the three tenia coli of the sigmoid colon and ending at the anal canal. Generally, the rectum is about 150 mm (15 cm) in length that ends at the anal canal, and in contrast the length of the colon is about 5 feet (1.5 m). The anal canal begins a few centimeters proximal to the classic dentate line and ends at the anal verge. The anal canal is about 5 cm in length (2). The rectal epithelium is lipoidal in nature, and in the absence of fecal matter, the rectum contains a very small amount of fluid that is nonbuffered in nature. In general, rectal fluid "pH" is about 7, but it varies because of its low buffering capacity when medicated drugs or chemical ingredients are administered. The general blood supply in the anorectal region is fairly rich, and therefore drug delivery through the mucous membrane or epithelial lining is well established. The lower, middle, and upper hemorrhoidal veins surround the rectum. Only the upper vein passes into the portal system, thus drugs absorbed into the lower and middle hemorrhoidal veins will bypass the liver. In the sense that at least a portion of the absorbed drug from the rectum may go directly into the inferior vena cava bypassing the liver. This is one of the advantages of rectal drug delivery system over the oral route.

Rectal Absorption of Drug

Variables affecting the absorption of drug from the rectum to the systemic circulation depend on several factors, such as drug solubility in the vehicle, drug pK_a, pH of the drug product, degree of ionization, lipid solubility, and anorectal physiology. The drugs that are dissolved in either fat or PEG base would melt or dissolve in the rectum, and hence partition through the epithelial membrane and get absorbed in the systemic circulation. Prior to absorption, the drugs that are suspended in the vehicle matrix would be released into the rectal fluid as particles and then would become available for partitioning and absorption.

Therapeutic Use
A drug may be administered in the suppository form and its effect may be either local or systemic depending on the intended therapeutic use. A category of drugs is available in the suppository form that may be used for treating local conditions of rectum, vagina, or urethra. The examples include antifungal agents, hormones, astringents, anesthetics, steroids, emollients, antioxidants, and contraceptives. A wide variety of drugs are also available for the systemic effect, for example, analgesics, antispasmodics, tranquilizers, sedatives, and antinauseatic and antibacterial agents.

SUPPOSITORY BASES AND THEIR CHARACTERISTICS
In the past decades, the use of the various suppository bases was based on their ease of availability rather than the scientific rationale. Theobroma oil, generally known as cocoa butter, is a naturally occurring ingredient that has a long history of use as the base for suppositories. The USP lists the following ingredients as usual and frequently used suppository bases: cocoa butter, hydrogenated vegetable oils (Type-1 & 2) also called hard fats, glycerinated gelatin, polyethylene glycol and their mixes with various molecular weights, and the fatty acid esters of polyethylene glycol.

An ideal suppository base from the physicochemical point of view may be described as follows:

1. It is nontoxic and nonirritating to the rectal tissues or mucus membrane.
2. It melts at or near rectal temperature (36°C ± 1°C).
3. It does not react with the drugs or any other ingredients of the formulation.
4. It does not show polymorphism.
5. It is stable at various temperatures.
6. It does not become rancid.
7. After cooling, the suppositories are released from the mold without cracking and brittleness.
8. It is stable in various storage conditions, for example, 5°C to 36°C for fat-type bases and up to 40°C to 50°C for PEG base.
9. It is nonsensitizing and nonallergic during its handling and use.
10. The interval between the melting and solidification point is narrow.

The other desirable chemical factors for a fat-type base can be described as follows: (*i*) its acid value is less than 0.02, (*ii*) its iodine value is less than 7, (*iii*) its saponification value is between 200 and 245, (*iv*) its hydroxyl value is between 5 and 60, (*v*) its solidification point is between 20°C and 30 °C, (*vi*) its melting range is between 30°C and 37°C, and (*vii*) its water number is between 20 and 30. Although, it is highly desirable to have an ideal suppository base, it is almost impossible to find one that has all the desirable characteristics.

A brief description of all the desirable chemical factors is described below (3).

- *Acid value*: The number of milligrams (mg) of potassium hydroxide needed to neutralize the free acids in one gram (1 g) of oil, fat, wax, or similar organic complex composition. The formation of free acids in the fats and oils is due to the hydrolysis of fatty esters due to the exposure to heat, light, chemical reaction, or bacterial contamination. It is observed that upon aging, fats and oils have a tendency of liberating free acids.
- *Iodine value*: The numbers of grams (g) of iodine absorbed by 100 g of fats or oils in a specific condition. It is the quantitative measurement of the proportion of unsaturated fatty acids present, both free and as esters, that has the property of absorbing iodine.
- *Hydroxyl value*: The number of milligrams (mg) of potassium hydroxide equivalent to the hydroxyl content of one gram (1 g) of the fats or oils. The hydroxyl value of the hard fat is important in prescreening the base during preformulation studies. In general, bases with low hydroxyl values have less plasticity and thereby they may become brittle during rapid or shock cooling.
- *Saponification value*: The amount of milligrams (mg) of potassium hydroxide required to neutralize the free acids and saponify the esters contained in one gram (1 g) of fat or oil.
- *Solid Fat Index* (SFI): This can be described as the percentage of the solid glycerides in the fat mixture at a certain temperature. The SFI is important for some of the fat bases if it becomes necessary to control the brittleness of the product by reducing the differential between mold temperature and mass temperature for trouble free molding. SFI can be determined using dilatometry, nuclear magnetic resonance, and differential thermal analysis.
- *Solidification point*: This is the temperature and time it takes to solidify a melted base. If the interval between melting range and solidification point is 10°C or more, the time taken to solidify the base may need to be shortened to facilitate efficient filling operation.
- *Water number*: The amount of water in grams (g) that can be incorporated in 100 g of fat or oil. The amount of water in the oil or fat base can be increased, however, by addition of suitable surface-active agents.

A brief description of some suppository bases is as follows:

Cocoa Butter

Theobroma oil or cocoa butter is a naturally occurring base that has been used for centuries. It is mainly a triglyceride, however, 40% of its fatty acid content is unsaturated. As a naturally occurring material from plant kingdom, its content and composition of fatty acids varies, and therefore a lot-to-lot variation is observed. Also, the place of its origin and the manufacturing treatment affect its quality. Cocoa butter exhibits polymorphism (i.e., existence of different crystalline forms) mostly because of the relatively high fraction of unsaturated fatty acid in its content. Four crystalline states can be described as follows.

1. The α form melts at around 24°C when the natural cocoa butter is melted and rapidly cooled to 0°C.
2. The β′ form melts at about 28°C to 31°C when melted cocoa butter is slowly cooled at 18°C to 23°C temperature.

3. By volume contraction, the unstable β′ form slowly changes into stable β form that melts at around 35°C. This particular β form of the cocoa butter is the most stable form and is used for the suppository preparations.
4. The γ form melts at 18°C. This type of cocoa butter is obtained by pouring melted (20°C) cocoa butter into a container and then rapidly cooling it at the deep-freeze temperature (3).

The formation of the various crystalline forms of cocoa butter depends on the speed and total time of heating and cooling processes. Prolonged heating above the critical melting temperature (36°C) may cause the formation of different unstable crystals of cocoa butter, but gradual cooling and then storing it at room temperature can form the stable β form within 1 to 4 days. During cooling, addition of a few stable crystals of cocoa butter prevents the formation of unstable form, and the addition of crystals in this manner is referred to as the "seeding" process (4). Cocoa butter may melt at higher than the normal room temperature, and because of the high amount of unsaturated fatty acid content it may become rancid during its storage. Therefore, cocoa butter should be stored in a cool, dry place.

Hard Fat
Hard fat, also called hydrogenated vegetable oil, is mainly composed of triglycerides, and mostly saturated fatty acid and esters. Hard fat can be produced from various vegetable sources, such as coconut and palm kernel oils, which are modified by esterification, hydrogenation, and fractionation to obtain products with desirable quality and melting ranges. Details of the manufacturing process of hard fat or hydrogenated vegetable oil can be obtained from literatures. Hard fat bases have great advantages over cocoa butter due to their excellent chemical stability and versatility. Hard fat does not show polymorphism similar to cocoa butter, and therefore, melting, molding, and cooling process is relatively simpler. Hard fat–based suppository is less prone to rancidity, and therefore can be stored at room temperature. Some of the drugs, for example, phenol, chloral hydrate, and creosote significantly lower the melting point of cocoa butter, and therefore, to address the problem a suitable hard fat base with higher melting point may be added to the formula. Also, inclusion of the suitable higher melting point waxes helps prevent softening of the suppositories.

Various hydrogenated vegetable oil (hard fat) or semisynthetic glyceride suppository bases are currently available from a variety of manufacturers. A short list of these oil-based bases is shown in Tables 2–4.

Polyethylene Glycol (PEG)
This base is of relatively recent origin compared to the other fat-type bases. PEG is water-soluble and has advantage over the fat bases with melting range close to the rectal temperature. Problems associated with handling, storage, and shipping are also considerably simplified. Polyethylene glycol polymers of varying molecular weights are available. The USP lists various grades of PEG. The grading is based on their molecular weight and viscosity range (5). Some of the commonly used polyethylene glycols, which may be considered for use as suppository base, are described in Table 5.

TABLE 2 Various Hard Fat Bases from Stepan (United States) for Compounding and Manufacturing of Suppositories

Product name	Composition	Melting range (°C)	Iodine value (g I_2/100 g)	Hydroxyl value (mg KOH/g)	Saponifi-cation value (KOH/mg)	Water number
Wecobee-M	Higher melting fractions of coconut and palm kernel oil	33.3–36	< 3	N/A	238–248	30–40
Wecobee-S	Same as above	38–40.5	<4	N/A	236–246	30–40
Wecobee-FS	Same as above	39.4–40.5	<3	N/A	236–248	30–40
Wecobee-R	Same as above	33.9–35	<4	N/A	236–246	30–40
Wecobee-W	Triglycerides	31.7–32.8	<4	N/A	242–252	30–40

TABLE 3 Various Hard Fat Bases from Sasol (Germany) for Compounding and Manufacturing of Suppositories

Product name	Composition	Melting range (°C)	Iodine value (g I₂/100 g)	Hydroxyl value (mg KOH/g)	Saponification value (KOH/mg)	Water number
Witepsol E75	Triglyceride of saturated vegetable fatty acids with monoglycerides	37–39	< 7	5–15	220–230	45
Witepsol E76	Same as above	37–39	<3	30–40	220–230	N/A
Witepsol E79	Same as above	36–38	<7	25–35	220–230	N/A
Witepsol E85	Same as above	33.9–35	<4	N/A	220–230	45
Witepsol H5	Same as above	34–36	<2	<5	235–245	N/A
Witepsol H12	Same as above	32.3–33.5	<3	5–15	235–245	100
Witepsol H15	Same as above	33.5–35.5	<3	5–15	230–245	100
Witepsol H19	Same as above	33.5–35.5	<7	20–30	230–240	N/A
Witepsol H32	Same as above	31.0–33.0	<3	<3	240–250	N/A
Witepsol H35	Same as above	33.5–35.5	<3	<3	240–250	N/A
Witepsol H37	Same as above	36.0–38.0	<3	<3	225–245	N/A
Witepsol H175	Same as above	36.0–38.0	<3	5–15	225–245	N/A
Witepsol H185	Same as above	38.0–39.0	<3	5–15	220–235	N/A
Witepsol S51	Same as above	30.0–32.0	<8	55–70	215–230	N/A
Witepsol S52	Same as above	32.0–33.5	<3	50–65	220–230	N/A
Witepsol S52	Same as above	33.5–35.5	<3	50–65	215–230	N/A
Witepsol S58	Same as above	31.5–33.0	<7	60–70	215–225	N/A
Witepsol W25	Same as above	33.5–35.5	<3	20–30	225–240	N/A
Witepsol W31	Same as above	35–37	<3	25–35	225–240	N/A
Witepsol W32	Same as above	32.0–33.5	<3	40–50	225–245	N/A
Witepsol W35	Same as above	33.5–35.5	<3	40–50	225–235	N/A
Witepsol W45	Same as above	33.5–35.5	<3	40–50	225–240	N/A

TABLE 4 Various Semisynthetic Glycerides or Hard Fat Bases from Gattefosse (France) for Compounding and Manufacturing of Suppositories

Product name	Composition	Melting range (°C)	Iodine value (g I_2/100 g)	Hydroxyl value (mg KOH/g)	Saponification value (mg KOH/g)	Acid value (mg KOH/g)
Suppocire A	Semi-synthetic glycerides (multi-purpose vehicle)	33–35	<2	20–30	224–246	<0.5
Suppocire B	Same as above	36–37.5	<2	20–30	220–244	<0.5
Suppocire D	Same as above	36–38	<2	20–30	210–232	<0.5
Suppocire AS2	Same as above	35–36.5	<2	15–25	224–246	<0.5
Suppocire BS2	Same as above	36–37.5	<2	15–25	224–246	<0.5
Suppocire AT	Same as above	35–36.5	<2	27–37	220–244	<0.5
Suppocire AIM	Semi-synthetic glycerides (low reactivity vehicle)	33–35	<2	<10	231–255	<0.2
Suppocire AM	Same as above	35–36.5	<2	<10	228–252	<0.2
Suppocire BM	Same as above	36–37.5	<2	<10	226–250	<0.2
Suppocire CM	Same as above	38–40.0	<2	<10	224–246	<0.2
Suppocire AIML	Semisynthetic glycerides (low reactivity vehicle compatible with large amount of powders)	33–35.0	<3	<10	228–252	<0.5
Suppocire AML	Same as above	35–36.5	<3	<10	228–252	<0.5
Suppocire BML	Same as above	36–37.5	<3	<10	225–249	<0.5
Suppocire NAI 25A	Semi-synthetic glycerides (high hydroxyl value for large scale production with shock cooling and/or large amount of powders)	33.2–35.2	<2	<30	230–250	<0.5
Suppocire NA 35	Same as above	34–36.0	<2	30–40	225–240	<0.3
Suppocire NAI 50	Same as above	33.5–35.5	<2	38–48	218–242	<0.5
Suppocire NAS 40	Same as above	34.5–35.5	<3	40–50	225–240	<0.3
Suppocire NAS 50	Same as above	33.5–35.5	<3	40–50	225–235	<0.3
Suppocire NAS 55	Same as above	33.5–36.5	<3	50–65	215–230	<1.0
Suppocire NAIS 90	Same as above	34–36.0	<1	80–100	215–245	<1.0
Ovucire WL-2944	Semisynthetic glycerides (base for vaginal suppositories)	32.5–35.5	<8	43–63	215–235	<0.5
Ovucire 3460	Same as above	32.5–34.0	<7	60–70	215–225	<1.0

TABLE 5 Various Polyethylene Glycols (PEGs) for Compounding and Manufacturing of Suppositories

Nominal average molecular weight (centistokes)	Viscosity range (centistokes)	Nominal average molecular weight	Viscosity range (centistokes)
1000	3.9–4.8	5000	170–250
1450	25–32	5500	206–315
2000	38–49	6000	250–390
2400	49–65	6500	295–480
3000	67–93	7000	350–590
3350	76–110	7500	405–735
3750	99–140	8000	470–900
4000	110–158		
4500	140–200		

Polyethylene glycols with molecular weight of less than 1000 usually occur in liquid form. If necessary, liquid PEGs can be used in combination with the other higher molecular weight grades of PEG to obtain the desired melting point of the suppository and/or release of the drug from the PEG matrix. In general, PEGs are chemically stable in air and in solution, and do not support microbial growth. During the manufacturing of PEG-based suppositories on a larger scale, and especially when molten base is heated and kept above 50°C for a long period of time, caution should be taken to prevent oxidative degradation (6). To prevent oxidation, the use of suitable antioxidants alone or in combination is recommended. Polyethylene glycol bases are also noted to be incompatible with salicylic acid, benzocaine, quinine, silver salts, tannic acid, aminopyrine, ichthammol, and sulfonamides.

Water-Dispersible Bases

Several nonionic surfactants have been developed that are chemically related to the PEG and can be used as suppository bases. Examples of such surfactants are polyoxyethylene sorbitan fatty acid esters and polyoxyethylene stearate (5). The other PEG-related chemical ingredient is polyethylene glycol monostearate (PGMS) that can be used as the base. Commercially, these surfactant bases are available from various manufacturers. A water dispersible base offers broad drug compatibility, nontoxicity, and nonsensitivity, and it does not support microbial growth. These bases are also suitable for storage at relatively elevated temperature.

Glycerinated Gelatin Base

Drug substances may be incorporated into glycerinated gelatin base by adding 20% gelatin and 10% water into 70% glycerin. Glycerinated gelatin suppositories are hygroscopic in nature and should be stored in a well-closed container, preferably at a temperature below 35°C (5). Glycerinated gelatin base gets less attention from the formulators because of the availability of varieties of other more suitable bases.

SUPPOSITORY FORMULATION

Suppositories are formulated based on their intended therapeutic use, such as for local action or systemic absorption. The formulation of suppositories mostly intended for the rectal use either for local action or systemic absorption has been discussed. Consideration of physicochemical properties of the API (Active Pharmaceutical Ingredient) is a key aspect of the formulation studies. Factors that may directly influence the formulation aspects of a rectal suppository include the physical and chemical properties of a drug such as the solubility, pH, pK_a, and polymorphism. Other factors such as physical and chemical compatibility of the drug with the primary base and with the other excipients, temperature sensitivity of the API, the selection of preservatives, and the drug–drug interaction (if two or more drugs are incorporated in the same formula) also play an important role.

Drug Solubility

A variety of drugs may be used in the suppository dosage form. Note that a drug may be of hydrophilic or hydrophobic in nature. For example, Promethazine HCl, Phenylephrine HCl, Pramoxine HCl, Ephedrine sulfate, Quinine hydrochloride, Sodium barbital, Chloral hydrate, Morphine sulfate, Caffeine, and Trimethobenzamide HCl are reported hydrophilic in nature. Note that Prochlorperazine, Acetaminophen, Bisacodyl, Miconazole nitrate, Indomethacin, and Ergotamine tartrate are practically insoluble in water, and therefore are considered hydrophobic. To become available for rectal absorption, a drug must be released from the suppository. The drug release may be achieved via melting of the suppository in the rectum or by dissolution in the rectal fluid. If a drug has more affinity for the lipid or the PEG base and if its concentration is very low in the suppository, then the drug molecule may remain within the suppository matrix for all practical purposes and thereby may not become available to the surrounding absorption sites. Drugs that have limited affinity for the fat-type base and are at or near the saturation concentration are more readily available for absorption sites (7).

pH

Some drugs are pH sensitive, and they show poor stability either in alkaline or acidic environment. In that case the base and the excipients must be selected such that the pH of the drug product remains within the desired level, for example, indomethacin degrades in alkaline medium, but is stable in a weakly buffered medium with a pH of 4 to 5. Drugs that are stable in a wide pH range (e.g., 3–8) are comparatively easier to formulate. During formulation it is important to study the effect of pH on the final product, especially when stability is a matter of concern.

Bases and Excipients

Suppository bases should be selected based on the review of the nature of the drug, and also the intended mode of action. Excipients may interact and can affect the critical quality attributes of the final formula, including the chemical stability and bioavailability of the drug. Excipients play a wide variety of functional roles in the suppository and therefore, the following aspects should be reviewed. An excipient

- may change solubility of the drug;
- may change bioavailability of the drug;

- may increase or decrease stability of the API;
- can help the active drug maintain preferred polymorphic form;
- maintains desired pH of the formulation;
- acts as an antioxidant, emulsifying agent, a binder, or a disintegrant;
- prevents aggregation of drug particles;
- helps prevent sedimentation of API and other excipients in a suspension medium; and
- Can provide coating on drug particles and/or other ingredients for delayed release.

Composition and characteristics of excipients can influence the safety and effectiveness of drugs, and therefore, can influence the bioavailability or bioequivalency of the product. Excipients can also play a role in stabilizing the product by preventing degradation of APIs or other excipients. Note that they can also play a role in destabilizing a product, its purity or effectiveness. When a suppository is too soft it is advisable to add a substance to harden it. A suitable suppository hardener can be white wax, carnauba wax, fatty alcohol, etc. These hardeners may also raise the melting point and affect the drug release from the matrix. It is postulated that the spreading capacity of the molten base, the surface area of the rectal mucous membrane in contact with molten mass, and the physicochemical properties of the drug and excipients affect the intensity of the therapeutic action. Use of a suitable surfactant may increase the spreading capacity of the molten base and hence increase the drug absorption.

Riegelman and Crowel showed that the use of a surfactant may increase the absorption, or it may also cause poor absorption of the drug (8). Therefore, the addition of a surfactant to the suppository formulation needs a careful consideration including the extensive in vitro investigation using a variety of discriminatory dissolution media preferably in the preformulation studies.

Polymorphism
This phenomenon is often characterized as the ability of a drug substance to exist in two or more crystalline phases that have different arrangements and/or conformations of the molecules in the crystal lattice (9). Polymorphs can show different chemical and physical properties, such as melting point, apparent solubility, optical properties, and chemical reactivity. These properties either directly or indirectly can impact on a product's quality-related parameters, such as the stability, in vitro release, and bioavailability. During the formulation stage, a formulator must take polymorphic drugs into consideration to prevent unwanted poor stability, poor dissolution, and failed bioavailability. By the laboratory experiments, scientists must identify the desired polymorphic form that will serve the best interest when identity, purity, safety, stability, and bioavailability are considered. In the New Drug Application (NDA), the innovator must inform and include all the relevant information concerning polymorphism of the new drug molecule to the Food and Drug Administration (FDA). For the generic drug manufacturer, FDA may refuse to accept and/or approve an Abbreviated New Drug Application (ANDA) if the application contains insufficient information to show that the drug substance is the "same" as the reference listed drug. Although, FDA has approved some generic products, which contain drug substances with different physical form when compared to the respective reference listed drug, for

example, Famotidine, Ranitidine, and Warfarin. An ANDA applicant is required to prove that the proposed generic drug product meets the standards of identity, quality, and purity, exhibits the satisfactory long-term stability, and it is also bioequivalent.

Temperature Sensitivity
Some drugs may be temperature sensitive or with the particular combination of adjuvants they may become heat sensitive. If a drug product is heat sensitive, studies should be conducted in a laboratory at various temperature settings to understand the effect of temperature and processing parameters at different temperature ranges. During manufacturing, caution must be taken to limit the exposure of the drug product to a temperature that has already been identified as risky prior to its scale up. Large scaled-up batches must be subjected to concurrent stability evaluation studies at least annually to evaluate the finished product quality.

Preservatives
Several antimicrobial agents are available that can prevent microbial growth and preserve the quality and integrity of the product. It is advisable to verify the solubility of all the intended preservatives, their compatibility in the formula, the confirmation of their preservative effectiveness at the selected concentration, the need for the combination of two or more preservatives for their synergistic effect, their initial and latter compatibility with the active drug and the intended excipients, and finally their tolerance in the rectum. Some of the commonly used preservatives are methylparaben, propylparaben, and butylparaben that can be used alone or in combination. Suitable antioxidants are also recommended to prevent oxidation of the drug and enhance stability.

Drug–Drug Interaction
When two or more drugs are incorporated in a single product, studies should be carried out to rule out any drug–drug interaction. The formulator should keep in mind that the effectiveness of both drugs is equally important when therapeutically indicated. Suppositories containing a combination of Trimethobenzamide and Benzocaine or Ergotamine Tartrate and Caffeine must impart the therapeutic effectiveness of the individual components when administered to a patient for a therapeutic purpose. Also, these combination drugs must show acceptable physical and chemical stability throughout the shelf life.

Solving Problems Associated with Formulation
During the formulation development, some unexpected problems may arise, and formulators should be on the lookout for that. It is important to keep in mind that a safe and stable drug product is not only essential, but is also required. Use of water as solvent to dissolve drugs or excipients, and then addition of this water phase in to the fat base should be avoided, if possible. Water accelerates the oxidation process of fats and oils that can destabilize the final product. If the water content is very low, and if it evaporates, the dissolved drug or excipient may then crystallize. Water content in the suppositories should be analytically checked to confirm the presence of the desired level of water in the finished product. A suitable antifungal preservative or combination of preservatives is

recommended if the formulation requires incorporation of water. Drug absorption from water-containing suppositories is enhanced only if an oil-in-water (o/w) emulsion exists with more than 50% of the water in the external phase (10). Suppositories made of glycerinated gelatin can lose moisture in a very dry climate and can absorb moisture under high humidity conditions. Polyethylene glycol (PEG) bases are also hygroscopic, but the degree of hygroscopicity of PEG depends on the chain length of the molecule. the shorter the molecular chain length (e.g., 400–2000), the higher is the possibility of hygroscopicity; the longer the molecular chain length (e.g., 3000–8000), the lower is the potential of hygroscopicity.

Fat bases show rancidity that results from auto-oxidation, which then leads to the subsequent decomposition of unsaturated fats into the low- and medium-molecular-weight (C_3–C_{11}) saturated and unsaturated aldehydes, ketones, and acids. The lower the content of unsaturated fatty acids in a suppository base, the greater is its resistance to developing rancidity. In other words, the higher the content of saturated fatty acids in a suppository base, the greater is its resistance to developing rancidity. In general, rancidity begins with the formation of hydroperoxides in the presence of oxygen. The peroxide value is an indicator that can be used to evaluate the formation and extent of rancidity of the base. The peroxide or active oxygen value is a measure of the iodine liberated from an acidified solution of potassium iodide, which is the so-called "peroxide oxygen" of the fats (4). To prevent auto-oxidation and thereby the rancidity of the fat base, a suitable antioxidant(s) is recommended. Several antioxidants are available and should be selected based on the nature of the base, drug, and other ingredients in the formula. The following is a list of antioxidants that are commonly used in the suppository preparation:

- Butylated Hydroxytoluene (BHT)
- Butylated Hydroxyanisole (BHA)
- Propyl Gallate and Gallic Acid
- Ascorbic acid and its esters
- Ascorbyl Palmitate
- Tocopherol (Vitamin E)
- Phenols
- Tannins or Tannic acid

Not all antioxidants are suitable for all formulations. Therefore, the compatibility of the selected antioxidants with the base(s), drug(s), and other excipients should be evaluated before finalizing the formula.

IN VITRO DRUG RELEASE STUDY

The active drug(s) must be available for absorption from the suppository. Before initiating any in vivo studies, a formulator should understand and evaluate the drug release rate and release pattern by conducting in vitro dissolution studies in the laboratory. Several FDA approved dissolution apparatuses are available to conduct the in vitro studies, for example, USP Apparatus-1 (basket system), Apparatus-2 (paddle system), Apparatus-3 (reciprocating cylinder system), etc.

A careful evaluation of the drug properties and the excipients is necessary before designing dissolution studies. In general, rectal pH is about 7 ± 0.2, which is considered neutral pH. The amount of rectal fluid is only a few milliliters (mL) and virtually has no buffer capacity; as a consequence, the dissolving drug(s) or even excipients can have determining effect on the existing pH in the anorectal area. Schanker found that the weaker acids and bases are more readily absorbed than the stronger acids and bases (12). It was also observed that nonionized drugs are more permeable through the colonic lumen than the highly ionized ones. Thus, changing the pH of the anorectal area to obtain more nonionized version of the drug can enhance the absorption of a drug. If the drug release is very quick or very slow then an adjustment of product pH can achieve the desired release and absorption rate. Schanker also showed that weak acids with a pK_a below 4.3 and weak bases with a pK_a below 8.5 are usually readily absorbed. Highly ionized drugs or compounds are poorly absorbed. Riegelman and Crowell have demonstrated that one of the rate-limiting steps in drug absorption is the diffusion of the drug to the site on the rectal mucosa at which absorption occurs. The diffusivity is influenced not only by the nature of the drug product, such as presence of a surfactant or other ingredients, but also by the physiologic state of the colon, that is, the amount and chemical nature of the fluids and solids present in the colonic lumen.

In general, the in vitro studies are conducted in a controlled environment or at an ideal condition where maximum drug release is expected within a stipulated time frame. Most of the dissolution studies are conducted at $37°C \pm 0.5°C$, but it should be recognized that although the internal human body temperature is about $37°C$, a patient's rectal temperature might vary. Especially the rectal temperature may vary from $36°C$ to $38°C$ depending on biological and physiological conditions. Therefore, in vitro drug release studies should be conducted at various temperature settings to rule out any formulation deficiencies. Once the formula is finalized, a specific temperature setting and fixed parameters should be selected to set up the regular Quality Control release criteria.

Coben and Lieberman summarized various physiologic factors that affect the absorption of drugs from the anorectal area, such as pH, colonic contents, lack of buffering capacity, circulation, and biological state. The principal physicochemical characteristics of drugs affecting membrane absorption are the water/lipid partition coefficient and degree of ionization (4).

Particle size plays an important role when drug particles are suspended in the dose matrix; better absorption rate is achieved when micronized particles are used compared to the macro particles. Use of micronized drug particles becomes irrelevant when drug is dissolved or solubilized in the suppository vehicle. In vitro drug release study is a great tool not only for evaluating drug release rate, but also for determining the qualitative and quantitative composition of the formulation that ultimately establishes the quality of the final product.

IN VIVO STUDY

The main objective of developing a medicated suppository is to deliver a drug through the anorectal area and achieve its local or systemic therapeutic action. The effectiveness of a drug may significantly depend on its formulation, and the physical and biological condition of the patient. In general, a suppository intended for systemic therapeutic effect must release its drug at the intended site of delivery, the drug needs to be absorbed, the absorbed drug would then go to the site of action and exert its therapeutic action. To achieve this objective, several formulas should be prepared and tested to select the one that has the most desirable drug absorption profile, and which is also therapeutically effective and safe.

An in vivo study requires a Study Design (Protocol) to properly conduct the study and obtain maximum information in reference to drug's pharmacokinetic and pharmacodynamic activities. In the case of a New Drug Application (NDA), the sponsor and/or applicant must provide all the pertinent pharmacokinetic and pharmacodynamic results to FDA for evaluation. An NDA applicant must provide bioavailability (BA), effectiveness, and safety profile data for approval. The abbreviated new drug (ANDA) applicant is not required to show effectiveness versus placebo and safety data, rather the sponsor must provide pharmacokinetic data to FDA to show bioequivalency (BE) compared to the referenced listed drug (RLD). For ANDA approval the safety and efficacy data can be referred to from the published literature or the data gathered by the brand developer and be used as a cross reference for the abbreviated drug product.

In general, therapeutic effectiveness data may be obtained from blood or serum concentration of the drug; the metabolic clearance information is obtained form the renal clearance or other related pharmacokinetic parameters as established by the innovator or the FDA. Some of the commonly used pharmacokinetic terms are summarized for general reference.

- C_{max}: Peak serum concentration of drug
- T_{max}: Time to reach peak serum concentration of drug
- T_{half}: Terminal elimination half-life
- $AUC_{0-\alpha}$: Area under the serum concentration curve
- AUC_t/AUC_r: Ratio between test versus reference
- T_{lag}: Lag time for modified release products, if present
- CL_{renal}: Renal clearance of drug as metabolite or other forms

Any proposed in vivo pilot or pivotal clinical endpoint study requires preapproval from an Institutional Review Board (IRB) Committee that reviews the study protocol prior to the study. The IRB committee reviews all the possible safety issues involved in the study design, and the relevant ethical issues. An in vivo study engages human subjects either as healthy volunteers or as actual patients. In studies involving clinical endpoints, it is advisable to have the proposed study protocol reviewed and commented upon by the FDA prior to the study.

MANUFACTURE OF SUPPOSITORIES

Suppositories can be made by various means, namely, hand molding, compression molding, and pour molding. Hand and compression moldings are relatively old methods, and the availability of modern machines almost obsoletes those processes. Hand-molded suppositories may be available by special orders and on an as needed basis. These are usually prepared and dispensed by the pharmacists (Figs. 2 and 3).

Manufacturing of a suppository batch should be predesigned to achieve the high-quality product. Manufacturing parameters should be well controlled within a specific range that was determined earlier either by small-scale laboratory batches or large scaled-up batches. Depending on the nature of the

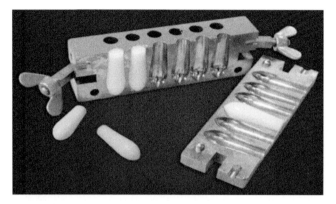

FIGURE 2 Mold for filling by hand. *Source*: Courtesy of G & W Laboratories.

product(s), the following process parameters should be identified, optimized, and established:

1. Maximum exposure time of the suppository base and/or base and excipient mix at a particular temperature setting.
2. Maximum temperature during the manufacturing of the batch with an allowable temperature range.
3. Optimum mixing time to obtain uniform content uniformity, especially when the drug is suspended in the melted bulk.
4. Minimum mixing time and temperature to dissolve a drug completely in the vehicle, if drug is dissolved as part of the manufacturing procedure.
5. Minimum temperature that needs to be maintained to avoid premature congealing of the melt during the molding, filling, and cooling operation.
6. Cooling temperature and cooling rate (degrees/minute) to congeal and form a solid suppository.

The previously mentioned parameters represent some of the basic process guidelines that are recommended during the manufacturing of a suppository batch.

Suppositories are usually manufactured in a stainless steel vessel, and the size of the vessel depends on the total volume of the melt. The manufacturing tank should be jacketed and be able to run cold water, hot water, or hot steam as the process requires. The processing tank, usually, is equipped with a suitable homomixer that can produce enough high sheer to mix all of the ingredients of

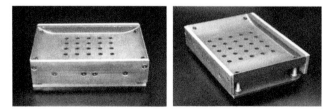

FIGURE 3 Mold for filling by hand. *Source*: Courtesy of G & W Laboratories.

the batch. The tank may be equipped with an agitator or scrapper for mixing and for better heat transfer to avoid local overheating or cooling. This also ensures proper mixing of the bulk melt throughout every corner of the mixing kettle. Circulation and recirculation of the melted base during processing is also recommended, especially when the drug and the other ingredients are suspended in the melted base. Maintenance of the designated temperature of the melt is critical when the drug itself or the combination of drug and excipient degrade in the presence of excess heat. Therefore, extra precautions may be needed during manufacturing of the heat-sensitive product.

Pour Molding

The advances in automatic suppository-making machinery have improved the entire molding operation. The modern molding operation can be described as a four-step process that includes pouring, cooling, sealing, and removing. In general, suppository melt is poured into an empty aluminum or plastic shell, then it is cooled at a desired cooling rate, the shell is then sealed to maintain product integrity, and then it is removed for final packaging.

Machine Molding

Several suppository molding machines are available, and these are highly sophisticated machines including computer-control to control the entire molding operation. These machines are sturdy, precise, and designed to achieve high production volume. Selection of a suitable suppository molding machine needs careful consideration involving the following features:

- Should be easy to operate with computer-controlled visible panel.
- It should form perfect empty suppository cavities from aluminum foil (by punch and matrix system) and from plastic films (by thermo-forming process).
- It should have the capability to seal each cavity or shell around the outline of the preformed cavity to ensure uniform sealing and also allow easy extraction of each suppository from the pack with the "peel-apart" opening system.
- It should have a volumetric filling pump that can accurately fill the required amount of melted bulk in each suppository mold cavity.
- It should have an adequate number of cooling chambers with automatic temperature controllers to congeal the suppositories at a desired cooling rate and temperature.
- It should have an operative system that is optimized to have minimum waste during the mold filling operation.
- It should produce suppositories in a continuous band that allows any number of suppositories to be cut (e.g., 2, 4, 5, 6, or 10 suppositories per strip) as needed.

The machine should be capable of sealing various kinds of aluminum foils (duplex and triplex) and various types of thermoforming plastic films (PVC, PVC/PE, PVC/PVDC/PE).

The molding run can vary from 5000 to 30,000 suppositories per hour of continuous operation depending on the nature of the product and the machine (Figs. 4 and 5).

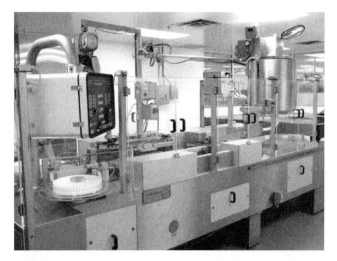

FIGURE 4 Sarong suppository filling and molding machine. *Source*: Courtesy of G&W Laboratories.

PROBLEMS ENCOUNTERED WITH THE MOLDING OPERATION

Brittleness

Suppositories may become brittle when cooling temperature and cooling rate are not optimized—chilling or shock cooling is an example. Shock cooling causes brittleness due to the sudden contraction of the fat base, which has very little elasticity. Hydrogenated oils are usually high in stearate content, and in the presence of other solutes, they become more prone to brittleness. Gradual cooling of the hard fat is recommended, and the temperature differential between the melted base and the cooling temperature should be as narrow as possible.

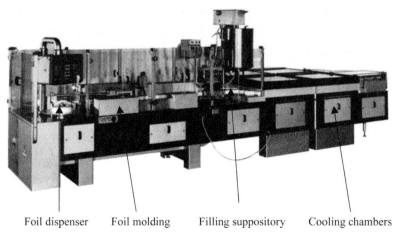

Foil dispenser   Foil molding   Filling suppository   Cooling chambers

FIGURE 5 A full frontal view of the Sarong suppository filling and molding machine.

Addition of a small amount of surfactant, monoglyceride, or a humectant will impart plasticity and thereby prevent brittleness. Natural cocoa butter consists of relatively more unsaturated fatty acids, and therefore, cocoa butter suppositories are quite elastic and do not fracture readily.

Filling Volume and Density of the Melt
Usually, oil-type bases are less dense than water and have lower specific gravity. The fill volume in a specific empty shell must be calculated before the molding operation. Knowledge about the suppository weight can be obtained from a given mold volume and density of the final product formula. The active drug and other required inactive ingredients should be added in such a manner that each suppository contains the exact amount of drug and excipients, as required. It should be remembered that density of the base alone should not be the sole criterion for calculating suppository weight even though volume of the mold is not variable. Processing factors such as the temperature of the melted bulk formula during the mold-filling step, the loss of water (if present in the formula) during the filling run, the lack of adequate weight control during processing, the lack of homogeneity of the melted bulk, and any wear and tear of the mold may also affect the weight of the final product.

Volume Contraction
Volume contraction usually occurs after cooling of the hot melted base in the metal mold cavity, aluminum shell, or in the plastic shell. A properly congealed suppository is fairly easy to remove from the mold without the use of any mold release additives or lubricant. In some instances, a contraction hole may form at the flat end of the suppository when temperature gradient between the cooling chamber and the suppository melt is quite large. Temperature setting of the cooling chambers should be such that the temperature difference between chambers is minimal, and this can be achieved by gradual cooling in stepwise manner. In high volume production, for example, glycerin suppository using standard open molds, wherein adequate control of temperature may not be possible, the mold is slightly overfilled with the melt and the extra mass is scraped off after congealing. This type of filling process eliminates the contraction hole in the suppository. For a fat- or oil-based suppository, the initial cooling chamber temperature should be close to the temperature of the melted bulk, and then stepwise cooling is recommended to avoid a contraction hole.

Mechanical Problems
The suppository filling and molding machines have made significant advances, but mechanical problems are an integral part of the molding operation regardless of the quality of the machine. The following factors can influence the molding process. These factors should be identified and rectified when a batch is running in the machine:

- Proper adjustment or alignment of the mold and the die plate.
- Aluminum foil or plastic film should not snap during operation.
- Adjustment of filling weight in each individual shell within ±2% of the target weight.

- Cooling chamber fails to maintain the desired temperature setting. This would result in either soft or brittle suppositories.
- Lack of appropriate heat-sealing of the aluminum foil or plastic film to maintain the integrity of each wrapped suppository.

These are some of the problems associated with the mechanical failures and which may result in suppositories with poor quality.

PROCESS OPTIMIZATION AND VALIDATION

In the FDA regulated industry, optimization of the manufacturing process and successful process validation are some of the integral parts of the product development project that ensure consistently reproducible good-quality product for years. During the early product development stage and later in the scale-up stage, manufacturing process parameters should be evaluated and determined based on the physical observation and chemical stability data of the finished product. As needed, a number of R&D batches should be made to establish good processing steps that show good physicochemical stability of the product with aging. The definition of validation as quoted from the FDA's guidance for industry is as follows: "A procedure to establish documented evidence that a specific process or test will consistently produce a product or test outcome meeting, its predetermined specifications, and quality attributes. A validated manufacturing process or test is the one that has been proven to do what it purports or is represented to do. The proof of process validation is obtained through collection and evaluation of data, preferably beginning with the process development phase and continuing through the production phase. Process validation necessarily includes process qualification (the qualification of the materials, equipment, systems, facility, personnel), but it also includes the control of the entire process for repeated batches or runs" (14). The word validation signifies a well-documented process that is reproducible and robust. Exercise of process validation requires several documents that may include the following, but not limited to

- Standard Operating Procedures (SOPs).
- Validation protocol designed for extensive bulk and finished product sampling, testing of samples per specifications, microbiological testing, the required test procedures, and proposed specifications as per the protocol.
- Manufacturing batch record that provides details of the manufacturing process and the controls.
- Cleaning Validation (CV) protocol.
- Stability protocol and the storage condition(s) for the finished goods.

The batch record should include the manufacturing formula and the details of the manufacturing steps including the order of addition of ingredients, mixing time, temperature ranges, the equipment and machines, and speeds of the machines that provide sheer and stress during mixing. Any special instruction or precaution must be clearly written and highlighted on the batch record and/or in the protocol so that everyone associated with the batch can follow the protocol as written. A process validation exercise usually involves production of three process validation batches. The exercise is not complete until a process validation report is issued, which includes the executed batch production records and the stability data.

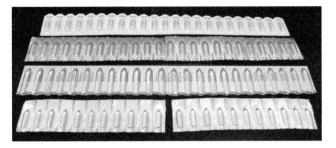

FIGURE 6 Aluminum and PVC foils for wrapped suppository. *Source*: Courtesy of G & W Laboratories.

PACKAGING OF SUPPOSITORIES
Packaging and labeling of suppositories is the final step of producing a finished product. In general, suppositories are individually wrapped with aluminum foil or plastic film (Fig. 6). Glycerin suppositories are usually available as unwrapped, but in a tightly closed plastic containers. The U.S. FDA regulates all packaging material and its components including metallic foils (e.g., aluminum), plastic polyvinyl shells (PVC), jars, labels, cartons, etc. In general, drug manufacturers are required to submit all the packaging information including the stability data of the finished product in the material that comes in direct contact with the product such as the proposed metallic foils, plastic shells, jars, caps, etc.

Packaging Problems
Packaging materials are subject to the atmospheric exposure and also come in direct contact with the drug substance, and therefore, several issues have been reported. Generally encountered problems are leaching, corrosion, permeation, and sorption. These are discussed as follows:

Leaching: Leaching can be described as the dissolution of a component material (e.g., plastic) into the drug product. In general, compounds leached from a plastic material are polymer additives, especially when they are in direct contact with the suppositories. High temperature and humidity can usually accelerate the leaching process. To prevent or minimize leaching or chemical reaction with the product, packaging compatibility studies must be conducted with the various packaging material to ensure acceptable quality of the finished product for the shelf life.

Corrosion: Corrosion of a metal can be described as erosion by chemical reaction between the product and the metal. In general, gradual destruction of a metal takes place due to a chemical process such as oxidation of the metal or by the action of a chemical on metal. For example, chloral hydrate corrodes aluminum shells when chloral hydrate comes in direct contact with the metal. In this case, a suitable plastic film can be selected based on the laboratory stability study data. The packaging studies of the drug product should start as soon as the preformulation studies are completed.

Permeation: Permeation can be described as a process of diffusion that starts from one side and progresses to the other side. This phenomenon is more common in the plastic material because of its composition, which also includes additional polymers, plasticizers, lubricants, fillers, and pigments. In addition, exposure to heat and humidity plays an important role and can increase the permeation process of the unwanted chemical into the drug product. Rugged plastic suppository shells that can provide a superior moisture barrier, minimum gas permeation, greater temperature resistance, as well as protection from the oxidative decomposition occurring due to light and heat are highly preferred.

Sorption: Sorption is a process wherein the drug product binds with the polymer molecules. Both the absorption and adsorption of a chemical to a polymer can be considered as the sorption process. The pharmaceutical excipients, especially the colorant or pigment, can easily be adsorbed or absorbed by the plastic. In selecting packaging components, careful consideration must be given to the physical and chemical nature of the final formula and the plastic polymer.

TESTING OF SUPPOSITORIES

All marketed suppository batches require testing to ensure the quality of the finished product. Analytical test methods must be validated and well documented to assure that every batch meets the standard that was defined during the R&D and scale-up batches. Specifications of the active drug ingredient(s) can be based on the selected compendium such as United States Pharmacopeia (USP), British Pharmacopeia (BP), European Pharmacopeia (EP), or Japanese Pharmacopeia (JP), etc. In some instance compendial specifications are not available, and in that case in-house specifications must be set and implemented.

Physical Tests

Several physical tests have been identified for suppositories, and they may be considered in the following manner:

- *Description and appearance*: The nature of suppository (e.g., waxy or nonwaxy type) should be identified and described in the quality control analysis report. The appearance or texture of the suppository should be identified, for example, smooth surface, etc.
- *Color*: Color of the finished suppository should be provided to confirm homogeneity of the batch. Also, if the color is intense, there should be a limit on the intensity of the color. This may be determined using the color Gardner chart, which will ensure same color and shade of the finished drug product from batch to batch.
- *Shape and size*: These two physical characteristics are important to confirm use of the correct suppository mold that was originally selected for the filling and molding operation. For example, a suppository can be torpedo shaped, bullet shaped, or may be of some other shape; but this should be appropriately identified for a product.
- *Weight*: Suppository weight must be confirmed by the quality control laboratory because the final weight of the suppository can influence the assay results of the API, preservatives, and antioxidants, if any.
- *Hardness (break point)*: The suppository hardness is an important factor if the suppository base has a tendency to form a polymorph and does not solidify

completely at its intended temperature, for example, cocoa butter suppository and also a suppository with a large amount of powder material.

- *Density*: Sometimes filling and molding operations may become difficult if the density or specific gravity of the final product is too low. If a wrong base and/or a wrong surfactant is used, the suppository mold then may not hold the required amount of melt in the cavity of the mold.
- *Melting point* (*melting range*): There are several different categories of suppository bases currently available in the market (Tables 2–4). One of the main differences among the bases is the melting point, or more precisely the melting range. Melting range of the suppository base plays an important role in selecting the best possible vehicle for a drug candidate. This may directly or indirectly also influence the melting point of the final product.

Chemical Tests

- *Assay*: The Active Pharmaceutical Ingredient (API) in a suppository must be tested and verified for its claimed strength, potency, and purity. The test method must be stability indicating and validated if intended to be used for stability studies. As per the recent FDA recommendation, the test methods for all preservatives should be validated.
- *Dissolution*: In vitro drug release of suppositories is a very common practice and FDA encourages all manufacturers to develop a robust and rugged dissolution method. This will help ensure drug release and bioavailability of the dosage form up to its expiration date. In general, various dissolution test methods can be employed relevant to the physicochemical nature of the drug product. The commonly used dissolution apparatuses are described below.
  1. USP Apparatus-1 (Basket System)
  2. USP Apparatus-2 (Paddle System)
  3. USP Apparatus-3 (Reciprocating Cylinder System)

The details of the above-mentioned dissolution apparatuses are well explained in the United States Pharmacopeia.

- *pH*: The pH of the suppository may play an important role if relevant to the chemical nature of the drug or to the ingredients used in the formulation. If pH is a critical factor, then a test procedure should be incorporated in the routine Quality Control tests.
- *Other Tests*: If a formula contains a suitable stabilizer other than the preservative, for example, an antioxidant, then a quantitative Quality Control test may be incorporated as part of the total physicochemical specifications.

REFERENCES

1. Mullins JD. Medicated Applications—Suppositories. Remington Pharmaceutical Sciences, 16th ed., Chapter 87. Easton, PA: Mack Publishing Company, 1980:1530–1533.
2. Trudel J. Anatomy and Physiology of the Colon, Rectum, and Anus. Minneapolis, MN: Division of Colon & Rectal Surgery, University of Minnesota.
3. Jenkins Gl, Christian JE, Hager JP. Quantitative Pharmaceutical Chemistry, 5th ed. New York: McGraw-Hill, 1957:269–288.
4. Coben LJ, Lieberman HA. Suppositories. The Theory and Practice of Industrial Pharmacy, 3rd ed. Philadelphia, PA: Lea and Febiger, 1986:564–588.

5. U.S. Pharmacopeia, USP-27/NF-22; General Information 'Pharmaceutical Dosage Forms', pp. 2586–2587.
6. Rowe RC, Sheskey PJ, Weller PJ. Handbook of Pharmaceutical Excipients 'Polyethylene Glycol', 4th ed. Chicago, IL: American Pharmaceutical Association, 2003:454–459.
7. Allawala NA, Riegelman S. The release of antimicrobial agents from solutions of surface-active agents. J Am Pharm Assoc Sci Ed 1953; 42:267–275.
8. Riegelman S, Crowell WJ. The kinetics of rectal absorption. II. The absorption of anions. J Am Pharm Assoc Sci Edition 1958; 47(2):115, 123–127 (1958).
9. Grant DJW. Theory and origin of polymorphism. In: Brittain HG, ed. Polymorphism in Pharmaceutical Solids. New York: Marcel Dekker, Inc., 1999:1–34.
10. Giacomini G, Mascitelli E. Sommistrane dei Farmaci per Via Rettale. Gitti Ed., Milano, 1954.
11. Schanker LS. Absorption of drugs from the rat colon. J Pharmacol Exptl Therapy 1959; 126:283.
12. Food and Drug Administration, Center for Drug Evaluation and Research. Guidance for Industry "Non Sterile Semisolid Dosage Forms", May 1997, SUPAC-SS, CMC-7.
13. The United States Pharmacopeial Convention Inc., USP 29/NF-24, 12601 Twinbrook Parkway, Rockville, MD 20852.

6 Nasal and Inhalation Drug Products

John Bell

Stewart Erl Associates, Loughborough, U.K.

Paul Flanders

Mylan, London, U.K.

INTRODUCTION

The reader should refer to previous publications in this series for an introduction to generic product development (1,2) giving overall regulatory guidance to the U.S. generic market. The following commentary is an introduction to regulation of generic products in the European Union (EU). These processes are under constant development and the initiatives of the International Conference on Harmonisation including the Common Technical Document may lead to dramatic changes in regulatory applications for generic products.

SELECTION OF A GENERIC PRODUCT

The EU differs in a significant number of ways to the U.S. market and the differences influence choice of a generic target.

Unlike the U.S. market no specific incentives are offered to the generic manufacturer in Europe comparable to the Hatch–Waxman 180 day exclusivity period for the first achiever. Commonly, several generic products will be launched at patent expiry within days of each other. This generates intense price competition with rapid falls in price levels, which has led generic organizations to seek specialist products requiring less common skills in an attempt to escape the rapid price de-escalation at first launch. Oral and nasal inhalation products are one such class in Europe.

Politically, the European landmass currently includes a European Union with 27 Member States. At the beginning of 2009, there are a further 3 "Applicant" States seeking Membership, and 19 other nonaligned States, generating a complex political and regulatory system (3). The Common Market States have established a European Medicines Evaluation Agency (EMEA) sited in London along the lines of the U.S. Food and Drug Administration (FDA), with an increasing role in the regulation of medicines. However each Member State retains sufficient authority to impose its own interpretation of the guidances and data with national requirements for approval of pharmaceutical products including generics; the remaining nonaligned States continue to operate their own regulatory systems.

LEGISLATIVE AND REGULATORY ISSUES

The development of the European Economic Community, or now European Union, is complex with roots stretching back centuries. It was established in 1958 by the Treaty of Rome with a Parliament in Strasbourg and an Executive sited in Brussels. Within the EEC all medicinal products must be evaluated by a relevant competent authority and approved before they may be marketed. The same levels of quality, safety, and efficacy should be demonstrated by all medicinal products and in all Member States of the Common Market area.

GENERIC DRUG APPROVAL

The approval process for generic products in Europe is considerably more complex than in the United States. Instead of a single authority and one procedure, that is, FDA and an Abbreviated New Drug Application (ANDA), there is one major authority, the EMEA, with three principal procedures and a large number of national regulators imposing their own local interpretations of what is required. This makes access to the European generic market fragmented and uncertain.

THREE AUTHORIZATION PROCEDURES

Until the end of 2005, a marketing authorization for a pharmaceutical product in more than one country in the European Union was applied for through one of two procedures: either the "Centralized Procedure" or the "Mutual Recognition Procedure" (MRP). A third, the "Decentralized Procedure" came into force with a newly revised EU Pharmaceutical Directive at the end of 2005.

The Centralized Procedure

The Centralized Procedure (CP) is administered by the EMEA in London. It consists of a single application that when approved, grants marketing authorization for all markets within the European Union. This procedure is available to all new or so-called "innovative" pharmaceuticals and is obligatory for certain therapy areas and biotechnology-derived products. It is now also open for generic applications when 10-year "Data Exclusivity Periods" (see later) granted to originator products authorized through this procedure begin to expire. There are also six year "Data Exclusivity Periods" that will not be observed under CP.

Mutual Recognition Procedure

Until the end of 2005, authorizations for generic medicines were normally applied for through the Mutual Recognition Procedure (MRP). Under MRP, the assessment and marketing authorization of one Member State, the "Reference Member State," should be "mutually recognized" by other "Concerned Member States." The MRP is set out in Directive 2001/83/EC, as amended by Directive 2004/27/EC, and further guidance is given in the Notice to Applicants, which forms Chapter 2 of the Rules Governing Medicinal Products in the EU (4). The MRP involves a national phase with the rapporteur Member State and subsequent release of an assessment report with a mutual recognition phase with the Concerned Member States.

Generic medicinal products are submitted as abridged applications and the applicants are not necessarily required to re-perform preclinical and clinical trials. Even so, the documentation and data required are extensive and very specific. Generics applications typically include chemical–pharmaceutical data and the results of bioequivalence studies, which demonstrate the quality and the "essential similarity" of the product. For information concerning the safety and efficacy of the molecule, the regulatory agencies are referred to the data that was established in the originator product's application for authorization. This is only possible once the Data Exclusivity Period has expired on that dossier (see later and Article 10 of Directive 2001/83/EC, as amended by Directive 2004/27/EC).

The Decentralized Procedure

The Decentralized Procedure (DCP) came into operation in late 2005. It is applicable in cases where an authorization does not yet exist in any of the Member States (MS). Identical dossiers will be submitted in all Member States where a marketing authorization is sought. A Reference MS, selected by the applicant with the agreement of the MS, will prepare draft assessment documents within 120 days and send them to the Concerned Member States. They, in turn, will either approve the assessment or the application will continue into arbitration procedures. The new DCP will involve Concerned Member States at an earlier stage of the evaluation than under the MRP in an effort to minimize disagreements and to facilitate the application for marketing authorization in as many markets as possible. The MS must justify grounds for not recognizing approvals, for example, serious risk to public health. A new conciliation committee has been established to facilitate greater discussion ahead of full CHMP (Committee for Medicinal Products for Human Use) arbitration. One disadvantage is that while under MRP, applicants may not be able to withdraw from any MS once the DCP has been entered.

Generics of medicinal products that were authorized through the Centralized Procedure will have the option of applying through either the Centralized or the Decentralized/Mutual Recognition Procedures.

Summary of Product Characteristics

The Summary of Product Characteristics—or SmPC—constitutes one of the major hurdles facing a generic medicine's application for authorization. The SmPC is the information that accompanies the product (a more detailed version of the patient information leaflet included in the box with medicines dispensed).

Until 1998, originator companies submitted applications for marketing authorization on a national basis, which gave rise to differing assessments of the same data from one country to another due to differences in local medical practices and originators made differing applications for a product in the various countries. As a result, the dosage, uses, warnings, etc., often vary between Member States. The generic applicant, however, must now introduce the same application file with the same SmPC in all Concerned Member States, requiring harmonization where none existed previously.

In some cases, Member States are unwilling to accept any difference between the SmPC of the originator and the generic for their national market, and the generic applicant is either forced to withdraw the application from one or more countries in order to save the others or else be forced into costly and time-consuming arbitration proceedings. In other cases, harmonized generic authorizations may be achieved by establishing what is known as "horizontal harmony," but the resulting SmPC differences between generic and originator

SmPCs, known as "vertical disharmony," are often considered unacceptable to national authorities. As a result, the generic medicine will not be included in lists allowing substitution or reimbursement in those countries, which in turn limits, or even blocks, sales of the lower-priced product.

Contested Applications
If disagreements between the Concerned Member States over the acceptability of an application remain unresolved at a certain point, the issue is automatically referred to the EMEA for arbitration under Article 29 of Directive 2001/83/EC (as amended). The relevant Committee's opinion is then communicated to the European Commission for its final decision. From late 2005 there has also been a 60-day intermediary step offering further opportunity to reach a compromise, before the formal arbitration procedure is initiated.

In practice, however, this recourse has not proven very effective and has received only limited use, largely due to its cost to the generic company in terms of finance, human resources and, most importantly, the delay to market in other Member States. Generics companies are, as a result, often forced to withdraw their MRP applications from problematic Member States to achieve approval in others. This ultimately leaves no avenue for that company for marketing in those countries where the application has been withdrawn.

Patents
Patent protection for pharmaceutical products has increased substantially since the 1980s and the EU currently has the highest level of market protection for pharmaceuticals in the world. Patents have been available for biotechnology products since 1998. The old 15-year patent regimes were replaced with mod-ern 20-year product patents in the early 1990s. This 20-year protection can now be increased by up to five more years through a Supplementary Protection Certificate (SPC). SPCs were introduced in 1992 to compensate originator companies for the time and cost of developing registration data.

In addition to this 25-year protection, additional patents for 20 year periods are regularly granted to pharmaceutical companies for new uses, e.g., indications, dosages, polymorphs, and changes in formulation. These patents provide additional years of market monopoly for often insignificant changes, providing little or no added therapeutic value to patients.

Strategic Patenting ("Evergreening")
Pharmaceutical originators practice "total product strategies" or "lifecycle maximization" by seeking to obtain as many patents as possible during the development and marketing cycle, as well as to extend them for new uses of established products, or to add on to the time-lag between patent grant and public health approval. Originators erect "picket fences" or families of dozens of patents around a single product, covering numerous aspects of the product such as

- basic composition, including new or alternative compounds;
- method of treatment, including new use of known compounds, different dosing, and therapies in combination with other drugs;
- synthesis of API;
- formulation and drug delivery;

- prodrugs releasing active ingredient;
- substances resulting from metabolism in body;
- different crystalline or hydrated structures;
- gene-markers showing response to drug therapy; and
- new delivery systems such as patches for administering the drug.

The complex science and the extent of coverage of pharmaceutical patents therefore introduce considerable risk of infringement into product development, making professional legal and scientific advice essential. There are many genuine disputes over patent validity in comparing inventions, requiring specialist court interpretation.

The business risk of infringement therefore arises in any drug or medical device development. For generics, it arises particularly because a generic medicine is defined as being identical to a branded drug in terms of active principle and having the same pharmaceutical form, safety level, and therapeutic effect.

Further Protection

Originator pharmaceutical companies in the EU also enjoy, in addition to patent protection, a separate period of regulatory data exclusivity during which the regulatory authorities are not allowed to refer to the data on file for an originator drug in order to process an application for marketing authorization for a generic medicine. The European Commission has recognized the importance of allowing drug originators time to recoup R&D investment by harmonizing this protection period. The European Union has reviewed the pharmaceutical legislation (Directive 2001/83/EC and Regulation 2309/93). After a protracted codecision procedure beginning in July 2001, the European Commission, European Parliament, and European Council agreed a position that was adopted by the Parliament in December 2003 and formally agreed by Council in March 2004.

The final outcome has been hailed as reaching a fair compromise between the various interests, as attested to by the fact that no one group seems fully satisfied with the results. The revised law contains a number of important advances toward making lower-priced generic medicines more readily available to European patients and healthcare systems. It also increased the market protection granted to the originator pharmaceutical companies to help them recover their investment in researching and developing new treatments.

Bolar Provisions

The "Bolar exemption" is a U.S. policy that allows generic manufacturers to prepare production and regulatory procedures before patents expire, so that products can be ready for sale as soon as the patent ends, rather than having to go through the lengthy preparatory process only after the patent period is over. In order to market a medicinal product, a manufacturer must first obtain regulatory approval by conducting clinical tests and trials to prove that the product is safe and effective. Producers of generic medicines are able to use the original manufacturer's approval if they can demonstrate that the generic version is bioequivalent to the approved medicine. However, the generic producer runs the risk of patent infringement if they conduct clinical trials on a patented product before the patent has expired. The "Bolar" exemption overcomes this risk by exempting these necessary studies, tests, and trials from constituting patent infringement.

The "Bolar exemption" was a small part of a comprehensive reform of the current EU pharmaceutical legislation proposed by the European Commission in July 2001. This comprised three proposals: a regulation on marketing authorizations and the functioning of European Medicines Evaluation Agency (EMEA), a directive on medicinal products for human use, and a directive on veterinary medicinal products.

Within the two directives on medicinal products are proposals to allow the "Bolar exemption." These can be found in Article 10, paragraph 5 of the directive on medicinal products for human use, and Article 13, paragraph 6 of directive on veterinary medicinal products. The text of these paragraphs is as follows:

> Conducting the necessary studies and trials with a view to the application of paragraphs 1, 2, 3 and 4 (paragraphs 1–5 on the Veterinary Medicinal Products Directive) and the consequential practical requirements shall not be regarded as contrary to patent rights or to supplementary protection certificates for medicinal products.

Agreement on the regulation and the directives was reached between the European Council, Commission, and Parliament on December 18, 2003, and the Council adopted them in March 2004.

The United Kingdom implementing legislation came into force on the October 30, 2005, the latest date for its adoption in the EU, and amends Sections 60(5) and 60(7) of the U.K. Patents Act 1977.

Data Exclusivity and Market Protection

Data Exclusivity guarantees additional market protection for originator pharmaceuticals by preventing health authorities from accepting applications for generic medicines during the period of exclusivity. It guarantees additional market protection for originator pharmaceuticals by preventing health authorities from accepting applications for generic medicines during the period of exclusivity. The effective period of market exclusivity gained by the originator company is the period of data exclusivity (currently 6 or 10 years) plus the time it takes to register and market the generic medicine - a further 1 to 3 years.

Data Exclusivity was introduced in 1987 to compensate for insufficient product patent protection in some countries. However, strong product patents are now available in all EU Member States, and the rules on data exclusivity have been changed in new EU pharmaceutical laws adopted in 2004.

Use of Originator Data

Data exclusivity prevents regulatory authorities from assessing the safety and efficacy profile of a generic application for a period of time beginning from the first marketing approval of the originator product. Generics applications cannot use data from the originator registration file. They are approved on their own merits, using their own development data, under the same EU requirements as the originators. However, since generics contain well-known, safe, and effective quality substances, unnecessary animal testing and clinical trials on humans performed by the originator companies are not repeated. Instead, regulatory authorities evaluate the generic application against the originator documentation on file - but only after the period of data exclusivity has expired. This assessment is carried out internally by the authorities. In no instance is the originator's research data released or disclosed to the generics producer or anyone else.

Data Exclusivity is not Data Protection
Data exclusivity has nothing to do with protecting research data. Long after the data exclusivity period has expired, the originator documentation remains protected by copyright laws and other legal provisions. Data exclusivity merely extends the originator company's market monopoly by not allowing the authorities to process an application for marketing authorization for a generic.

Previous Legal Framework
Under Directive 2001/83/EC, EU data exclusivity laws guarantee market protection for originator medicines for either 6 or 10 years. Data exclusivity extends for six-years after European marketing authorization is granted in Austria, Denmark, Finland, Greece, Ireland, Portugal, Spain, Norway, and Iceland and is the period adopted by the 10 new Member States during their negotiations for EU accession. Ten-year periods of exclusivity are operated in Belgium, Germany, France, Italy, Luxembourg, the Netherlands, Sweden, and the United Kingdom. A ten-year period is also granted to an originator gaining marketing approval through the Centralized Procedure.

New Legal Framework
New EU pharmaceutical legislation adopted in 2004 has created a harmonized EU eight-year data exclusivity provision with an additional two-year market exclusivity provision. This effective 10-year market exclusivity can be extended by an additional one year maximum if, during the first 8 years of those 10 years, the marketing authorization holder obtains an authorization for one or more new therapeutic indications, which, during the scientific evaluation prior to their authorization, are held to bring a significant clinical benefit in comparison with existing therapies. This so-called 8+2+1 formula applies to new chemical entities (NCEs) in all procedures and to all Member States (unless certain new Member States are awarded derogations, which they can request following publication of the new law).

In practical terms, this means that a generic application for marketing authorization can be submitted after Year 8, but that the product cannot be marketed until after Year 10 or 11.

The revised legislation also provides a one-year data exclusivity provision for products switching from "prescription-only" to "over-the-counter" (OTC) status, on the basis of new preclinical or clinical data. The law also grants one-year data exclusivity for any new indication for a product that can demonstrate well-established use. This latter provision is noncumulative, that is, it covers only the use of the new indication and can only be used once.

Because of the adamant opposition to this overall increase in data exclusivity from the current six-year countries - especially from the Accession countries, that had not agreed to this law in their accession agreements, that were not yet entitled to vote on it during the legislative process, and those that felt it would have a significant effect on their government medicines bill - an additional clause was inserted at the last minute making the law prospective. As a result, the new periods of data exclusivity will only take effect for reference products applying for marketing authorization after the new law is fully in effect. Therefore, the first generics applications under the 8+2+1-year data exclusivity period will not occur until late 2013.

Resources for Generic Submissions
General references covering books and articles, reports, and studies can be found at http://www.egagenerics.com/ (accessed February 27, 2008).

PRODUCT DEVELOPMENT
Advising on the development of generic products in this technology is fraught with difficulty. There is intense scientific and technical activity driving rapid development of the subject, matched by a parallel comprehension and regulation of the area. New approaches, for example, FDAs Pharmaceutical Quality Assessment System and its associated Process Analytical Technology and Quality by Design initiatives, continually distract the product developer. The area is characterized by a widening library of updated guidelines from regulators, for example, EMEA Guidelines on Quality and Clinical Documentation, the lobbying of industry groups, and inevitably, growing consumer interest. Airways diseases are unfortunately increasing in society, and the need for local treatment of the whole respiratory tract, already an attractive commercial market, has been dramatically increased recently by the emerging reality of systemic drug administration through the lungs.

The first development, marketing approval, manufacture, and supply of a nasal or inhalation product from an NCE is a challenging enough task, requiring considerable resources and skill. This task is equaled today by the problems posed in bringing to market a generic copy of the established innovator medicine and complicated in comparison with most other generic products since nasal and inhalation products have two elements, a device and a formulation. These must each be developed separately in a coordinated program to provide a product "essentially similar" (in Eurospeak) to the original.

Administration of actives into the nasal or respiratory airways uniquely in drug delivery systems requires the generation of an aerosol of appropriate characteristics. The definition of these characteristics has been exhaustively investigated, is dealt with extensively in many other places, and will not be elaborated in depth here (5,6). The broad consensus is that for local treatment of the lung airways particles of 1 to 5 μm Mass Median Aerodynamic Diameter (MMAD) comprises the therapeutic fraction of the inhaled aerosol, although the presence of submicron particles in the aerosol from a number of pMDI products has recently been documented (7) and their relevance questioned. Particles of this size range, that is, 1 to 5 μm, are sometimes referred to as "respirable" particles, but due to the imprecision of this description, the terms "Fine Particle Fraction" (FPF) and "Fine Particle Mass" (FPM) were introduced referring to those particles with an MMAD of <5 μm. The former term FPF is rightly falling out of favor since it can obscure the actual dose by mass of fine particles. In contrast, treatment of the nasal airways is less clear since large drops from a nasal dropper or coarse or fine sprays may all be effective and a major objective in development of nasal product sprays is to avoid the presence of the more aerodynamically stable 1 to

5 μm particles to inhibit penetration of the nasal spray into the deeper airways of the lung.

While faithfully matching the characteristics of the innovator product is the key to development of a successful generic product, the reader must constantly bear in mind changing standards, for example, assessment of nasal products is under scrutiny by Pharmacopoeial authorities, and the expectations of the Regulator with respect to dose content uniformity of an inhaler may well have changed since the original approval of the branded product more than a decade previously. Approved propellants for pressurized products have changed from the ozone destructive chlorofluorocarbon (CFC) types to the ozone benign hydrofluoroalkanes (HFA) materials, necessitating substitution of these important excipients. This latter development has created major challenges not only for the generic industry, but also for the originator companies that were forced to undergo the unusual experience of replicating their own products (8,9). Nontechnical barriers also exist, which are outside the control of the technologist; Levine et al. discuss the adverse reaction of patients to generic substitution of inhalers even though therapeutic equivalence had been properly established (10).

The technical work can commence when the target markets have been identified, account taken of relevant intellectual property and regulatory barriers in the territories of interest, adequate resources are available, and representative samples of the originator product obtained.

The generic formulator's approach is first to characterize fully the branded product. Guidance for this is the regulatory requirements that need to be met, for example, as detailed in the FDA draft Notes for Guidance (11), EMEA guidance (12), or other countries guidances, which spell out how the Regulator will evaluate data packages. The reader should note that in the U.S. market, the draft Guidance issued by the FDA in 1998 continues under review at the time of writing. It is being extensively modified to be more concise and less numerically focused with respect to end-product testing, but with more emphasis on fostering quality-by-design (QbD) approaches and no date is currently available for its reissue. The content of the draft has been extensively criticized by the industry cooperative IPAC-RS, which can be consulted at http://www.ipacrs.com/PDFs/IPAC_Final_Comments_on_CMC.PDF (accessed February 5, 2008).

Overlaid may be further general guidance's, for example, relating to the continuing transition from CFC propellants (13). The parameters required for the future generic product are, however, reasonably defined by these documents. This evaluation presents the generic manufacturer with their first technical challenge to develop methods suitable for analysis of the obtained representative samples of the branded product, and then eventually to further develop these for their own product. The reader should refer to a companion volume for general guidance on analytical method development (14). This product analysis is crucial to the project as it sets the development specification targets. Occasionally useful data may be available in the literature (15). The product characteristics established will determine performance of the product in the clinic including such elements as particle size distribution data, dose delivered to the patient, and dose content uniformity, all generated using available pharmacopoeial methods plus product stability in terms of the established and other yet-to-be-set parameters. The importance of establishing accurate baseline data cannot be overemphasized since without this information the generic developer stands no chance of success. The limited market success to date of generic (CFC free) beclomethasone dipropionate (BDP) pMDI products is related in part to the technical problems of replicating the originator product (16), in terms of particle size distribution. The originator product is formulated with CFCs generating an aerosol particle size distribution with MMAD ~3.6 μm; the generic copy using the new HFA propellants in which the active is soluble is much finer with MMAD ~1.1 μm (17) and questions of product safety and clinical equivalence arose, which continue to time of writing.

The analysis of the brand leader product must also include a breakdown of components of the device to cover valve and metering volume for pressurized products, container volumes, materials of construction, resistance to air flow, and other elements discussed later. It is of considerable advantage if identical device and formulation components can be obtained from the supplier and exhaustive enquiries in this direction must not be overlooked. The generic developer may also look for potential areas of improvement where commercial advantage may be obtained. This could be a longer shelf life, better dose uniformity, or ease of use and convenience over the brand.

Development of respiratory and nasal medicines today has become an extensive and complex challenge and all aspects of this work cannot be covered comprehensively in a single chapter. The development of oral and nasal generic products will be approached by first considering important characteristics of the delivery systems, move onto excipients and formulation, and consider in detail Pharmaceutical Development and the Final Product Specification. Aspects such as impurities, stability testing, manufacture, and validation of equipment will be discussed only briefly. The science and technology of pharmaceutical inhalers continues to develop and is discussed in an ever-increasing number of treatises (18–21), which will repay study by entrants to the field. These will provide a broad background helpful for the development of products widely recognized as among the most complex of modern drug delivery systems.

Use of In Vitro Data Alone to Substantiate Therapeutic Equivalence

An important development in the field of generic respiratory products is the potential use of in vitro data alone to establish similarity. There is no fundamental reason why equivalence in performance of mechanical devices such as inhalers should not be capable of establishment by in vitro determinations alone, comparable to the use of dissolution testing in solid dosage forms. Variation in the performance of inhaler devices is entirely due to variation in use by the patient provided the devices being compared are proved similar by comprehensive evaluation and are consistent in performance. It is the lack of confidence of Regulators and Clinicians in the current technology to provide "comprehensive evaluation" that is generating the need for increasingly complex and expensive clinical studies, which are the greatest single barrier to entry of generic copy products in the respiratory field. However, recent European Guidelines have cautiously begun to explore the way forward in this area, and the growing expertise and strength of respiratory technology in the generic industry should be encouraged to enlarge this initiative, understandably but indefensibly against the wishes of the innovator industry. The guideline is discussed below.

EMEA Guideline on the Pharmaceutical Quality of Inhalation and Nasal Products (12)

This Document identifies the characteristics of orally inhaled and nasal products that must be evaluated for clinically supported approvals in the European Common Market and Canada and extends this for the first time in an Appendix to generic products. This Appendix states that therapeutic equivalence to the innovator product must be substantiated by in vivo and/or in vitro studies, but in all cases the comparability must be established by stated parameters. Any differences beyond normal analytical variability must be accompanied by a rationale as to why the differences would not result in different deposition or absorption characteristics. The interpretation of such a statement has yet to become established in practice but provides a significant opportunity in the creation of a generic respiratory market. It should be recognized by the technologist that in the case of *clinical study*–supported applications, the copy generic batches used in the clinical studies are the key reference points in building a submission, whereas for *nonclinical* equivalence applications the copy batches used in the in vitro studies are key.

The critical parameters identified for in vitro establishment of therapeutic equivalence in the Guidance are

1. Complete individual stage particle size distribution profiles.
2. The above at a range of flow rates, if flow rate dependency is not excluded.
3. Delivered dose.

The Guideline recognizes that information from batches used in vivo will not be available and makes suggestions for areas where further data will be required.

1. Extractables/leachables
    A safety assessment could be made based upon a comparison of the generic versus the innovator product provided this is justified by the composition of the packaging. This could apply to use of identical valve components in pressurized inhalers where the generic technologist is able to establish that the same components are used. The same approach will be applicable in nasal and other liquid products. In the situation where alternative materials are used, a full identification and safety evaluation will have to be completed.
2. Delivered dose uniformity and fine particle mass over patient flow range
    In the absence of in vivo studies, the range of flow rates used in products sensitive to such variation, for example, dry powder inhalers, must be justified. This information can readily be obtained by simple studies in respiratory clinics. Studies using both the generic and innovator device should be carried out noting the European guidance comments precluding the use of variable innovator product as reference.
3. Particle/droplet size distribution, total drug delivered, and drug delivery rate
    Particularly applicable to continuous liquid spray products, information on the characteristics of the copy product from the commercial scale process is required compared to the key batches used for the establishment of in vitro equivalence. This is not unusual and replaces the usual requirement for comparison with the batches used in clinical studies.

4. Pharmacoepial Excipients

    Particularly applicable to dry powder devices, specification of excipients, for example, lactose must be linked to the characteristics of the excipient in the batches used to establish equivalence.

EMEA Guideline on the Requirements for Clinical Documentation for Orally Inhaled Products (OIP) Including the Requirements for Demonstration of Therapeutic Equivalence Between two Inhaled Products for Use in the Treatment of Asthma and Chronic Obstructive Pulmonary Disease (COPD) in Adults and for Use in the Treatment of Asthma in Children and Adolescents (http://www.emea.europea.eu/pdfs/human/ewp/4850108en)

Somewhat confusingly the Clinical Guideline includes many requirements that are pharmaceutical rather than clinical. However, it follows a similar pattern but is even more helpful, giving general considerations concerning the requirements for in-vitro characterisation and clinical documentation in respect of all inhaled dose forms and multi-strength products, and identifying nine criteria to be satisfied if therapeutic equivalence by in vitro means to a reference product is to be substantiated, together with a number of very useful elaborations on essential supportive data. There are indications that the problem of approval of a generic inhaler should be solely on the grounds of bioequivalence and not include the interchangeability argument. The nine criteria are as follows:

1. The product contains the same active substance (i.e., same salt, ester, etc.). This is an unambiguous requirement.
2. The pharmaceutical dosage form is identical. Together with criteria 4 and 7 below, there is a strong thread that the copy product should be capable of being used by the patient without further training causing associated staff costs, which simply detract from the economics of generic products.
3. "The active substance is in the solid state (powder, suspension); any differences in crystalline structure and/or polymorphic form should not influence the dissolution characteristics, the performance of the product or the aerosol particle behavior." The terminology of the draft guideline is not clear for this criterion, but it appears to mean that the active if in the solid state should dissolve at the same rate as the reference product. It is not clear if it is intended to exclude actives in solution, for which no substantiation is apparent to the authors.
4. Any qualitative and/or quantitative differences in excipients should not influence the performance of the product (e.g., delivered dose uniformity, etc.), aerosol particle behavior (e.g., hygroscopic effect, plume dynamic and geometry) and/or be likely to affect the inhalation behavior of the patient (e.g., particle size distribution affecting mouth/throat feel or "cold Freon" effect). This requirement appears to be a nonspecific catchall and is elaborated to suggest effects in quality or quantity of the dose delivered, plume geometry, and organoleptic effects. It is also backing up requirements 2 and 7 in contributing to seamless substitution of the copy product.
5. Any qualitative and/or quantitative differences in excipients should not change the safety profile of the product. It may be difficult to be certain that use of an alternative excipient has no effect on the active(s) solubility rate,

on absorption, or on local effects in the airways, and changes should only be made if absolutely essential. The consequences in terms of clinical and safety evaluation may be far reaching.

6. The inhaled volume needed to get sufficient amount of active substance into the lungs should be similar $(+/-15\%)$. In the case of nebulizer products, the inhaled volume used will be an effect of the device used and the performance is readily checked. Pressurized inhalers used with the standard actuator are unlikely to cause any difficulty. Introduction of a breath-actuated system will require careful checking to ensure that actuation takes place at the beginning of inhalation and not at high flow rates when measurable volumes of air will have been inhaled before release of drug. Whether such a variation would have any therapeutic effect is a situation that the generic formulator is advised to avoid. Dry powder inhalers generally pick up the drug at the beginning of inhalation, although capsule types, for example, Cyclohaler™ may sprinkle the contents throughout the inhalation cycle. The objective should be to use a similar mechanism to the reference product.

7. Handling of the inhalation devices for the test and the reference products in order to release the required amount of the active substance should be similar. This requirement is likely to generate considerable debate. Again pressurized inhalers with a standard actuator will comply readily, as will nebulizers. Dry powder inhalers may cause a major difficulty due to the wide variation in design.

8. The inhalation device has the same resistance to airflow (within $\pm15\%$). This point is mentioned later in this chapter in the discussion on dry powder inhalers.

9. The target delivered dose should be similar (within $+/-15\%$). Again this is unambiguous.

The Guidance gives helpful commentary on the detail of the data required from aerosol particle size analysis data. Inevitably, it is required to be from impactor/inertia analysis: the whole of the distribution must be examined and separated into not less than four classes or "groups of stages." Where flow rate dependency is present, that is, in dry powder devices, the flow rates used must be based on the patient population targeted and include the 10th and 90th percentile. Limits of $\pm15\%$ are suggested as maximum in vitro differences on each of the stage sets used and 90% confidence intervals are required.

While it appears that the European Regulator intends to be tough on compliance with the proposals, the generic designer will find the approach a welcome alternative to the relative imprecision, cost, time, and other problems of executing clinical studies.

Quality by Design (QbD) and Process Analytical Technology (PAT)

That quality should be designed and not tested into a product was given a new impetus in August 2002 when FDA outlined its initiative on Pharmaceutical cGMPs for the 21st Century (22). Subsequent guidance under the PAT framework (23) has further supported the move to make Quality by Design a reality. The intention is to acquire and develop greater knowledge of product and process characteristics such that quality is designed into products and away from quality assurance by end product testing, in turn creating a basis for a more flexible regulatory approach. The "Control Space" of a product would be positioned within a totally understood "Design Space" providing an enhanced understanding of the product characteristics. The concept and its ramifications as applied to respiratory products have begun to be explored and reports are being made in the literature and at relevant conferences currently. The European Pharmaceutical Aerosol Group (EPAG) presented their preliminary studies on its application to a pressurized inhaler recently (24) and identified the following six significant steps in the process:

1. Determine the critical performance characteristics of the product.
2. Identify the potential material attributes and process parameters.
3. Perform a risk assessment of these parameters.
4. Perform screening experiments to establish the "Explored Space."
5. Perform design space experiments to establish the "Design Space."
6. Determine the "Control Space."

The main driver for both innovator and generic technologists will be enhanced understanding of their product. To what extent the barriers, that is, the additional work required, will be outweighed by the benefits will only emerge over the next few years and the evolution of the process should be observed closely in the literature, at meetings, and by discussions with Regulators at appropriate times.

There are three main classes of products of current generic significance to be considered for drug delivery to the respiratory tract:

Pressurized metered dose inhalers
Dry powder inhalers
Aqueous droplet inhalers

PRESSURIZED METERED DOSE INHALER

The pressurized metered dose inhaler (pMDI) is a long established delivery system (25) and the reader should consult O'Callaghan and Wright for a general review (26), Smyth (27) and Purewal and Grant (28) for more detailed information, and Bell and Newman for a recent update (29). An update on the known physics of aerosol generation from pMDIs has been given by Versteeg and Hargrave (30). Copying an innovator product is materially helped by the observation that unlike DPIs, pMDIs are visually similar and are constructed and perform similarly. The basic packaging components are the canister, a pharmaceutical quality-metering valve, and a mouthpiece/actuator combination. Subsequent to the Montreal Protocol and propellant transition (31), the formulation in the canister comprises non-CFC propellants, selected from the hydrofluoroalkanes P134a and P227 (NB: The correct terminology for this propellant is 227ea but the generally used reference is 227), the active, and optionally, excipients. In addition to these core elements, there have been developed a range of "add-ons" including counters, devices to aid synchronization of dose generation with inhalation, and mechanisms to assist impaired patients to use the inhaler.

The generic designer must evaluate the total range offered by the original in developing the program of work and identify exactly what is to be developed.

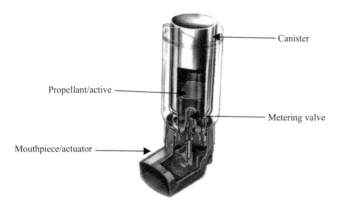

FIGURE 1   Pressurized metered dose inhaler.

Devices

The general form of the pMDI is shown in Figure 1. An aluminium canister is contained within a molded plastic actuator body, which allows canister insertion at one end and has a mouthpiece extension at the other. Active drug is suspended or dissolved in the propellant filled into the canister, which is then fitted with a properly crimped on metering valve. The stem of the valve fits into a shaped receiver inside the actuator. Operated in the orientation shown in Figure 1, downward pressure on the canister base forces the valve stem to slide into the valve housing, releasing a dose from an internal metering chamber through an atomizing nozzle into the mouthpiece where violent break-up of the jet occurs as the propellant explosively evaporates, resulting in the progressive formation of an aerosol cloud, which is inhaled by the patient.

Components

Canister

The container for the bulk product must primarily be capable of withstanding the high internal pressure generated by the propellant and provide an inert surface to resist physical or chemical changes to the contents. There are two containers in general use, aluminium monobloc and plastic-coated glass, with the former being most widely used. With a 50-μL valve, a 60 shot product will require a 5-mL container, a 120 shot an 8–mL, and a 200 shot a 12 mL.

Aluminium monobloc: Canisters are available in a variety of sizes and can be tailor made to the purchaser's specification. Coating of the internal wall of the container has entered the technology in recent years to combat chemical interaction between walls and contents and to reduce adhesion of suspended active, which would cause dosage variation. The use of can coatings must be carefully considered by the generic developer. Inspection of the brand leader products will demonstrate which strengths of the product use coated cans, often the lowest only, and chemical analysis will identify which coating is used. The developer has then to consider if the coating is required for the formulation under development and the expectations of the target market's Regulators. The patent situation with can coating also needs to be considered, as this varies from country to country. As newer can coatings become available and patents are successfully

TABLE 1 Materials of Construction of Metering Valves

Component	Material	Effect on performance
Metering chamber	Polyester, acetal	Dimensional stability, extractables, and leachables
Core extension	Polyester, acetal, stainless steel	Ditto
Inner core	Polyester, acetal	Ditto
Body	Polyester, acetal, nylon	Ditto
Washers/seats/gaskets	EPDM, nitrile, bromobutyl, chloroprene	Leakage, moisture ingress, extractables, and leachables
Spring	Stainless steel	Consistent valve movement
Ferrule	Anodized aluminium	-

challenged the choice available to the developer increases. The choice will also depend on supporting information such as chemical analyses, extractable data, and preclinical data to support the registration package.

If a coated canister is used then leachable and extractable data will need to be generated for preclinical assessment and subsequent license application.

Manufacturers of aluminium canisters include Presspart (www.presspart. com, accessed February 3, 2008) and 3M (www.3m.com/dds, accessed February 3, 2008).

Metering Valve
The pharmaceutical metering valve is a key component of the pMDI drug delivery system and has evolved over the past several decades into a sophisticated and remarkably reliable mechanism. Materials of construction are shown in Table 1 and current industry suppliers are shown in Table 2 with the source of valves used on a number of products. Valve selection must also take into account the patent landscape in each of the territories where the product will be marketed. For example a patent, EP 0708805B1, covering the use of EPDM in valves is still valid in some European countries. The development scientist should check the patent validity in the countries where the product will be marketed. Technology will have moved on since the innovator product was developed, and Regulators may be unwilling to accept the innovator dose variability. The generic technologist should bear in mind that as in all science, attention to detail is imperative (32).

Figure 2 shows the pharmaceutical metering valve in its simplest form; it is important to grasp the basic metering principles so that modifications introduced to the design can be understood.

The valve stem passes through the metering chamber with its distal end (exit) firmly fixed in the actuator housing and its proximal end immersed in the propellant drug mix. At rest the propellant/drug bulk mix in the canister is continuous with liquid in the metering chamber via a slot in the valve stem and consequently the mix fills the metering chamber from the canister bulk contents. When actuated by pressing on the canister base, the stem slides to close

TABLE 2 Metering Valves and Formulations Used for Various pMDIs

Product	Active(s)	Company	Valve supplier	Formulation
Airomir	Albuterol/ Salbutamol	3M	3M	134a + EtOH
Q-Var	BDP	IVAX	3M	134a + EtOH
Alvesco	Ciclesonide	Altana	3M	134a + EtOH
Allergospermin	Sodium Cromogly-cate/Reproterol	Asta	Bespak	134a + EtOH
Beclazone	BDP	Ivax	Bespak	134a + EtOH
Salbumol/ Salamol	Albuterol/ Salbutamol	Ivax	Bespak	134a + EtOH
Budiair	Budesonide	Chiesi	Bespak	134a + EtOH + nonvolatile
Beclojet	BDP	Chiesi	Bespak	134a + EtOH + nonvolatile
Atimos/Forair	Formoterol	Chiesi	Bespak	134a + EtOH + nonvolatile
Intal	SCG	Sanofi-Aventis	Bespak	227
Tilade	Nedocromil	Sanofi-Aventis	Bespak	227 + PVP + PEG + flavor
Berodual	Fenoterol/ Ipratropium	Boehringer	Bespak	134a + EtOH
Berotec	Fenoterol	Boehringer	Bespak	134a + EtOH
Atrovent	Ipratropium	Boehringer	Bespak	134a + EtOH
Salbutamol	Albuterol/ Salbutamol	Aldo-Union	Bespak	134a + EtOH
Salbutamol Evohaler	Albuterol/ Salbutamol	GSK	Valois	134a only
Flixotide Evohaler	Fluticasone	GSK	Valois	134a only
Seretide Evohaler	Fluticasone/ Salmeterol	GSK	Valois	134a only
Serevent Evohaler	Salmeterol	GSK	Bespak	134a only

off the metering chamber from the bulk and continuing movement releases the metering chamber contents through an orifice in the valve stem into the actuator and thence to the mouthpiece.

All pharmaceutical metering valves use these basic principles with various modifications to eliminate faults that have been identified during the past several decades of use, and the generic technologist must understand the background to changes made to this critical component of the delivery system. Historic problems that led to modifications include sealing problems (often due to the propellant characteristics), dose variation due to drainage from the metering chamber at rest (referred to as "loss of prime"), and problems resulting from unstable suspensions.

Sealing problems

There are three seals (also known as gaskets) in the metering valve all of critical importance to performance. A large seal closes gaps between the main valve

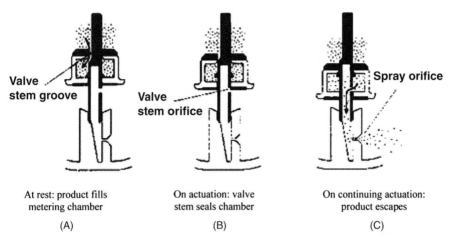

FIGURE 2 Pharmaceutical metering valve—how it works. From Ref 113.

body and the aluminium/glass canister rim. Two further small seals are used, one to close the distal (outer) exit of the stem and one the proximal (inner metering tank) end of the stem.

Seals are made of rubber, which has a complex composition, and the generic technologist must pay considerable attention to these components. Rubber seals are dimensionally unstable due to swelling under influence of the propellant, which unfortunately is also capable of extracting components of the rubber. The nature of these extractives has required the investment of much effort on the part of the industry. A report of work carried out at the Product Quality Research Institute provides valuable information (33). The consensus is that at least leachables (see later) must be identified both qualitatively and quantitatively in the innovator and generic products during development. It has been reported that leachables may have an effect on particle size of a suspension product in particular storage orientations (34). During manufacture of the inhaler, assembly dimensions must be accurately controlled to ensure accurate and precise sealing. The seals may be pre-extracted by the supplier with the propellant (and possibly other solvents) to be used in the product. The technologist should develop a close relationship with the designated supplier to understand exactly what is being done in order to design in-house extractive testing as well as to ensure a trouble free situation. The choice of materials for use as valve seals is slowly being extended. Each new material has its own benefits and issues, whether it is better compatibility with alcoholic formulations or reduced extractables and leachables. The selection of a valve seal material requires great care.

Dose variation due to drainage at rest—"Loss of prime"
The metering chamber of the pMDI fills during the return to rest by the valve arrangement discussed previously. Since the valve chamber is in direct communication with the bulk at rest, emptying can also occur depending on the orientation of the canister. If active drains from the chamber in varying amounts,

the result is a variable dose when the inhaler is actuated. This is the reason why pMDIs are recommended to be operated for one or two non-inhaled ("waste") shots prior to use after a period of nonuse. The phenomenon, which has become known as "loss of prime," was first reported by Cyr (35) when it was identified by the Canadian Health Protection Branch during the development of generic salbutamol pMDIs. Technologists in the industry were already aware that analytical methods for the demonstration of consistent dosing in vitro required considerable care.

Innovation in valve design is currently occurring to ensure that consistent dosing from pMDIs in the clinical context is a more robust process. It does not appear to have been recognized of significance clinically in use of β_2 adrenergics, possibly because patients might titrate out bronchostriction with an additional "puff" or dose (36); however, the increased use of other drugs, for example, steroids where there is no immediate feedback to the patient of their effect demands improved technology. Quantifying the loss of prime is required under draft and final guidances from various health authorities and should be reported in the license application and reflected in the patient instructions. The introduction of counters on pMDIs is likely to draw patient's attention to the impact of "wasted" shots, increasing the demand for better technology (37–39).

Problems with unstable suspensions

Formulators working on pMDIs up to the 1990s when the CFC propellants were in use were able to develop stable suspensions by utilizing suitable surfactants, notably sorbitan trioleate (Span 85), first used (40) in the 1950s. Oleic acid was later introduced into the first suspension formulations of salbutamol by Sir David Jack and his group at Allen & Hanbury. These together with lecithin, sorbitan monooleate (Span 80), and combinations of the three CFC propellants P11, P12, and P114 served to develop many products. The introduction of the ozone-benign hydrofluoralkane propellants HFA 134a and 227 has posed a serious problem in product development due to the lack of solubility of the established excipients. A very large number of claims for alternative excipients are recorded in the patent literature from the mid-1990s onwards (41), as a frantic search ensued but to date no significant advances have appeared.

HFA-based pMDIs have now appeared in most markets; data given by Bagger-Jorgensen et al. (42) drawn from an IPAC-RS database indicate that the variation in delivered dose of HFA suspension pMDIs is significantly greater than the CFC suspensions replaced (RSD HFA:CFC = 9.6:6.4). The development of data demonstrating satisfactory dose uniformity is a challenge to in vitro analytical methodology.

This increased problem of suspension instability exaggerates loss of prime, since there is an enhanced likelihood of active deposition both in the metering chamber and canister during nonuse periods. This can lead to higher and lower doses than label claim at the first dose if the canister is not adequately shaken and waste shots fired. This situation in the relatively newly marketed HFA formulations makes the task of the generic formulator no easier.

Important points to look for with a metering valve are

- Shot weight reproducibility should be better than ±5%.
- Leachables should be minimized, of low toxicity, and at consistent levels.

- High uniformity of delivered dose, desirably all at nmt ± 20% of target.
- Low drug deposition and absorption on to valve surfaces.
- Low leakage of propellant.
- Low moisture ingress via the valve.
- Materials inert to chemical interaction with drug/excipients/propellant.
- Reliable supply chain.

Suppliers of metering valves include Consort Medical (formerly Bespak) (www.consortmedical.com, accessed February 3, 2008)), Valois (www.valois.com, accessed February 3, 2008), and Neotechnic (www.3m.com/dds, accessed February 3, 2008).

Actuator/Mouthpiece

The mouthpiece of the standard pMDI serves a dual purpose, also acting as the actuation mechanism as illustrated in Figure 1. The stem of the metering valve is inserted in a tight-fitting housing inside the actuator body. In use the base of the canister is pressed with the thumb to overcome the resistance of the valve mechanism, usually about 40 N, allowing the metered dose to be released forward into the mouthpiece and inhaled by the patient. The mouthpiece should be sufficiently long to be comfortable for the patient to grasp with their lips. All current pMDIs are used in a valve down orientation since the metering valve operates without use of a dip tube; in use the canister is therefore spatially vertically above the mouthpiece and the geometry of the system is arranged to ensure that patients can use the device comfortably, tilted away from the face.

Mouthpieces are usually constructed of high-density polyethylene or polypropylene with suitable identifying coloring materials. In the development phase, close attention must be paid to drug deposition on the mouthpiece during dosing caused by the impingement of spray from the explosive evaporation of the propellant. About 10% of the metered contents are usually retained by the actuator, and during the development phase, close observation of this parameter must be maintained since it directly affects the delivered dose and variation can indicate changes in components.

The physical forms of inhaler actuators may be subject to design registrations or other intellectual property protection, and the generic technologist should initiate suitable checks during the product development phase.

The actual effect of the actuator on the aerosol delivered to the patient is a matter of some contention.

Purewal and Grant (43) raise a number of issues about which the technologist should be aware including the influence of the actuator spray orifice diameter and taper, and the "dead volume" including the valve stem into which the propellant is admitted and first begins to change state from liquid to gas. Chiesi have used valve orifice variation as one component to modify the particle size distribution of their proprietary Modulite™ formulations (44).

Important points to observe in actuator manufacture are

- Strict dimensional control over the orifice diameter and circularity.
- Strict control over the molding process and polymer formulation.
- Plume alignment with the mouthpiece to minimize and ensure consistent deposition.
- Absence of flex on actuation.

- Adequate support for the canister to prevent lateral forces on the valve.
- Absence of flash or other molding defects that could affect plume aerosolization or direction.

Spacers

Spacers are means of containing in a reservoir the therapeutic aerosol cloud generated by an inhaler so that it can be inhaled by the patient a few seconds later. Introduced as a simple approach to overcome the need for synchronization of inhalation with actuation of the pMDI inhaler, spacers have turned out to be a surprisingly complex add-on (45–47). The generic pMDI should always be investigated, initially by in vitro means, to confirm that there is equivalence of performance with the innovator inhaler in use with its spacer. One approach to this is to determine the half-life of the aerosol cloud in the spacer device compared to the generic; keeping in mind that infant respirations using a spacer with a pMDI are intermittent and shallow, the stability of the cloud up to at least 30 seconds should be studied. If recommended for use in a particular patient population, for example, pediatrics, its use may have to be validated in that population and its suitability supported by appropriately designed clinical studies. The requirement should be discussed with the Regulator, especially in Europe where if in-vitro equivalence cannot be demonstrated, then clinical studies are mandatory as is spacer development for standard pMDIs designed for children. An investigation of the effect of the recommended cleaning procedure should also be made. Within Europe the generic manufacturer is able to recommend use of its product with a spacer owned by an innovator organization, but has no control over future changes to the device. It is usually necessary to identify an available specific spacer for use with the generic pMDI such as the Trudell Aerochamber (48). While spacer devices have demonstrable advantages, it is uncertain how many patients actually use them consistently; however, their incorporation into the generic clinical program must be considered either within the phase III program or as a separate study.

Breath operated systems

The coordination of dose delivery with the act of inhalation by the patient has long been suspected as a problem in consistent use of pMDIs (49,50). Passive dry powder inhalers, that is, devices that rely on the patient's inhaled air stream for operation, are inherently inhalation coordinated creating the particle cloud during inspiration, while in contrast pMDIs generate an aerosol in a burst, independent of the patient's inhalation effort. The availability of bronchodilator "rescue" drugs in convenient pocket/handbag/sports bag friendly robust pMDIs over many decades has established the pMDI as a global favorite system with patients, but at the same time effectiveness and popularity of this particular medication may have obscured the problem of lack of coordination. At the present time while coordination is intuitively desirable and increasingly important with the entry of other actives, for example, steroids, to the market, the clinical evidence in favor of breath-actuation (51,52) is very strongly indicative rather than definitive. A number of devices developed to overcome coordination failure are available in the market, and their use is increasing. Examples are shown below (Fig. 3). It is noteworthy that the EasiBreathe device is uniquely provided with a special spacer add-on, the Optimizer. Since breath-operated pMDIs eliminate the

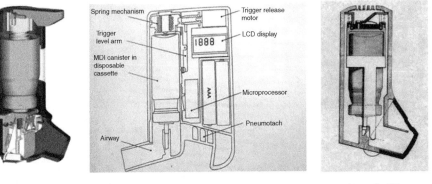

Autohaler ™   Smartmist ™   EasiBreathe ™

FIGURE 3 Complex breath-operated actuators. From Ref. 114.

synchronization problem and the Optimizer serves both to slow the fast moving jet and remove coarse particles from the aerosol reducing deposition in the upper respiratory tract (53), it fulfils the function of a spacer device. This approach has yet to be fully exploited in pMDI delivery systems.

The use of a breath-actuated system represents an opportunity for the technologist to differentiate their generic inhaler from an innovator standard pMDI to the advantage of the patient. Clinical studies using a pMDI invariably include significant training, coaching, and monitoring of patients in the use of the conventional press and breathe devices, a standard which is only likely to be maintained in every day use by a minority of patients. Breath-actuated systems will maintain the discipline of correct administration of doses throughout inhaler life and therefore their use should be encouraged. Clinical experience to date is strongly indicative of the potency of breath-operated devices on pMDIs, and it is probable that only the very high cost of large double-blind placebo-controlled confirmatory trials inhibit their widespread adoption. The recent design and introduction of low-cost systems, for example, using a kinked plastic tube, the K-valve (54), shown in Figure 4, and currently well advanced in development, should open up the desirable aim of universal replacement of the standard ski-boot form with modern breath-operated actuators. There is no reason why generics should not consider taking the lead in this patient friendly development.

Counters

A draft Guidance from FDA (55) in 2003, reinforced by encouragement from EMEA (56) in 2006, has spurred industry to develop counting systems for pMDIs. A comprehensive review of the topic has been given by Bradshaw (57) who noted that the primary purpose is to protect patient safety. While the principal needs of the Regulator are count reliability and display method, the generic technologist must consider access including intellectual property ownership, costs and marketing, combined to give the patient ease of use, and robustness similar to the innovator product, which may not in fact include a counter since the current regulatory approach is not retrospective. Generic products have an opportunity

Autoversion. Dose loaded as mouthpiece is opened and released on inhalation

The metered dose is discharged into a kinked plastic tube and then released by straightening of the tube during inhalation.

Manual version. Dose loaded as canister pressed and released automatically on inhalation

FIGURE 4 The K-valve and K-halers. From Ref. 114.

to add value to their copy product relative to the innovator by inclusion of a low-cost mechanism.

As analyzed by Bradshaw, counters are either direct (rely on an active event of firing such as sound, temperature, or pressure change) or indirect (rely on canister movement/thumb pressure). The patent field for the first type appears to be relatively free but is very crowded for counters of the second type. The available indirect counters consist of two types, pressure operated (utilizing the "press" employed to operate the canister), or displacement operated (using the canister movement to work a mechanism).

Direct counters may be the least likely to fail being linked to the actual delivery of a dose but so far are technically difficult to achieve. Indirect counters are simpler to devise and it is in this technology that all the available devices are found at present. At time of writing only one organization has introduced a counter to the market, but in view of the proposal for all new pMDIs to have at least one in development, it can be expected that they will become a standard item. Available information on counters is given in Table 3.

FIGURE 5 Helix Counter. An indirect displacement counter. Mechanism attached to counter. Variety of formats. From Ref. 114.

TABLE 3 Sources of pMDI Counters

Organization	Type	Source
DCA		www.dca-design.com (accessed February 3, 2008)
Valois	Displacement, numeric, color	www.valois.com (accessed February 3, 2008)
Cambridge Consultants	Displacement, numeric, color	www.cambridgeconsultants.com (accessed February 3, 2008)
Trudell	Range of displacement types	www.trudellmed.com (accessed February 3, 2008)
Clinical Designs (Helix, Fig. 5)	Displacement with 1 count resolution and color at end of life	Info@clinicaldesigns.com (accessed February 3, 2008)
Bang & Olufsen Medicom	Displacement, numeric	www.medicom.bang-olufsen.com/ (accessed February 3, 2008)
Bespak	Displacement, numeric	www.consortmedical.com (accessed February 3, 2008)

Formulation development

Materials

Propellants

There are two propellants approved for use in pMDIs, hydrofluoralkane 134a and hydrofluoralkane 227. Both are pharmaceutical grades of widely used industrial refrigerants and are usually known as HFCs or HFAs. These two were selected in the early 1990s as nonchlorine containing molecules with suitable properties for extensive safety evaluation by an international group of pharmaceutical companies acting as the International Pharmaceutical Aerosol Consortium (IPAC) who funded programs to evaluate HFA 134a (IPACT-1) and HFA 227 (IPACT-2).

Modern HFA formulations are constructed very differently to the previous chlorofluorocarbon products (CFC 11, CFC 12, and CFC 114), which most commonly were binary or tertiary mixtures of the three allowing blends to be constructed that delivered optimal formulation properties with respect to solvency, density, and boiling point. With HFAs the choice is far more restricted and there is no HFA replacement for CFC 11, which with a boiling point of 8°C could be handled unpressurized in cold rooms and found extensive use as a dispersant for suspension formulations. This forced alternative formulation approaches where ethanol is used to replace CFC 11 and more use is made of pressurized manufacturing vessels. Thus, two rather similar HFAs are available, nearly always unblended and generally with the use of ethanol acting as a cosolvent or as a dispersing agent.

Some key properties are listed in Table 4, compared to a traditional CFC blend.

HFA 134a tends to be the default propellant. As a simpler molecule, with less fluorine atoms and being associated with a large-scale technical refrigerants business, it has major cost advantages over HFA 227. However, HFA 227 can be required if a denser propellant medium or a higher boiling point is needed.

Testing: Both HFAs are fully tested by the suppliers and will normally be accompanied by comprehensive Certificates of Analysis. A pMDI manufacturer

TABLE 4 Propellant Properties

	HFA 134a	HFA 227	P11:P12 = 30:70
Formula	CF_3CFH_2	CF_3CFHCF_3	$CFCl_3/CF_2Cl_2$
Boiling point ($^{\circ}$C)	-26.1	-16.5	~-5
Liquid Density (kg/m^3 at 25°C)	1207	1387	1350
Solubility of water in liquid propellant (% w/w, at 25°C)	0.11	0.061	~ 0.006
Solvent properties	Weakly alcohol-like	Weakly CFC-like	A strong, nonpolar, solvent
Flammability	Nonflammable	Nonflammable	Nonflammable

will normally be expected to perform at least basic ID testing (such as identity by infrared), but Regulators vary in their requirements. Some suppliers have arrangements with independent laboratories to facilitate separate "acceptance" testing, as required in the United States.

Purity: While there is a range of minimum purity specifications across different suppliers, it is true to say that all HFAs are considerably purer than the medical CFCs used to be, in some cases by orders of magnitude.

Manufacturers: There are three international manufacturers and suppliers of the medical HFAs. The leading supplier is INEOS Fluor, followed by DuPont and Solvay.

GMP standards: Although these gases are excipients, the regulatory authorities treat them as though they were actives, in terms of both documentation review and GMP standards required. The suppliers have fully followed this requirement, and these gases are some of the tightest controlled excipients available to the industry.

HFA 134a. HFA 134a can be considered from a vapor pressure viewpoint as a CFC 12 replacement—a propellant in the true sense of the word. However, its other physical properties are very different and are largely responsible for the extensive reformulation effort that is needed to produce a viable 134a formulation. Key properties include

Density: Significantly lighter than the CFCs, this can add to problems with suspension stability, with behavior like claying intensified.

Polarity: This is a polar molecule and is miscible in all proportions with ethanol—a very widely used adjuvant in many 134a formulations. This can allow for a medium polarity active that has been traditionally handled as a suspension (e.g., beclomethasone dipropionate) to be formulated as a solution of 134a and ethanol.

However, the main problem that arises is the enhanced solubility of, and affinity for, water, which is 18 times that of CFC 12. Water can diffuse into filled aerosols through elastomeric seals, giving serious problems with stability particularly in high humidity storage. Also, traditional surfactants are not soluble in pure 134a, although oleic acid is sparingly soluble in ethanol/134a mixtures.

Safety: The toxicology of HFA 134a has been studied intensively, both by the manufacturer and the pharmaceutical industry. The key studies were carried out by two groupings of pharmaceutical companies. The most widely quoted study was by IPACT-1, which established the safety of a benchmark 134a specification in these applications. The European CPMP (Committee for Proprietary Medicinal Products) published the specification and further announced that no further submissions on propellant toxicology would be needed unless it did not conform to the specification. This approach tends to be followed in many other countries, with the exception of the United States, where it may still be necessary to join the relevant IPACT group to gain access to the original data.

Specifications: All HFA 134a specifications from the main manufacturers comply with both the IPACT-1 specification and an FDA draft guidance specification. Neither HFA has a published finished monograph in any of the Pharmacoepia as yet, so these two sources are universally taken as benchmarks.

HFA 227. With its higher boiling point and greater density, HFA 227 slightly resembles some of the old CFC blends. It is usually used alone in formulations, although there has been occasional discussion about the possibility of a blend with 134a. However, the advantages of such an approach seem to be slight, and no product has yet emerged onto the market based on such blends.

Density: Its greater, CFC-like density has been used to stabilize difficult high-loaded suspensions, such as sodium cromoglycate.

Polarity: It is significantly less polar than 134a, but moisture uptake is still a threat and, again, traditional surfactants have little solubility. It is not usually used in blend with ethanol, as this would only make it more 134a-like.

Safety: The critical toxicology studies were carried out by the consortium IPACT-2, which again justified a minimum purity specification. Unlike the IPACT-1 work, this specification has not entered the public domain, but suppliers have nonetheless managed to provide compliant product by simply going to extremes of purity.

Suppliers: www.ineosfluor.com (accessed February 3, 2008), www.solvay-fluor.com (accessed February 3, 2008), and www2.dupont.com/Dymel_Pharmaceutical/en_US/ (accessed February 3, 2008).

Other excipients
There are very few other excipients used in pMDIs due to the twin challenges of safety and suitability. Oleic acid, sorbitan monooleate, sorbitan trioleate, and phosphatidyl lecithin cholate were commonly used in CFC formulations up to 2% but are all relatively insoluble in the HFA propellants. Ethanol finds more frequent use particularly with steroid products to both enhance solubility and provide a manufacturing aid. However, ethanol can be used to enhance the solubility of oleic acid in the formulation. Other excipients such as polyvinyl pyrolidone and polyethylene glycol have been used, although the patent situation is complex. A number of formulation patents exist in the United States whose application in Europe is uncertain.

Formulation options
There are two types of formulation in current use, suspensions and solutions. The first product that appeared in the market in pMDI form in 1956 was a solution of isoprenaline sulfate in CFC propellants using a high proportion of ethanol as a cosolvent. The wide particle size characteristics of this product became clear when inertial impingers were introduced into the technology (58) and the suspension approach was preferred to provide better aerosol particle size control. However, when HFA propellants became available in the 1990s, it was found that steroid actives were sufficiently soluble to allow the use of solution formulations and a significant reduction of particle size was achieved. Thus, both solution and suspension formulations are found in HFA products today. Aspects of both formulation types are heavily covered by patents in many countries and the generic developer must be fully aware of the situation for the target markets.

Suspensions. This approach is suitable for drugs with negligible solubility in HFA propellants such as salbutamol/albuterol, salmeterol, and fluticasone. Chemical stability will be less of a problem, but physical stability has emerged as a major problem. Good dispersion of the suspension is essential to obtain a uniform dose and the absence of satisfactory dispersants is a continuing difficulty. Ostwald ripening is always a possibility and this will require careful evaluation during development. Excipient free suspension formulations have been developed, which are described in the patent literature (59).

Solutions. This approach is available for actives that are soluble in HFA propellants/ethanol mixtures such as beclomethasone, ciclesonide and flunisolide (60). Inevitably, chemical stability may be more of a problem but drug content uniformity is excellent, enabling the use of a weighing procedure in some determinations where it can be justified. The particle size of the inhaled aerosol requires careful investigation, as it is likely to be fine although methods have been developed to manipulate this parameter (61). It is worth noting that clinical studies on some steroid solution formulations are still contentious (62).

Pharmaceutical Product Development
An extensive range of performance studies must be carried out during the development of new formulations of pressurized inhalers and these will be discussed in this section, followed by a review of the Final Product Specification, before considering manufacturing methods briefly.

Initially, analytical methods should be developed, which will be fully validated with the final product formulation, after which small numbers of formulations are prepared in canisters using one of the techniques described later. Glass vials with a protective coating are commonly used at this stage to allow viewing of trial suspension and solution formulations. The performance of the chosen formulations should then be evaluated against a number of criteria—and the innovator product—using Pharmacoepial methods (USP/Ph Eur), where applicable. This will eventually enable a draft Final Product Specification to be prepared. When sufficient confidence in a formulation has been accumulated, a trial batch of several hundred canisters should be prepared using a draft documented procedure and evaluated with a view to obtain the information discussed below. Storage of the product in a variety of orientations should be undertaken at a range of conditions, with samples withdrawn periodically. It is wise to store samples of the innovator product for comparison purposes. With growing confidence further batches can be prepared.

Investigations must now be conducted to establish that the dosage form, formulation, manufacturing process, container closure system, microbiological attributes, and instructions for use are appropriate and result in acceptable product performance. Sufficient batches should be studied, often three, to take batch variability into account. Of particular importance are batches used in clinical studies, which should be very fully characterized.

A number of authors have reported on the evaluation process for several current molecules including flunisolide (63), salbutamol (8), salmeterol (9), and budesonide (15). A useful description of the early formulation assessment process is given by Robins and Brouet (64) who developed a budesonide HFA pMDI. Important aspects of development are given below with commentary on the different items.

Drug substance. Unless the drug substance is to be dissolved in propellant, for example, beclomethasone dipropionate, the particle size and particle size distribution of the micronized active must be fully characterized. Laser light scattering techniques are in common use for the raw material and apparatus is readily available (65). The total particle size range should be well defined and acceptance criteria set. The product used in pivotal clinical studies is key and underpins the specification. A variety of information may be required by Regulators such as particle morphology, solvates and hydrates, polymorphs, amorphous forms, solubility profile, moisture and/or residual solvent content, microbial quality, dissociation constants, and specific rotation. These will all form part of the API raw material specification.

Physical characterization. If the drug substance is to remain in the solid state in a suspension formulation, it should be characterized with respect to crystallinity to ensure that any changes during storage are anticipated. Beclomethasone undergoes changes when suspended in CFC propellants, although to date there are no reports of similar effects in HFA propellants. Changes have been reported with budesonide postmicronization (66). Processing, including micronization, may affect the crystallinity of materials and should be studied as part of the product development.

Chirality. If the active is chiral, studies should be carried out to determine any potential for racemization. These will lead to clarity on the need for achiral or specific assay procedures to control the presence of the enantiomer.

Fill weight. The range of fill weights should be determined and specified and reassurance obtained that the minimum fill weight justifies the claims for the number of delivered doses to be claimed, taking into account factors such as leakage rate, filling machine tolerances, ullage and overages for factory, and patient priming.

Extractables and leachables. The reader should refer to the earlier discussion under metering valves and sealing problems. "Extractables" are materials that can be removed from packaging materials and components by use of solvents under extreme conditions to represent the worst scenario of undesirable materials that might enter the medicine. "Leachables" are what is actually extracted by the formulation during the claimed shelf life of the product. It is necessary to identify at minimum the leachables and possibly the extractables, conduct safety assessments where necessary, and set limits. The innovator product should be examined similarly and the results compared if appropriate. Reference should be made to the territory Regulators to determine current thinking on the topic. At the time of writing, the studies conducted by PQRI (33) and EMEA Guidance (12) are key sources of intelligence. A preclinical assessment of the resultant data must be carried out and used to aid in the justification of the specification.

Delivered dose. The delivered dose is the quantity of drug delivered per actuation from the pMDI actuator to the patient. The metered dose is the quantity of drug per actuation in the pMDI metering chamber, usually taken as the quantity that can be collected directly from the valve stem. The current trend is to measure the delivered dose by a Pharmacoepial apparatus (USP or Ph Eur) or suitably validated alternative, and the purpose of the test is to evaluate the uniformity of the doses delivered throughout the life of the product. With this in mind, attention should be paid to any instructions to be given to the patient with respect to shaking or orientation of the inhaler. In Europe, a total of 10 doses are collected at the beginning (after any priming shots), middle, and at the end (e.g., 3, 4, 3) of a number of canisters and assayed separately. A standard is set by the Pharmacoepias (USP, Ph Eur), for example, all individual shots are within 20% and the mean within 15% of the label claim. The same approach is used to determine within and between canister and batch variation. A second tier is set to include more containers if a few results outside the first limits are found. An examination of doses fired after the label claim has been reached should also be made and the tail-off determined, unless a lockout device is fitted as part of a counter mechanism. Such devices are not favored on asthma "rescue" drugs, for example, salbutamol/albuterol. Standards proposed by FDA (11) are contentious following an investigation by IPAC-RS (http://www.ipacrs.com/PDFs/IPAC_Final_Comments_on_CMC.PDF, accessed February 5, 2008) and alternative approaches are being considered. The generic technologist should determine the current position before evaluating data since it is not only possible that standards have changed since the innovator product was developed but the overall statistical approach may be different.

Fine particle mass. The critical particle size and particle size distributions of medical aerosols are determined by inertial impaction techniques, which allow chemical determination of the mass of active at various size ranges. Details of apparatus and methods are now available in the USP (67) and Ph Eur (68) to which reference should be made. The Fine Particle Mass (FPM) representing the active dose is considered to be the mass of active particles below 5 μm aerodynamic diameter and can be obtained by pooling the appropriate stages of the multistage impactor for a single-dose determination if necessary. Even with this approach, determination of the FPM for single doses can be challenging. If impractical for low-dose drugs, a justification should be constructed and discussed with the Regulator. The FPM should be determined throughout the container life and for the tail-off portion. An evaluation of any spacer device to be used must be carried out to determine the effect on FPM. Spacers (69) are likely to increase the amount of fine particles inhaled as propellant evaporates.

They also remove large particles by impaction or settling mechanisms, otherwise destined to deposit in the upper respiratory tract. The determination of the suspended cloud half-life over 30 seconds will give a useful picture of events in the spacer. The study should take into account any instructions to be given to the patient, for example, recommended delay before inhaling or use of tidal breathing. Cleaning of spacers should also be investigated. The occurrence of static electricity generated by vigorous cleaning procedures can change dramatically the available dose, requiring an investigation and definition of appropriate procedures. The information obtained from these studies should be reconciled with any clinical data.

Particle size distribution. The overall particle size distribution of the aerosol must be measured usually with an impactor according to the Pharmacoepial methods. The total range is important, not just the fine particles, since material impacting in the upper respiratory tract may enter the gastrointestinal tract and be absorbed leading to undesirable drug activity. A plot of cumulative percentage less than the impactor cutoff sizes leads to the Mass Median Aerodynamic Diameter (MMAD—note the word order is sometimes changed leading to AMMD), and Geometric Standard Deviation if the data indicate a uni-modal log-normal distribution. It has been recommended that a mass balance should be carried out on the data but this apparently simple procedure has recently been revealed as presenting yet another possible problem (42) in the technology. Regulators require the test to be based on a patient dose, despite the fact that some products, for example, formoterol are 6 or 12 µg per actuation. This can lead to problems in quantification.

Actuator deposition. Drug inevitably deposits on the actuator mouthpiece and should be accurately determined across a number of devices and batches. The value should be consistent since variation may indicate, for example, a molding problem. Together with Delivered Dose, the total value should be similar to the dose determined ex-valve.

Priming, re-priming, and shaking requirements. In the case of suspension products, the generic technologist must carry out careful evaluation of the copy product to ensure that directions can be given to the patient to ensure a correct and consistent dose. If required, the time of shaking and the number of priming shots needed to achieve label claim must be investigated including storing the product in various orientations and using different time intervals to define the need for re-priming. The product should be tested in this way during storage testing. Information given on the innovator patient information leaflet should be examined and it may be worth confirming the claimed procedures.

Cleaning requirements. The need for cleaning of the actuator must be investigated using dose delivered and aerosol particle sizing techniques under conditions replicating patient use with regard to dosing regimens and priming.

Low-temperature performance. The storage of pMDIs at low temperatures, for example, in a domestic refrigerator, will lead to a reduction of the vapor pressure of the propellant with a possible fall off in uniform dosing and fine particle generation; HFA 134a falls from 4.9 barg to 1.9 barg when the temperature is changed from 20°C to 0°C. Warming the canister in the hand is a simple method to restore the system if investigation reveals an unacceptable fall in FPM and/or Dose Content Uniformity.

Temperature cycling.   Fluctuations of temperature in storage and patient use can lead to a variety of effects in pMDIs and should be investigated thoroughly. During storage testing, samples from several batches should be cycled from high (e.g., 40°C) to a low temperature (e.g., −5°C or even to freezer temperatures) possibly via an anticipated recommended storage temperature (e.g., 20–25°C) with 12 to 24 hours at each point. Appearance, leak rate, weight loss, delivered dose uniformity, FPM, related substances, and moisture content should be examined.

Robustness.   Patients inevitably will find all possible ways to use and misuse the inhaler and its robustness must be challenged in the development phase—and in the final form if changes have been made during development. Tumbling in a rotary mixer and dropping in various orientations from a fixed height on to a hard surface are essential to check for weak points. If prototypes were used in any significant early evaluations, the performance of the final to-be-marketed device must be confirmed. Actuators made from single cavity molds should be examined for performance when manufactured on multicavity molds for particle size, dose delivery, and actuator deposition. Purewal (70) has dealt with the topic unintentional misuse and suggested test methods.

Device development report
Records should be kept of the development of the device and the various steps taken to confirm suitability for use documented. If standard valves, actuator, and canisters are used this is a relatively simple undertaking. However, if a novel breath-operated actuator, spacer, etc., is to be introduced, the significance of such records increases.

Microbiology.   Pressurized products generally are a hostile environment to microorganisms. In addition to testing routinely for contamination by standard Pharmacoepial procedures, a test should be carried out in the development stage by seeding product samples with representative microorganisms and confirming nonviability. It may be possible to justify nonroutine examination of batches from the information obtained.

Microscopy.   Examination of a sprayed slide has little to offer pMDI product development in the 21st century unless the active has a specific particle appearance that the technologist has reason to check or an investigation of the possible presence of foreign particulates is desired (71). However, microscopic examination can give a good indication of any particle coarsening mechanisms (crystal growth or agglomeration) in suspension formulations.

Water.   Use of HFA 134a in particular may lead to moisture increase and studies should be carried out to determine if moisture levels affect product performance. Careful attention to the elastomers used in valve sealing and discussion with suppliers may help to reduce any problem encountered.

Leak rate. Elastomer valve seals swell after propellant filling and newly filled batches are quarantined for 2 to 4 weeks after the filling process. After stabilization, the leak rate should be determined on a representative number of canisters from each of several batches by reweighing after storage at a range of temperatures. This profile can then be used to set a leak rate specification. Different approaches are used to remove leakers but in general canisters are heated to about 50 to 55°C for 2 to 10 minutes in a water bath or air blast and check weighed. Some workers use canisters with inserted temperature probes to confirm that 50°C to 55°C is reached.

Packaging. Current FDA draft guidance suggests that the color of the canister contents and inside of the canister and valve assembly should be examined as well as any changes monitored during the stability evaluation stage. This will require either use of a special penetrating needle device to release propellant or cooling of the canister to below −30°C and removal of the valve. The contents should then be allowed to warm to room temperature, allowing the propellant to evaporate before examination. Alternatively, the contents may be transferred to a clear container (glass or PET), a blank ferrule fitted, and the formulation is then examined for any changes.

Final product specification
The Specification for the Final Product represents the standards to which the marketed product will comply, encompassing the totality of the pharmaceutical development studies plus information obtained during storage testing. Again, it should be firmly based on the batches used to establish clinical equivalence with the innovator product. A complete description of the acceptance criteria and analytical procedures with sampling plans may be required in the submission detailing the number of samples tested, individual or composite samples specified, and number of replicate analyses per sample. The proposed validated test procedures should be documented in sufficient detail to allow validation by the reviewing agency. The following are the elements that should be considered:

Description: A description of the formulation, defining the active substance(s) and ingredients, including the FPM per actuation, and of the full product, including actuator and canister. It should include the visible components, their color, and any characteristics of their shape.

Identification: The FDA and EU Health Authorities require two independent identification tests and also a test for the counter ion if present. Chromatography, UV, and IR spectrophotometry represent choices.

Assay: The amount of identified active present usually expressed as weight per canister.

Impurities and degradation products: The standard required is that all impurities or degradation products above 0.1% must be identified and controlled. There is considerable documentation from the International Conference on Harmonization in this and related areas, which should be consulted (111).

Water content: If a limit for water content has been established to ensure product stability, it should be defined. Testing in the development stage may have shown that there is no problem and this will remove the need for a limit.

Delivered dose: The delivered dose used is increasingly the ex-actuator dose, the main exception being older products in some areas of Europe, which have traditionally used the ex-valve dose. It is calculated from the Delivered Dose Uniformity test.

Delivered dose uniformity: Compliance will have been investigated during the development phase and while the Pharmacoepias and Regulators are currently agreed on standards and apparatus, attention should be paid to changes in this area as discussed previously.

Fine particle mass: The FPM is accepted currently as the mass of active below 5 μm MMAD in size. It is determined by Pharmacoepial impactor methods and reported as actual mass not as a percentage of any other parameter, with upper and lower limits.

Leak rate: The leak rate determined during the development phase with limits is given.

Microbiology: The limit can be a Pharmacoepial limit or if the earlier studies justify no limit then the specification is omitted.

Leachables: The results of the detailed studies carried out in development on Extractables and Leachables will define the nature of the specification. The copy product should be at least comparable to the innovator.

Number of doses: The number of doses varies according to canister size and valve size but the number of doses must be stated.

Manufacture

There are four options to manufacture pMDIs:

Hand filling: this is the usual laboratory approach to prepare small numbers for preliminary trials.

- Weigh drug and excipients into individual canisters
- Crimp on valve
- Pressure fill or cold fill with required amount of propellant

Cold filling: this is a technique popular with the previous CFC propellants but less in use with HFAs.

- Formulation chilled to liquid state
- Filled as a liquid
- Valve crimped on
  The following techniques are the basis of large-scale manufacture:

Two stage filling

- Can purged with pure propellant
- Valve crimped on
- Formulation filled through valve
- Pure propellant filled through valve

Single stage filling

- Can purged with pure propellant
- Valve crimped on
- Formulation filled through valve

Suppliers of equipment include http://www.pamasol.com (accessed February 3, 2008).

When satisfactory information has been obtained, which may be utilized for obtaining authorization for clinical studies and the technologist is confident that the product is equivalent to the innovator product, the next step is to scale up the process and prepare large batches for long-term stability and possibly further clinical studies. Regulators favor clinicals from product made at the final site of manufacture and scale. At this stage, it is important to agree a program of work that provides full information for the eventual ANDA/Product License Application. The characteristics of the clinical trial batches will set the standards for the eventual marketed product.

If an experienced contract organization is not to be employed, a manufacturing process is devised and the process validated. Three batches are then prepared and appropriate numbers set up in an appropriate stability program, for example, see ICH Guidelines (72) and suitable numbers diverted for clinical studies. The recently issued ICH Guideline Pharmaceutical Development Q8 provides general advice on manufacturing process development. A specific example relating to inhalation products is in course of development at time of writing. Within Europe, inhaler manufacture is considered a nonstandard process and there is an expectation that process validation data is submitted at the time of license application. Further progress of the product is by the usual route through Regulators, manufacturing, etc.

DRY POWDER INHALER

A wide variety of forms of the dry powder inhaler (DPI) have evolved (73), with different internal component configurations, different means of storing the powder mass, measuring a dose, and causing particle aerosolization. Developing a generic DPI means either accessing an available developed device, or carrying out the total device development process of a new concept. This process can be surprisingly prolonged and indeed may be as long although not as costly as the evolution of an NCE.

Devices
All DPIs must provide the following four elements (Fig. 6).

Powder formulation
The active drug has to be present in the inhaled aerosol in particles of MMAD 1 to 5 μm. Such particles have a very high surface area to mass ratio, exhibit strong adhesive and cohesive properties, and are very difficult to handle both industrially in bulk and more particularly to redisperse into their ultimate fine particle form with the low energy input available from current portable DPIs. A considerable amount of research effort has been invested in developing an understanding of their properties, and techniques are at last becoming available offering the possibility of improved powders for inhalation. Their flow and dispersal properties significantly affect the performance of DPIs and the methods in use are discussed later.

Mechanism to meter the dose
The DPI may be either self-metering, that is, have an in-built mechanism for metering a dose from a bulk reservoir, for example, Turbuhaler^{TM}, Easyhaler^{TM},

or premetered, that is, carry out the metering in the manufacturing stage into a suitable container such as a foil pocket (Diskhaler™, Diskus™) or hard gelatine capsule (Handihaler™, Aeroliser™). The former approach, bulk reservoir, is likely to be less precise than the latter that can add to problems of Dose Delivered uniformity in the performance of the final product. Although there is continuing debate about the merits of the two approaches, the protection offered by foil packs to the active is at least superficially an attractive feature.

Mechanism to transfer the dose to the inhaled airstream
The mechanism may be active or passive. The active mechanisms shake or vibrate the metered dose into the inhaled air, for example, the Aeroliser™ pierces the gelatine capsule, then shakes it violently in a centrifugal airstream to sprinkle the active into the passing inhaled air (74). A similar mechanism is used in the Rotahaler™, Handihaler™, the newer Flowcaps™ (75) device, and in a more sophisticated form in the Spinhaler™. The passive mechanisms use impingement of the inhaled air onto the exposed premetered dose to lift it into the airstream, for example, Diskus™, Diskhaler™, Turbuhaler™, the newer Gyrohaler™ (76) and Xcelovair™, and the Exubra inhaler for systemic drug delivery (77).

Mechanism to disperse the dose into respirable particles
Virtually all current clinical DPIs use the energy of the inhaled air stream to separate the agglomerated 1 to 5 μm particles into individual particles by creating turbulence inside the device, for example, Turbuhaler™, and it is variation in this feature as well as the powder formulation that contribute to variation in the critical Fine Particle Mass of the administered aerosol.

Characteristics of the DPI device
The performance of a DPI can be evaluated in vitro with respect to dose delivered, particle size distribution, FPM, and drug content uniformity with accuracy and precision. Similarly, the shelf-life characteristics regarding stability of the components and performance over a prolonged period can be determined by well-established procedures. However, regulatory bodies continue to require clinical substantiation of performance of devices and this can be the most costly element in the development of a generic inhaler.

In the case of pressurized metered inhalers, this position might eventually be eroded. Discomfort of clinician's regarding equivalence of generic pMDIs may be based more on variation in patient use of the device than related to any variations in the product that could not be quantitated, once identified, by in vitro means. However, in the case of DPIs there is evidence of more complex relationships (78). Variations in air-flow resistance, point of dose injection, and formulations, for example, may all influence the entry, penetration, and distribution of drug particles in the lung airways. High air-flow resistance of an inhaler

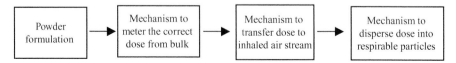

FIGURE 6   Basics of a DPI.

might be construed as leading to higher shear and turbulence in the internal airstreams of the device leading to greater dispersion of agglomerates and higher FPMs. Srichana et al. (79), however, using lactose/salbutamol mixes showed that a Diskhaler™ with one-third the air-flow resistance could deliver approximately twice the FPM of a single capsule Inhalator™ both operated at 90 L/min, the difference being attributed to formulation. Such effects are currently under scrutiny (80) in many laboratories. Bondesson et al (81) have, equally counter intuitively, demonstrated that dose delivery late in the breath can increase dry powder aerosol penetration into the lungs. It cannot currently be excluded that such technical variation between devices might lead to detectable change in clinical performance of generic DPIs. A further contribution to the topic was made at a Conference in May 2005 (82) when the issue of whether DPIs, now appearing extensively in Europe, could be used interchangeably was discussed. The consequences of variation in the increasing number of devices are causing concern to physicians. Price et al. (83) have noted that the approval on the basis of regulations designed to safeguard quality of dry powder inhalers does not mean that devices are interchangeable. Clinical trials to establish equivalence may not take into account factors in patients behavior or variations in patient inhaler technique that may affect use of devices in real-life situations. The cost savings of using generic devices may be offset by the cost of demonstrating to the patient how to use the new device and the additional physician time to address concerns and management costs if disease control is affected. The availability in the literature of studies to establish the patient-friendly characteristics of many individual devices are basically unhelpful; Nielsen and coworkers praised the characteristics of the Diskhaler™ in 1997 (84): within a few years a study by Diggory et al. (85) concluded that most elderly people cannot use the device, and treatment was unlikely to be effective unless the delivery system was improved.

The development of successful DPI generic products continues as a challenge for technologists, clinicians, and Regulators.

Product development
The general problems involved in development of a DPI have been reviewed by Newman and Busse (86). The Pharmaceutical Development and Product Specification of a generic dry powder inhaler here will center on the European market where the entry of generic DPIs is commonplace. In the United States market there is yet to be approval of a generic DPI, which may underscore a difference in approach of the two Regulatory areas (87).

Device selection
There is no generic physical form of the DPI analogous to the basic pMDI. Historically, many organizations have developed DPI devices in anticipation of replacing the pMDI when it came under siege in the 1990s. Certain of these have progressed to multicavity mold stage, are in production, and the first step is to determine the availability of a suitable device. Table 5 shows some devices that might be investigated. A number of organizations have licensed the Innovata Clickhaler™ to market generic products; the Novoliser™ and ISF inhaler are also among generic DPIs in Europe, and Ranbaxy have recently launched a range of DPI products in the United Kingdom in the Orion Easyhaler™. As discussed above it may be wise to select initially a device with similar characteristics to the innovator product (Fig. 7), although it must be clearly understood that

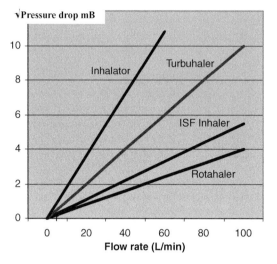

FIGURE 7 Examples of pressure drop and inhaled air flow relationship for four DPIs.

neither inhaler airflow resistance alone nor the inhalation flow rate achieved by the patient can predict how efficiently a DPI will deliver a drug to the lungs.

The requirement for a dose counter must be considered. A counter is essential on the self-metering devices since the patient usually is not able to determine how much treatment remains. The premetered devices may require a counter if the prepackaged dose is not visible, for example, Accuhaler™. If the premetered

TABLE 5 Some Possible Generic Devices

Device	Source	Location
Cyclohaler™	Plastiape	http://www.plastiape.it/
Easyhaler™	Orion	www.orionpharma.com
Pulvinal™	Chiesi	www.chiesigroup.com/
Next™		
Novolizer™	Viatris	http://www.medapharma.de/patient/novolizer/
Taifun™	LAB	taneli.jouhikainen@labinc.fi
Aspirair™	Vectura	www.vectura.com
Eclipse™	Sanofi-	http://www.sanofi-aventis.com/
	Aventis	
Flowcaps™	Hovione	www.hovione.com
Turbospin™	PH & T	www.phtpharma.com
Directhaler™	Direct Haler	www.direct-haler.com
	A/S	
ODPI	Otsuka	www.otsuka.com
Gyrohaler™	Vectura	www.vectura.com
AcuBreathe™	Respirics	www.respirics.com
Microdose™	3M	www.microdose-tech.com/
Oriel	Oriel	www.oriel.com
Clickhaler™	Innovata	www.innovata.co.uk
Dispohaler™	AC Pharma	www.ac-pharma.de
MIAT	MIAT	www.miat.it

product uses separately packaged hard gelatine or HPMC capsules, for example, Flow Caps™, Eclipse™, which can be readily determined, then a counter is unnecessary. The newly available Vectura Gyrohaler™ is understood to use a strip pack in a container that is translucent and appears to avoid the need for a counter mechanism, thus lowering manufacturing costs.

Formulation development
Materials
Excipients
The range of excipients for use in DPI formulations, like pMDIs, is again limited due to safety and suitability. The most extensively used is lactose, which apart from a small number of lactose intolerant patients has proved satisfactory for 50 years. Other materials, for example, glucose, magnesium stearate, leucine, and mannitol have been explored (88) and may find future application.

In order that powders can be mixed effectively to form a bulk that can be handled on filling machines and yet be readily disaggregated by the patient's inspiratory effort, particle characteristics such as size, shape, rugosity, and possibly electrical charge are important. The factors influencing the use of lactose have been subject to considerable investigation (89,90) particularly focusing on the presence of "active sites" on lactose crystals that can be inactivated by lactose fines or other materials leading to varying degrees of active drug release from blends. This work provides the generic technologist with powerful means to tailor the copy product to mirror the performance of the innovator product.

Formulation options
The active powder for inhalation has to be prepared in the form of individual particles in a size range of 1 to 5 μm. The usual process used for this is air jet milling ("micronizing") and this process produces powders with a spread of particle sizes. This procedure is best carried out for generic products by using the services of a contractor (91) organization, although the active substance supplier may provide ready milled substance. The specification for the material will be Pharmacoepial with an additional particle size clause, for example, 99% w/w <10 μm, 95% w/w 1–5 μm. The specification will of necessity be required to match that of the innovator product and achieving this can be a prolonged exercise.

The active fine powder at this particle size inevitably has very poor flow properties due to electrostatic and Van der Waal forces causing interparticle cohesion and adhesion to other surfaces. It is difficult to handle and only the early Turbuhaler™ has ever attempted to utilize micronized active alone, which was considered leading edge technology at the time but a challenging project to meet dose uniformity standards.

Four techniques are currently known for developing formulations for DPIs.

The simplest approach initially used in the 1940s in Abbott's Aerohaler™ and refined later in the Intal™ Spinhaler™ (92) was to incorporate sufficient coarse particle size flow aid with good flow properties to enhance the flow properties of the micronized active. Lactose is used as this coarse flow aid excipient almost exclusively in current formulations.

When a high proportion of active is present (e.g., 50% w/w in Intal™), such blends are found to be difficult to manufacture with consistent physical properties, and this led to the development of a pelletization process (93) later utilized by Astra in its Turbuhaler^{TM} formulations (94). More recently (95) the potential of spray drying (96), supercritical fluid (97), and "airy particulate" (98) technologies are being investigated. This latter advance is particularly innovative, based upon particles incorporating a honeycomb structure providing geometrically large particles of relatively small aerodynamic size, reported simultaneously to have improved flow properties plus a suitable size for inhalation. At the present time no products are available in the market based on the spray drying, supercritical fluid or "airy particulate" (99) technologies, all of which are subject to considerable intellectual property rights. Clark and York (100) have reported uninspiring data from an investigation on supercritical fluid derived powders; Sakagami and Byron (101) have recently reviewed the potential of respirable microspheres and concluded that critical and unique challenges including toxicological evaluation have to be overcome. The generics of the near future are most likely to be based upon the lactose blend or pelletization approaches and only these two techniques will be considered.

Lactose blend: Early products in the field introduced the idea of mixing the micronized lumpy, cohesive, difficult-to-handle active with coarse lactose having inherently good flow properties to produce a blend with suitable handling characteristics. A suitable lactose specification might not be more than 10% below 30 μm and not more than 30% greater than 70 μm, determined by a suitable technique such as an Air Jet Sieve or laser diffraction. This limits the finer particles that inhibit powder flow and the very coarse ones that are uncomfortable in the oropharynx on inhalation. Much subsequent work has been carried out on the characteristics of lactose that have an influence on the active fine particle content of the aerosol cloud generated by the inhalation device and this knowledge provides ways to tune the formulated product to match the innovator product. Techniques are becoming available for preparing mixtures on the small scale (102).

Pelletization: Pelletization of medicaments is described in the patent (93) and in general literature (103), and the initial approach is to utilize the techniques described if available. The objective is to agglomerate the fine active drug particles into loose rounded pellets that have adequate flow properties to enhance handling but are sufficiently loosely bound to disperse into discrete respirable particles when injected into the inhaled air stream and exposed to the turbulent disruptive forces of the inhaler device.

Pharmaceutical product development

As with pMDIs, an extensive range of performance studies must be carried out during development of a generic DPI product.

A device has to be selected and this is usually on the basis of contacting various organizations, for example, as in Table 4, and considering the mix of device appearance and characteristics, technology/experience available, terms, etc., as is usual in the generic industry. It is unwise to choose a device at too early a stage in its development, for example, pre-single cavity molds, since the path to approval can be surprisingly long and tortuous, particularly if the technology of injection molding and general engineering is foreign to the generic organization. Obvious features to look for in a DPI are

Patient use.

- Simple operation—ideal is open, breathe, close. Capsules are fiddly. Twisting, pressing, inserting, ejecting, etc., are to be avoided.
- Quickly taught/learned—this is a generic and excessive physician time to teach destroys the economics.
- Clear counter mechanism—large digits or obvious analogue presentation.
- Unobtrusive—patients do not want to advertise their use of the device.

Technical.

- Accurate metered dose—premetering is a good approach.
- Consistent FPM—no air-flow dependence (15 L/min upwards).
- No cleaning—not easy; at least ensure it is not critical.
- Total moisture protection, including exhaling through device.
- double dosing eliminated.

Although there appears to have been no serious difficulty in bringing generic DPIs to market in Europe with the current range of asthma medication, at the present time it would be wise to select a device with

- similar air-flow resistance,
- similar FPM over the whole flow range,
- similar dose injection point in the inhaled air stream (little data at present), and
- similar user handling characteristics

to the innovator product targeted.

The nature of the powder formulation to be employed will follow from what is already known about the performance of the device and batches of the product can then be prepared, or obtained. Analytical methods should be developed, after which a number of formulations are prepared using one of the techniques described earlier. The performance of the product should then be evaluated against a number of criteria—and the innovator product—using Pharmacoepial methods where applicable. This will eventually enable a target draft Final Product Specification to be prepared. When sufficient confidence in a formulation has been accumulated, a trial batch of several hundred filled devices should be prepared using a draft documented procedure and evaluated with a view to obtain the information discussed below. Storage of the product in a variety of orientations should be undertaken at a range of conditions including humidity variation, with samples withdrawn periodically. It is wise to store samples of the innovator product for comparison purposes. Include stresses that compact the powder, for example, vibration at a range of frequencies and amplitudes.

Investigations must now be conducted to establish that the dosage form, formulation, manufacturing process, container closure system, microbiological attributes, and instructions for use are appropriate and result in acceptable product performance. Sufficient batches should be studied, often three, to take batch variability into account. Of particular importance are batches used in clinical studies, which should be very fully characterized.

Drug substance and physical characterization
The comments under pMDIs are applicable.

Chirality
If the active is chiral studies should be carried out to determine any potential for racemization. These will lead to clarity on the need for achiral or specific assay procedures to control the presence of the enantiomer.

Fill weight
In the case of a reservoir device, it is necessary to demonstrate that the minimum fill weight chosen is sufficient to provide the labeled number of doses, at the same time ensuring that the final dose is able to meet the standard for delivered and fine particle dose. If a premetered dose device is used, Pharmacoepial limits for unit fill should be consulted but it will be necessary to confirm that adoption of the standards will ensure that the total and fine particle dose criteria are met.

Delivered dose
The delivered dose is the quantity of drug delivered from the DPI to the patient and is measured by a Pharmacoepial apparatus or suitably validated alternative. Note that a dose may be more than one inhalation. As with pMDIs the purpose of the test is to evaluate the uniformity of the doses delivered throughout the life of the product. With this in mind attention should be paid to any instructions to be given to the patient with respect to cleaning or particularly orientation of the inhaler when being loaded or used. In Europe, a total of 10 doses are collected at the beginning, middle, and end (e.g., 3,4,3) of a number of devices and assayed separately. A standard is set by the Pharmacoepias, for example, all individual shots are within 20% and the mean within 15% of the label claim. The same approach is used to determine variation within and between devices and batches. A second tier is set in the standard to include more devices, if a few results outside the first limits are found. An examination of doses fired after the label claim has been reached should also be made and the tail-off determined, unless a lockout device is fitted as part of a counter mechanism. Such devices are not favored on "rescue" drugs, for example, salbutamol/albuterol. Standards proposed by FDA (11) are contentious following an investigation by IPAC-RS (http://www.ipacrs.com/PDFs/IPAC_Final_Comments_on_CMC.PDF, accessed February 5, 2008) and alternative approaches are being considered. The generic technologist should determine the current position before evaluating data since it is not only possible that standards have changed since the innovator product was developed, but the overall statistical approach may be different.

Fine particle mass
The comments under pMDIs apply in part but the different characteristics of passive (i.e., using energy from the patient inhalation only) DPIs have to be considered. The FPM of early formulations and devices showed considerable dependence on inspiratory air flow rate, and subsequent investigations over decades have revealed a complex situation involving device resistance, formulation, device dispersing power, patient inspiratory effort, and effects on respiratory tract anatomy, which continues to absorb pharmaceutical scientists today.

The FPM must not only be investigated throughout the life of the product, including tail off, but use of a range of air flow rates to determine the consistency of the fine particle dose. The EMEA Guidance (12) suggests that the range of flow rates selected should be related to information obtained during clinical studies or published data for the device and the range of flow rates studied should be from the 10th to the 90th percentile. It may be necessary to determine if there is an effect from relative humidity of the air on the FPM determined; if this is found it will be necessary to control the humidity of any impactor sizing apparatus air supply. The characteristics of impactors at varying airflow rates are well understood and determinations although tedious are routine. Asthma patients have less difficulty inhaling against a resistance than generally thought but vary in the volume of air inhaled and their ability to exhale. DPI devices all have different resistances so the Pharmacoepial approach to standardize matters is to use a constant pressure drop across the impactor and use a constant volume of air. The analytical method represents a significant challenge as the low dose of some compounds (6 μg) makes accurate recovery from the impactor and subsequent sample preparation difficult.

Particle size distribution
As with pMDIs, the overall particle size distribution of the delivered aerosol must be measured usually with an impactor according to the Pharmacoepial methods. Similarly, the total range is important, not just the fine particles, since material impacting in the upper respiratory tract may enter the gastrointestinal tract and be absorbed leading to undesirable drug activity. The coarse lactose flow aid usually employed predominantly deposits in the upper tract, carrying retained drug particles. A plot of cumulative percentage less than the impactor cutoff sizes leads to the Mass Median Aerodynamic Diameter (MMAD; note the word order is often changed leading to AMMD), and Geometric Standard Deviation if the data indicates a unimodal log-normal distribution. It has been recommended that a mass balance should be carried out on the data but this apparently simple procedure has recently been revealed as presenting yet another problem (42) in the technology.

Actuator deposition
During development studies, the distribution of drug on the DPI device should be investigated. Any variation in Delivered Dose may be due to some change in the device, which has changed the amount of drug retained by the device and this can be more readily checked.

Shaking requirements
In the unlikely event that the device requires shaking an appropriate study should be carried out to support instructions to be given to the patient.

Cleaning requirements
The need for cleaning of the DPI must be investigated using dose delivered and aerosol particle sizing techniques under conditions replicating consumer use with regard to dosing regimens.

Temperature/humidity cycling
It is unlikely that temperature cycling per se will be cause problems but an associated effect on relative humidity can be critical (99). The product should be exposed to stress conditions, for example, 25°C/30% RH and 40°C/75%RH for at least three months and examined for delivered and fine particle dose. If the product uses hard gelatine or other capsules, the effects of low and high humidities on brittleness and opening should be examined. Comparison to the innovator product is always useful.

Robustness.
As for pMDIs.

Device development report
Records should be kept of the development of the device and the various steps taken to confirm suitability for use documented. In certain circumstances, e.g., a re-useable device, CE marking of the device may be required in Europe.

Microbiology
In addition to testing routinely for contamination by standard Pharmacoepial procedures, a test should be carried out in the development stage by seeding product samples with representative microorganisms and examining viability particularly relative to moisture levels.

Microscopy
Examination of the powder formulation has little to offer for DPI product development unless it has a specific particle appearance that the technologist has reason to check or an investigation of the possible presence of foreign particulates is desired.

Water
Moisture levels can affect drug dispersion in powder formulations and a study should be carried out to determine any effect on delivered or fine particle dose. Any effect on microbial characteristics has been mentioned previously.

Final product specification
As with pMDIs, the Specification for the Final Product represents the standards to which the marketed product will comply, encompassing the totality of the pharmaceutical development studies plus information obtained during storage testing. Again it should be firmly based on the batches used to establish clinical equivalence with the innovator product. A complete description of the acceptance criteria and analytical procedures with sampling plans may be required in the submission detailing the number of samples tested, individual or composite samples specified, and number of replicate analyses per sample. The proposed validated test procedures should be documented in sufficient detail to allow validation by the reviewing agency. The following are the elements which should be considered:

Description
A description of the formulation, defining the active substance(s) and ingredients, including the FPM per actuation, and of the full product. It should include the visible components, their color, and any characteristics of their shape.

Assay
The amount of identified active present. This might be as weight per device if contained in a self-metering or enclosed system, or the weight in individual dosage units if separately packaged.

Impurities and degradation products: The standard required is as for pressurised inhalers, all impurities or degradation products above 0.1% being identified and controlled. Documentation from the International Conference on Harmonization should be consulted (111).

Water content
If a limit for water content has been established to ensure product stability, it should be defined.

Delivered dose
The delivered dose used is the ex-device dose. It is calculated from the Delivered Dose Uniformity test.

Delivered dose uniformity
Compliance will have been investigated and achieved during the development phase, and while the Pharmacoepias and Regulators are currently agreed on standards and apparatus, attention should be paid to changes in approach in this area as discussed previously.

Fine particle mass
The FPM is accepted currently as the mass of active below 5 μm MMAD in size. It is determined by Pharmacoepial impactor methods and reported as actual mass not as a percentage of any other parameter, with upper and lower limits. The flow rates and pressure drops utilised will require specification.

Microbiology
The limit can be a Pharmacoepial limit or if otherwise it will need explaining.

Number of doses
The number of doses as established during Pharmaceutical Development must be stated.

Stability
An appropriate Stability program should be devised (72) and carried out.

Manufacture
Initially, validated analytical methods should be developed, after which small batches of the blend will be made up. The performance of the product should

then be evaluated against the draft specification using Pharmacoepial methods, where applicable. When sufficient confidence in the product has been accumulated, a trial batch to fill several hundred devices should be prepared using a draft documented procedure and evaluated against the Specification. Storage of the product should be undertaken at a range of conditions, with samples withdrawn periodically. It is wise to store samples of the innovator product for comparison purposes.

When satisfactory information has been obtained, which will be utilized for obtaining authorization for clinical studies, and the technologist is confident that the product is equivalent to the innovator product, the next step is to scale up the process and prepare large batches for long-term stability and clinical studies. If an experienced contract organization is not to be employed, a manufacturing process is devised and the process validated. Three batches are then prepared and appropriate numbers set up on stability according to ICH guidelines and used for clinical studies.

At this stage, it is important to agree a program of work that provides full information for the eventual ANDA/Product License Application. The characteristics of the clinical trial batches will set the standards for the eventual marketed product.

AQUEOUS DROPLET INHALERS (ADI) OR NON-PRESSURIZED METERED DOSE INHALERS (NpMDI)

There are two types of ADIs.

The first is the widely available plug-in or battery operated "nebulizer," which utilizes an air pump to generate respirable droplets from an aqueous solution or suspension of the active. Nebulizers exist in a wide variety of physical forms with their own unique characteristics and introduce active product from either a bulk container or more usually today a unit dose container, and it is the latter which is the market for a generic product. The active is contained in a plastic unit manufactured under sterile conditions in a form-fill-seal machine. The technology is essentially as for parenteral products, either solutions or suspensions and will not be considered further.

The second type is a sophisticated ADI, sometimes referred to as a "soft mist" inhaler (114), which has appeared in recent years as an alternative to use of propellant driven and powder devices. These devices, extensively patented and requiring complex manufacturing arrangements, are yet to carve out and define their market and as yet do not attract generic attention for locally applied therapy in the airways.

NASAL PRODUCTS

The market for nasal administration products has increased substantially in recent years and is ahead of the inhalation market in introducing systemic products.

Since the introduction of nasal squeeze bottles decades ago, the range of nasal dispensing systems has grown tremendously in terms of diversity. The disadvantages of the early nasal devices—pipettes, drops, and squeeze bottles—has

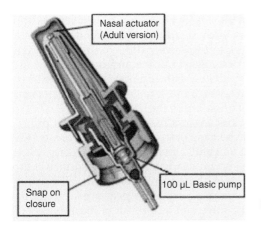

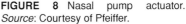

FIGURE 8 Nasal pump actuator. *Source*: Courtesy of Pfeiffer.

been superseded in terms of dose accuracy, resistance to contamination, and efficiency in terms of a proper coverage of the nasal mucosa.

Today's nasal dispensing systems use mechanical pump principles (Figs. 8 and 9). There are different kinds of pump mechanism on the market, but essentially they all follow more or less the same basic principle. Pressure is applied to a closed chamber filled with a liquid formulation or suspension, which is ejected through a nozzle to cause the liquid jet to break up. A current trend is to develop preservative free formulations. In a traditional multidose formulation, the preservative takes care of a possible contamination issue but an unpreserved formulation has to rely on the integrity of the primary packaging. Basically, two options can be considered to combat microbial contamination.

The first option can be to include a preservative either in the formulation or into the pack component(s), which are in contact with the formulation. Most

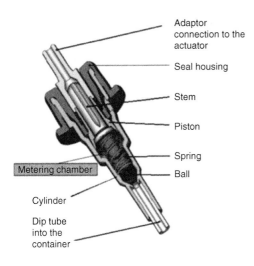

FIGURE 9 Nasal pump actuator metering unit. *Source*: Courtesy of Pfeiffer.

common for the latter is the implementation of silver ion–releasing materials into the device. The preservative activity of such an arrangement will depend on a favorable ratio of container surface area to the surrounding volume. Additionally, the efficacy depends on the microbiological burden and the nature of the contamination. Such systems must be validated with appropriate challenge tests. These are chemically protected systems, which mean microorganisms can enter during use and contaminate the formulation inside the nasal actuator.

The alternative is a preservative free system based on mechanical principles. Here a mechanically protected dispensing system seals directly behind the product exit. This mechanical barrier inherently prevents any microorganisms from entering the formulation. A small spring-loaded pin seals the orifice at the very end of the nasal actuator. As soon as the patient actuates the dispensing system, the pressure inside the pump increases. When the pressure in the pump system becomes higher than the spring force and as a result of incompressibility of the liquid formulation, the pin will retract and release the metered dose. Depending on the dispensed dose volume, the velocity of the particles expelled from the orifice is between 50 and 70 km/hr. Subsequently after the dispensing act the pressure in the volume chamber drops and the pin will reseal the orifice. To cope with the volume replacement by entering air, an embedded filter will prevent contamination reaching the formulation. Whether a depth filter or an absolute microbiological filter is used depends on the choice of design of the delivery system developer. A depth filter comprises of pore sizes that are large enough for microorganisms to penetrate; however, the travel distance is long enough to ensure the microorganisms, which adhere to dust particles, are retained.

An absolute microbiological filter consists of a thin membrane with pores small enough to hold back any microorganisms to enter the system. Usually, 0.2 μm membranes are applied. An absolute filter might bear some advantages as nonadhered bacteria and bacteria spores are more efficiently retained.

Unit- and bidose systems

In addition to the multidose systems described, unit- and bidose systems are available, designed to administer one or two doses into the nasal cavity. These generally find application for the increasing number of compounds administered via the nasal route acting systemically and requiring a fast onset of action. Particularly demanding therapies that require accurate and precise performance from the delivery system will most likely employ these single/dual-use disposable systems. Already established in the market are calcitonin formulations to treat osteoporosis and various drug products to treat migraine or offer pain relief. The latest activities focus on nasal administration using disposable bidose systems. These devices are also considered efficient for the administration of vaccines where it is important to cover the mucosa in both halves of the nasal cavity. The increasing numbers of systemic acting products represented by proteins or peptides are demanding new application principles for different dosage forms. Such products are increasingly proteins and peptides, which are more stable in the dry and solid state.

Dry powder dispensers for nasal administration offer the advantage of maximum product safety and ease of use for the patient.

The administration of controlled substances for pain management requires that safety aspects have to be seriously considered. The idea behind these sophisticated devices is to offer a maximum of mobility and safety to the patient compared to other traditional more invasive parenteral routes. The alternative nasal route can also be attractive in terms of reducing the overall cost of a therapy. Lockout systems are available that electronically regulate the administration profile in terms of time and dosing. During the nonoperating mode the actuation of the device is blocked. Additionally, it is possible to record patient specific data.

With the growing attraction of nasal administration, the regulatory authorities are paying increasing attention to this route. Of key importance is the physical performance of a nasal spray. The pivotal parameters are the accuracy of the dose volume, the spray pattern or spray angle, droplet size distribution, and consistent dosing performance. In particular, it is important to avoid the generation of particles small enough to reach the lung. Current guidelines intensively regulate these devices and sometimes formulation related parameters. Similarly, plume geometry, characterized by determination of the spray angle and the spray pattern, and the relevant deposition site in the cavity itself. The nasal cavity can be described as two elongated cavities, which are separated by the nasal septum: the shape of the nasal cavity varies from individual to individual, and the interaction between device performance and anatomical diversity makes precise regulation of spray pattern and spray angles difficult.

The APIs available today include steroids such as mometasone, fluticasone, betametasone, and budesonide for local action and sumatriptan, zolmitriptan, buserelin, calcitonin, and desmopressin, intended to produce systemic action. Within these examples, there are different formulation approaches such as solutions, suspensions, preservative free and preserved, unit dose, and multiple doses from 2 to 200 actuations. As an innovator all formulation options are available; however, for the generic manufacturer, there is a need to match innovator performance, however good or bad, and remain outside the relevant intellectual property. To add to the constraints, it would not be surprising if the regulatory requirements had changed since the innovator product was launched usually by being more demanding. Excellent guidances are available from the principle Regulators (12,104,105). The innovator companies will add to the development hurdles by lobbying health authorities to impose even more requirements. For all the challenges the currently marketed nasal products represent an attractive market to generic development houses.

There are three types of products to be considered:

Metered dose nasal spray
Pressurized metered dose nasal spray
Dry powder insufflators

The technology of the latter two follows that of orally inhaled pMDIs and DPIs, which have been dealt with in detail in other sections. The nasal section will focus on metered aqueous droplet products.

Metered Dose Nasal Spray (MDNS)
Devices
The two elements of the MDNS are the pump and the container.

Pump.   The pump contains components that are responsible for metering, atomizing, and delivering the formulation to the user. It should repeatedly spray discrete, accurate, small doses of the formulation in the desired physical form. Development of the pump in conjunction with the formulation is critical since formulation characteristics can influence (106) the delivered spray from variation in a number of variables. The characteristics of nasal generation are dependent on a combination of actuation force, viscosity, rheological properties, surface tension, and the pump. To these can be added speed of actuation, hold, and return times. The design of the pump—inlet channels, and swirl chambers—and compatibility of the components with the formulation should be examined before undertaking critical clinical, bioequivalence, and stability studies. Partial or multiple dosing should be eliminated by the design and as with all multidose products, Regulators are currently urging a counter mechanism if the patient is unable readily to ascertain how much medicine remains.

    The presence of elastomers in the pump design will raise the question of leachables entering the product during storage, and reference should be made to the discussion under pMDI metering valves.

    Manufacturers of pumps include www.pharmaceutical-technology.com, www.valois.com/html/pharmacie/cadre.asp, and www.consortmedical.com/ (all accessed February 3, 2008).

Container.   The container is usually a bottle of dark glass to provide protection against actinic radiation.

Pharmaceutical Product Development
An extensive range of performance studies must be carried out during development of new formulations of nasal sprays and these will be discussed in this section, followed by a review of the Final Product Specification, before considering briefly manufacturing methods.

    Initially, analytical methods should be developed, after which small numbers of formulations are prepared based on the formulation of the innovator product. The performance of the product should then be evaluated against a number of criteria—and the innovator product—using Pharmacoepial methods where applicable. There is currently increasing activity in this area and enquiries should be made regarding new methodology with the secretariats. This will eventually enable a draft Final Product Specification to be prepared. When sufficient confidence in a formulation has been accumulated, a trial batch of several hundred units should be prepared using a draft documented procedure and evaluated with a view to obtain the information discussed below. Storage of the product in a variety of orientations should be undertaken at a range of conditions, with samples withdrawn periodically. It is wise to store samples of the innovator product for comparison purposes

    Investigations must now be conducted to establish that the dosage form, formulation, manufacturing process, container closure system, microbiological attributes, and instructions for use are appropriate and result in acceptable product performance. Sufficient batches should be studied, often three, to take batch variability into account. Of particular importance are batches used in clinical studies, which should be very fully characterized. Both FDA and the EMEA have published useful Guidelines in this area (107,108).

Drug substance and physical characterization
Unless the drug substance is to be dissolved in solvent, for example, water, the particle size and particle size distribution of the active must be fully characterized. Laser light scattering techniques are in common use for the raw material, and apparatus is readily available. The total particle size range should be well defined and acceptance criteria set. The product used in pivotal clinical studies is key and provides the basis for the specification. A variety of information may be required by Regulators such as particle morphology, solvates and hydrates, polymorphs, amorphous forms, solubility profile, moisture and/or residual solvent content, microbial quality, dissociation constants, and specific rotation. The usual physical attributes, for example, color, appearance, identification, moisture, residue on ignition, assay, impurities, melting range, etc., will all form part of the active raw material specification and further detail will be found in the Regulatory Guidances.

Chirality. If the active is chiral, studies should be carried out to determine any potential for racemization. These will lead to clarity on the need for achiral or specific assay procedures to control the presence of the enantiomer.

Prime. The problem of a metered dose emptying wholly or partially from the metering chamber before actuation exists with MDNS as with pMDIs and the required time interval and number of priming shots must be established with a suitable experimental design. Orientation, time, and storage conditions, for example, shaken in a pocket or handbag, can all affect dose content uniformity (Spray Content Uniformity in FDA speak). The key information to establish is the approximate interval that can pass before the drug product should be reprimed and the number of sprays required, and the effect of time and orientation on this interval. In the case of suspension products, the generic technologist must carry out careful evaluation of the copy product to ensure that directions can be given to the patient to ensure a correct and consistent dose. If required, the time of shaking and the number of priming shots needed to achieve label claim must be investigated including storing the product in various orientations and using different time intervals to define the need for repriming. The product should be tested in this way during storage testing. Information given on the innovator patient information leaflet should be examined and it may be worth confirming the claimed procedures.

Temperature cycling. Fluctuations of temperature in storage and in patient use can lead to a variety of effects and should be investigated thoroughly. During storage testing samples from several batches should be cycled from high (e.g., 40°C) to a low temperature (e.g., −5°C or even down to freezer temperatures) possibly via an anticipated recommended storage temperature (e.g. 20–25°C) with at least 24 hours at each point. Appearance, color, clarity, weight loss, delivered dose uniformity, droplet size distribution, related substances, and moisture content should be examined and the results may disclose effects on a suspension, or pump action. FDA Guidance recommends 12-hour cycles with temperatures ranging from freezer values ($-10°C$ to $-20°C$) to $40°C$ for four weeks. Sterility testing should not be overlooked if a microbial barrier is in use in the container system.

Dose proportionality.   When a variety of drug strengths are offered a study should be carried out to compare the dose, uniformity of dose and particle size of the family of products, and the integrity of the range justified.

Particulate matter.   A standard for the presence of particulate matter in nasal sprays is not at present a consideration in Europe, but the FDA guidance requires a standard in the Final Product Specification. Foreign particulates can enter the product from a variety of sources including formulation container and closure components, and methods for evaluation should be developed. Solutions can be filtered and examined microscopically or particle counters employed. Of particular importance is to establish whether there is any change during storage of the product under conditions of time, temperature, and stress, since if there is no change only a batch release test will be required. If particulates are encountered it will be necessary to identify and eliminate them from further consideration.

Cleaning instructions.   The need for cleaning must be investigated using dose delivered and aerosol particle sizing techniques under conditions replicating consumer use with regard to dosing regimens and priming. This information will form part of the labeling.

Device robustness.   Patients inevitably will find all possible ways to use and misuse the product and its robustness must be challenged in the development phase—and in the final form if changes have been made during development. Tumbling in a rotary mixer, vibrating, and dropping in various orientations from a fixed height on to a hard surface are essential to check for weak points. If prototypes were used in any significant early evaluations, the performance of the final to-be-marketed device must be confirmed. If multiple use of the pump with replacement formulation is allowed, the validity of the exercise should be established with dose content uniformity and particle size confirmation throughout the life.

Device orientation.   The angle at which a metering device is refilled and/or used can sometimes affect the dose delivered and a study should be designed to investigate any effects. The use of dose delivered and particle size are again key.

Dose delivered (or spray content) uniformity.   The delivered or sprayed dose is the quantity of drug delivered from the pump to the patient. The minimum number of sprays should be used according to the sensitivity of the analytical methods available, Regulators preferring a single spray or two at maximum if that is the recommended dose. This may have to be negotiated. There is currently no pharmacoepial apparatus, and a modification of the arrangement described for collection of delivered dose for inhalation products may be useful. The purpose of the test is to evaluate the uniformity of the doses delivered throughout the life of the product. With this in mind attention should be paid to any instructions to be given to the patient with respect to shaking or orientation of the device. In Europe, the EMEA Guidance suggests standards "consistent with the pharma-coepia" and the Ph Eur provides a standard (monograph 0676 revised in 2006) using collection of 10 individual doses. A second tier extends to 20 more units if not more than 3 doses are >25% but <35%, allowing a maximum of 3 >25% but all <35%. Whilst there is no suggestion of determining doses from the beginning and end of use the most recent edition of the Pharmacopoeia should be consulted. The same approach can be used to determine within and between units, and batch variation. An examination of doses fired after the label claim has been reached could also be made and the tail-off determined, although no regulators appear to be concerned with this standard. The FDA standard is to take 1 shot at the beginning and 1 shot at the end from each of 10 units (after priming), total 20 shots, with standard 20% variation and 15% for the mean. If 3–6 are >20% but <25%, tier 2 is invoked with an extra 20 units and an additional 40 shots. Compliance is not more than 6/60 are >20% and none are >25%, and the means of both beginning and end doses separately are not more than 15% of label claim. Although standards proposed by FDA (11) for inhalation products have been challenged, there does not seem to have been similar commentary from the nasal products sector on the standard. The generic technologist should confirm the current position before evaluating data since it is not only possible that standards have changed since the innovator product was developed, but the overall statistical approach may be different. The current guidances already permit alternative approaches that provide equal or greater assurance.

Particle size distribution. The overall particle size distribution of the spray must be measured usually with a laser light scattering apparatus. The total range is important, particularly the fine particles, which might penetrate into the lower respiratory tract leading to undesirable drug or additive activity. It has been recommended that a mass balance should be carried out on the data but this apparently simple procedure has recently been revealed for certain respiratory products as presenting a possible problem in the technology. In the case of suspension products it may be necessary to utilize microscopic methods of size evaluation to examine agglomeration or crystal growth.

Spray pattern and plume geometry. In Europe, little credence is currently afforded to spray pattern and plume geometry in respiratory and nasal products. IPAC-RS have strongly criticized their use (109) in the FDA draft Guidance of 1998 for respiratory products commenting on the multiple sources of any located variation, which in their view could be better controlled by more targeted approaches. However, FDA has retained the requirement for investigation in the final Nasal Spray Guidance of 2002. Spray pattern is determined by spraying from the nasal pump on to a collection surface that might be a TLC plate or a filter paper. The spray deposited is then visualized with a suitable reagent and examined. Variables in the procedure might be the number of sprays, distance from the pump, and its orientation to the collection surface. Acceptance criteria must be developed to describe the shape and size of the pattern and ratios of the longest and the shortest axis are suggested. Plume geometry can be investigated by a variety of procedures, for example, using high-speed photography; a new instrument, the Varidose™ (110) may also prove useful in this area. Details of the employed technique are required by the Agency including visualization methods, exposure time, and orientations.

Preservative effectiveness and sterility maintenance. If the nasal spray contains a preservative, it is necessary to demonstrate the effectiveness of the system during development. Suitable procedures are available in the USP and Ph Eur. It is particularly important to demonstrate that the effectiveness is maintained at the end of the claimed shelf life and the first batch stored is valuable to demonstrate this. Products that use a pump with a microbial barrier and therefore no preservative system should demonstrate sterility at the beginning of stability, during testing, and at the end.

Final Product Specification

The specification for the final product represents the standards to which the marketed product will comply, encompassing the totality of the pharmaceutical development studies plus information obtained during storage testing. Again it should be firmly based on the batches used to establish clinical equivalence with the innovator product. A complete description of the acceptance criteria and analytical procedures with sampling plans may be required in the submission, detailing the number of samples tested, individual or composite samples specified, and number of replicate analyses per sample. The proposed validated test procedures should be documented in sufficient detail to allow validation by the reviewing agency. Limits should be set where appropriate for manufacture and end of shelf life. The following are the parameters which should be considered:

Description. A description of the formulation, defining the active substance(s) and ingredients, including the dose per actuation, and of the full product including pump and container components. It should include the visible components, their color, and any characteristics of their shape and size. Note if the product is colored either initially or as a result of degradation a limit test will be needed.

Identification. The FDA requires two independent identification tests and also a test for the counter ion if present. Chromatography, UV, and IR represent choices. FDA have elaborated the case of requirements for a chiral drug substance suggesting that an achiral assay procedure is acceptable only if racemization has been shown to be insignificant.

Assay. The amount of identified active present often expressed as % weight per unit volume together with the target fill volume. The procedure should be a stability indicating and limits set for manufacture and end of shelf life.

Impurities and degradation products. The standard required is the usual one with all impurities or degradation products above 0.1% to be identified and controlled. There is significant documentation from the International Conference on Harmonization in this and related areas, which should be consulted (111).

Preservatives and stabilizing excipients. A number of additives may have been used in the nasal formulation, for example, benzalkonium chloride, EDTA, surfactants, and assay procedure for these are required together with limits for manufacture and end of shelf life.

Delivered dose. Referred to as "Pump Delivery" by FDA Guidance, the delivered dose used is the shot weight metered by the pump. Provided the manufacture of the pump dimensionally is properly controlled and accurate, there should be no problem in meeting a limit of 15% of target weight for individual shots and 10% for the mean of 10 shots. If very small doses are metered, for example, 20 μL, the criteria may have to be extended and justified.

Delivered dose uniformity. Again FDA use a different term "Spray Content Uniformity." Compliance will have been investigated during the development phase and compliance with the necessary standards determined.

Spray pattern and plume geometry. There is no current requirement in Europe for a standard for either of these. However, in the United States, full details of methods with proposed standards are required.

Particle size distribution. The standards identified in the development phase provide the basis for the standard. As noted the proportion less than 10 μm is important and should be similar to the innovator product.

Particulate matter. A standard identified in development should be included.

Microbiology. During Pharmaceutical Development, it will have been demonstrated that the product does not support the growth of microorganisms and that microbial quality is maintained throughout the claimed life. The standard proposed in Pharmacoepial Forum (112) for multidose nasal sprays is

Total aerobic microbial count	NMT 100 cfu/mL
Total combined yeasts and molds count	NMT 10 cfu/mL
Absence per milliliter	*S. aureus*, *P. Aeruginosa*, bile tolerant Gram-negative bacteria

A note is made that these standards may not be acceptable to FDA and also draws attention to USP chapter <61> that an error factor of 2 may be applied to such acceptance criteria.

Leachables. The results of the detailed studies carried out in development on Extractables and Leachables will define the nature of the specification.

Number of doses. The number of doses should be defined. This might be covered in defining net contents and any losses occurring during storage should be considered. Data obtained during storage testing with the product in various orientations will assist in the definition.

Viscosity, tonicity, pH.   If relevant to the formulation, standards and methods are required for these parameters.

Manufacture

The manufacture of nasal sprays draws on the techniques of parenteral production and reference should be made to this area. Initially, analytical methods are developed for the product, after which small batches of the solution or suspension will be made up. The pump device will be chosen after considering the characteristics of the innovator product and discussions with pump suppliers; similarly, the container. The performance of the assembled product should then be taken through the various Pharmaceutical Development steps discussed and evaluated against a draft specification using pharmacoepial methods where applicable. When sufficient confidence in the product has been accumulated, a trial batch of several hundred devices should be prepared using a draft documented procedure and evaluated against the draft Specification. Storage of the product should be undertaken at a range of conditions, with samples withdrawn periodically. It is wise to store samples of the innovator product for comparison purposes.

When satisfactory information has been obtained, which may be utilized for obtaining authorization for clinical studies, and the technologist is confident that the product is equivalent to the innovator product, the next step is to scale up the process and prepare large batches for long-term stability and clinical studies. If an experienced contract organization is not to be employed, a manufacturing process is devised and the process validated. Three batches are then prepared and appropriate numbers set up on stability according to ICH guidelines and used for clinical studies. Use of commercial scale lots at the final site of manufacture is appreciated by regulators.

At this stage it is important to ensure a program of work that provides full information for the eventual ANDA/Product License Application has been executed. The characteristics of the clinical trial batches and information collected from development and storage testing will set the standards for the eventual marketed product.

ACKNOWLEDGEMENTS

The support of Dr. Tol Purewal of Consort Medical (previously Bespak), Dr. Tim Noakes of Ineos for the section on propellants, and Dr. Rene Bomber of Pfeiffer on nasal products is gratefully acknowledged.

REFERENCES

1. Shargel L, Kanfer I. Generic Drug Product Development: Solid Oral Dosage Forms (Drugs and the Pharmaceutical Sciences), vol. 143. New York: Marcel Dekker, 2005: 1–16.
2. Berry IR. The Pharmaceutical Regulatory Process, vol 144. New York: Marcel Dekker, 2004.
3. http://europa.eu.int/abc/governments/index_en.htm. Accessed February 5, 2008.
4. http://ec.europa.eu/enterprise/pharmaceuticals/eudralex/index.htm.   Accessed March 3, 2008.
5. International Commission on Radiological Protection. Human respiratory tract model for radiological protection. Ann ICRP 1994; 24:36–52.

6. Heyder J, Gebhart J, Rudolf G, et al. Deposition of particles in the human respiratory tract in the size range 0.005–15μ. J Aerosol Sci 1986; 17:811–825.
7. Crampton M, Kinnersley R, Ayres J. Sub-micrometer particle production by pressurised metered dose inhalers. J Aerosol Med 2004 (Spring); 17(1):33–42.
8. Cripps A, Riebe M, Schulze M, et al. Pharmaceutical transition to non-CFC pressurised metered dose inhalers. Respir Med 2000: 94(Suppl B):S3–S9.
9. Peyron ID, Britto IL, Bennison LB, et al. Development and performance of a new hydrofluoralkane (HFA 134a)-based metered dose inhaler of salmeterol. Respir Med 2005; 99(suppl A):S20–S30.
10. Levine M, Gaebel K, Spino M. A study of patient responses to a perceived change in salbutamol metered dose inhalers. J Generic Med 2005; 2(3):201–208.
11. FDA draft Guidance for Industry 1998. Metered Dose Inhaler (PMDI) and Dry Powder (DPI) Drug Products. Chemistry, Manufacturing and Controls. http://www.fda.gov/ohrms/dockets/ac/00/backgrd/3609b1j.pdf. Accessed March 3, 2008.
12. EMEA 2005 Guideline on the Quality of Inhalation and Nasal Products issued April 19, 2006. http://www.emea.europa.eu/pdfs/human/qwp/4931305en.pdf. Accessed March 3, 2008.
13. Replacement of CFC's in Metered Dose Inhalation Products. 75/318/EEC-Council Regulation No. 594/91 and Notes for Guidance on Requirements for Pharmaceutical Documentation for Pressurised PMDI Products. CPMP/QWP/2845/00.
14. Shargel L, Kanfer I. Generic Drug Product Development: Solid Oral Dosage Forms (Drugs and the Pharmaceutical Sciences), vol. 143. New York: Marcel Dekker, 2005:31–51.
15. Ganderton D, Lewis D, Davis R, et al. The formulation and evaluation of a CFC-free budesonide pressurised metered dose inhaler. Respir Med 2003; 97(suppl D):S4–S9.
16. Milanowski J, Qualtrough J, Perrin VL. Inhaled beclomethasone(BDP) with non-CFC propellant (HFA 134a) is equivalent to BDP-CFC for the treatment of asthma. Respir Med 1999; 93:245–251.
17. Leach C. Improved delivery of inhaled steroids to enlarge small airways. Respir Med 1998; 92(suppl A):3–8.
18. Moren F, Newhouse MT, Dolovich MD, eds. Aerosols in Medicine, 2nd ed. Amsterdam, The Netherlands: Elsevier Science Publishers, 1986.
19. Byron PR, ed. Respiratory Drug Delivery. Boca Raton, FL: CRC Press, 1990.
20. Hickey AJ, ed. Inhalation Aerosols. New York: Marcel Dekker, 1996.
21. Bisgard H, O'Callaghan C, Smaldone GC, eds. Drug Delivery to the Lung. New York: Marcel Dekker, 2002.
22. Pharmaceutical cGMP's for the 21st Century: A Risk Based Approach. http://www.fda.gov/oc/guidance/gmp.html. Accessed February 3, 2008.
23. Guidance for Industry. PAT—A Framework for Innovative Pharmaceutical Develeopment, Manufacturing and Quality Assurance. http://www.fda.gov/cder/gmp/patprogressreport.htm. Accessed February 3, 2008.
24. Bowles N, Cahill E, Haeberlin B, et al. Application of quality by design to inhalation products. Respir Drug Deliv Eur 2007:61–69.
25. Thiel CG. Program and Proceedings of Respiratory Drug Delivery, Vol 5. Buffalo Grove, IL: Interpharm Press, 1996:115–123.
26. O'Callaghan and Wright. The Metered Dose Inhaler. In: Drug Delivery to the Lung. Bisgard H, O'Callaghan C, Smaldone GC, eds. New York: Marcel Dekker, 2002:337–370.
27. Smyth HDC. The influence of formulation variables on the performance of alternative propellant driven metered dose inhalers. Adv Drug Deliv Rev 2003; 55:807–828.
28. Purewal TS, Grant DJW, eds. Metered Dose Inhaler Technology. Boca Raton, FL: CRC Press, 1998.
29. Bell J, Newman S. The rejuvenated pressurised metered dose inhaler. Expert Opin Drug Deliv 2007; 4(3):215–234.
30. Versteeg HK, Hargrave GK. Fundamentals and resilience of the original MDI design. Respiratory Drug Delivery 2006; 1:91–100.

31. Leach CL. The CFC transition and its impact on pulmonary development. Respir Care 2005; 50(9):1201–1208.
32. Berry J, Kline L, Naini V, et al. Influence of valve lubricant on the aerodynamic particle size of a metered dose inhaler. Drug Dev Ind Pharm 2004; 30(3):267–275.
33. http://www.pqri.org/workshops/leach_ext/pqrilandeworkshop.asp. Accessed November 28, 2009.
34. Berry J, Kline LC, Hart JL, et al. Influence of the storage orientation on the aerodynamic particle size of a suspension metered dose inhaler containing propellant HFA-227. Drug Dev Ind Pharm 2003; 29(6):631–639.
35. Cyr TD, Graham SJ, Li KY, et al. Low first-spray drug content in albuterol metered-dose inhalers. Pharm Res 1991; 8(5):658–660.
36. Smith IJ, Parry-Billings M. Inhalers of the Future? Pulm Pharmacol Ther 2002; 1–18.
37. Wilby M, Purkins G et al. Novel valve designs to eliminate loss of prime. Respiratory Drug Delivery X 2006:373–375.
38. Williams L, Velasco V, et al. SpraymiserTM Valve: Designed with Patient Use in Mind. Respiratory Drug Delivery 2006; 2:377–380.
39. Topliss P, Ward D, Southall J, et al. Pharmaceutical performance of a valve for metered dose inhalers (MDI's) designed to eliminate loss of prime. Respir Drug Delivery X 2006; 499–502.
40. Thiel CG. Dispensing device. 1957 US Pat No. 2,886,217.
41. Bowman PA, Greenleaf D. Non-CFC metered dose inhalers: The patent landscape. Int J Pharmaceutics 1999; 186:91–94.
42. Bagger-Jorgensen H, Sandell D, Lundback H, et al. Effect of inherent variability of inhalation products on impactor mass balance limits. J Aerosol Med 2005; 18(4):367–378.
43. Purewal T, Grant DJW, eds. Metered Dose Inhaler Technology. Interpharm/CRC, 1998:46–49.
44. Lewis DA, Meakin BJ, Brambilla G. New actuators versus old: Actuator modifications for HFA solution HFA's. RDD 2006; X:101–110.
45. Dubus JC, Dolovich M. Emitted doses of salbutamol pressurised metered dose inhaler from 5 different plastic spacer devices. Fundam Clin Pharmacol 2000; 14(3):219–224.
46. Newman SP. Spacer devices for metered dose inhalers. Clin Pharmacokinet 2004; 43(6):349–360.
47. Bisgaard H, O'Callaghan C, Smaldone GC, eds. Drug Delivery to the Lung. New York: Marcel Dekker 2002:389–420.
48. http://www.trudellmed.com/. Accessed February 3, 2008.
49. Larsen JS, Hahn M, Ekholm B, et al. Evaluation of conventional press and breathe metered dose inhaler technique in 501 patients. J Asthma 1994; 31:193–199.
50. Van Beerendonk I, Mesters I, Mudde AN, et al. Assessment of the inhalation technique in outpatients with chronic obstructive pulmonary disease using a metered dose inhaler or dry powder device. J Asthma 1998; 35:273–279.
51. Price D, Thomas M, Mitchell G, et al. Improvement in asthma control with a breath-actuated pressurised metered dose inhaler (BAI): A prescribing claims study of 5556 patients using a traditional pressurised metered dose inhaler (PMDI) or a breath actuated device. Respir Med 2003; 97:12–19.
52. Rau JL. Practical problems with aerosol therapy in COPD. Respir Care 2006; 51(2):158–172.
53. Hardy J, Jasuja A, Frier M, et al. A small volume spacer for use with a breath operated pressurised metered dose inhaler. Int J Pharm 1996; 142:129–133.
54. Patent number PCT/GB98/00770,(WO 98/41254)
55. FDA Guidance for Industry. Integration of Dose-Counting Mechanisms into MDI Drug Products. Rockville, MD: FDA Guidance for Industry, 2003.
56. EMEA/CHMP/QWP/49313/2005 Final. Guideline on the Quality of Inhalation and Nasal Products. EMEA/CHMP/QWP/49313/2005 Final, 12. Para 4.2.1.19. London.
57. Bradshaw DRS. Developing dose counters: An appraisal based on regulator, pharma, and user needs. RDD X 2006, 121–131.

58. Bell JH, Brown K, Glasby J. Variation in delivery of isoprenaline from various pressurised inhalers. J Pharm Pharmacol 1973; 25(suppl):32P–36P.
59. Patent No EPP 0990437.
60. Waugh J, Goa KL. Flunisolide HFA. Am J Respir Med 2002; 1(5):369–372.
61. Lewis DA, Ganderton D, Meakin BJ, et al. Modulite: A simple solution to a difficult problem. Respiration 2005; 72(suppl 1):3–5.
62. Derom E, Pauwels RA. Pharmokinetic and pharmacodynamic properties of inhaled beclometasone dipropionate delivered via hydrofluoroalkane-containing devices. Clin Pharmacokinet 2005; 44(8):815–36.
63. Waugh J, Goa KL. Flunisolide HFA. A J Respir Med 2002; 1(5):369–372.
64. Robins E, Brouet G. Formulation development of a budesonide HFA-MDI. RDD 2006; 3:931–934.
65. http://www.malvern.co.uk, www.sympatec.com. Accessed February 3, 2008.
66. Joshi V, Dwivedi S, Ward GH. Increase in the specific surface area of budesonide during storage post micronisation. Pharm Res 2002; 19(1):7–12.
67. USP 2003 United States Pharmacoepia. 26 <601>.
68. EP2005 European Pharmacoepia. 5th ed. and supplements.
69. Bisgaard H, Anhoj J, Wildhaber JW. Spacer devices. In: Bisgard H, O'Callaghan C, Smaldone GC, eds. Drug Delivery to the Lung. New York: Marcel Dekker, 2002:389–420.
70. Purewal TS. Test methods for inhalers to check performance under normal use and unintentional misuse conditions. Drug Delivery to the Lung 2001; XII:92–98.
71. D'Abreu-Hayling C. Quality by Design and Risk Management Approaches to Foreign Particles in OINDP. Respir Drug Deliv 2006; X:639–642.
72. International Conference on Harmonisation. http://www.ich.org. Accessed February 3, 2008.
73. Smith IJ, Parry-Billings M. The Inhalers of the Future? A Review of Dry Powder Devices on the Market Today. Pulmon Pharmacol Therap 2002:1–18.
74. Coates MS, Fletcher DF, Chan HK, et al. The role of capsule on the performance of a dry powder inhaler using computational and experimental analyses. Pharm Res. 2005; 22(6):923–932.
75. http://www.flowcaps.com/. Accessed February 3, 2008.
76. http://www.vectura.com/. Accessed March 3, 2008.
77. White S, Bennett DB, Cheu S, et al. Exubra: Pharmaceutical development of a novel product for pulmonary delivery of insulin. Diabetes Technol Ther 2005; 7(6):896–906.
78. Borgstrom L, Bisgaard H, O'Callaghan C, et al. Dry powder inhalers. In: Bisgard H, O'Callaghan C, Smaldone GC, eds. Drug Delivery to the Lung. New York: Marcel Dekker, 2002.
79. Srichana T, Martin GP, Marriott C. Dry Powder Inhalers: The influence of device resistance and powder formulation on drug and lactose deposition in vitro. Eur J Pharm Sci 1998; (7):77.
80. Young PM, Edge S, Traini D, et al. The influence of dose on the performance of dry powder inhalation systems. Int J Pharm 2005; 296:26–33.
81. Bondesson E, Bengtsson T, Borgstom L, et al. Dose delivery late in the breath can increase dry powder aerosol penetration into the lungs. J Aerosol Med 2005(Spring); 18(1):23–33.
82. Price D. Foreword: can dry powder inhalers be used interchangeably? Int J Clin Pract 2005; 59(suppl 149):2.
83. Price D, Summers M, Zanen P. Could interchangeable use of dry powder inhalers affect patients? Int J Clin Pract December 2005; 59(suppl 149):3–6.
84. Nielsen K, Okamoto L, Shah T. Importance of selected inhaler characteristics and acceptance of a new breath actuated powder inhalation device. J Asthma 1997; 34(3):249–253.
85. Diggory P, Fernandez C, Humphrey A, et al. Comparison of elderly people's technique in using two dry powder inhalers to deliver zanamivir: Randomised controlled trial. BMJ. 2001; 322(7303):49–50.
86. Newman SP, Busse WW. Evolution of dry powder inhaler design, formulation and performance. Respir Med 2002; 96:293–304.

87. Atkins PJ. Dry Powder Inhalers: An Overview. Respir Care 2005; 50(10):1312 [Hickey in discussion].
88. Lucas P, Anderson K, Staniforth JN. Protein deposition from dry powder inhalers; fine particle multiplets as performance modifiers. Pharm Res 1998; 15:562–569.
89. Hersey JA. Ordered mixing: a new concept in powder mixing practice. Powder Technol 1975; 11:41–44.
90. Kassem NM, Ganderton D. The influence of carrier surface on the characteristics of inspirable powder aerosols. J Pharm Pharmacol 1990; 42:11.
91. http://www.microntech.com/. Accessed March 3, 2008.
92. Bell JH, Hartley P, Cox JSG. Dry Powder Aerosols I: A new powder inhalation device. J Pharm Sci 1971; 60:1559–1564.
93. Bell JH. Pelletised Medicament Formulations. Patent No. GB 1,520,247.
94. Borgstrom L, Bisgaard H, O'Callaghan C, et al. Dry powder inhalers. In: Bisgard H, O'Callaghan C, Smaldone GC, eds. Drug Delivery to the Lung. New York: Marcel Dekker, 2002:424.
95. Chan HK. Dry powder aerosol delivery systems: Current and future research directions. J Aerosol Med Spring 2006; 19(1):21–27.
96. Chew NY, Shekunove BY, Tong HH, et al. Effect of amino acids on the dispersion of disodium cromoglycate powders. J Pharm Sci 2005; 94(10):2289–300.
97. York P, Kompella UB, Shekunov BY. Supercritical Fluid Technology for Drug Product Development. Drugs and the Pharmaceutical Sciences, Vol 138. New York: Marcel Dekker, 2002.
98. Dunbar C, Scheuch G, Sommerer K, et al. In vitro and in vivo dose delivery characteristics of large porous particles for inhalation. Int J Pharm 2002; 245(1–2):178–189.
99. Edwards D, Hanes J, Caponetti G, et al. Large porous particles for pulmonary drug delivery. Science 1997; 276:1868.
100. Clark AR, York P. SCF-Pulmonary pharmaceuticals: the uncoated truth. RDD 2006; X:317–326.
101. Sakagami M, Byron PR. Respirable microspheres for inhalation: The potential of manipulating pulmonary disposition for improved therapeutic efficacy. Clin Pharm 2005; 44(3):263–77.
102. Egen M, Zensi A. Mixing process for powders on low gram-scale. Respir Drug Deliv X 2006; 503–505.
103. Yang TT, Kenyon D. Use of an agglomerate formulation in a new multidose dry powder inhaler. In: Dalby RN, Byron PR, Farr SJ, Peart J, eds. Respiratory Drug Delivery VII. Raleigh, NC: Serentec Press, 2000:503–505.
104. FDA. Guidance for Industry. Nasal Spray and Inhalation Solution, Suspension, and Spray Products—Chemistry, Manufacturing and Controls Documentation. FDA, 2002.
105. Guideline on the Pharmaceutical Quality of Inhalation and Nasal Products. April 2006. EMEA/CHMP/QWP/49313/2005 Final. 24. London.
106. Dayal P, Sudhan M, Singh M. evaluation of different parameters that affect droplet size distribution from nasal sprays using the Malvern SpraytecTM. J Pharm Sci 2004; 93(7):1725–1742.
107. http://www.fda.gov/cder/guidance/index.htm. Accessed March 3, 2008.
108. EMEA/CHMP/QWP/49313/2005 Final.
109. ITFG/IPAC-RS Collaboration. CMC tests and methods technical team. A response to the FDA draft Guidance for Industry. May 2001. ITFG/IPAC-RS Collaboration.
110. http://www.varidose.co.uk. Accessed March 3, 2008.
111. http://www.ich.org/cache/compo/276–254-1.html. Accessed March 3, 2008.
112. Adjei L, Amann A, Blumenstein J, et al. Microbial testing for orally inhaled and nasal drug products. Pharm Forum 2005; 31(4):1258–1261.
113. Clark AR. Medical inhalers. Past, present and future. Aerosol Sci Tech 1995; 22:374–391.
114. Hochreiner D, Holz H, Kreher C, et al. Comparison of the velocity and spray duration of Respimat Soft Mist inhaler and pressurised metered dose inhalers. J Aerosol Med 2005; 18(3):261–263.

7 Locally Acting Nasal and Inhalation Drug Products: Regulatory and Bioequivalence Perspective

Gur Jai Pal Singh

*Division of Bioequivalence, U.S. FDA, Rockville, Maryland, U.S.A.

INTRODUCTION

The Federal Food, Drug, and Cosmetic Act (FD&C Act) is the basic food and drug law in the Unites States. It requires that "new drugs," as defined by the Act, be reviewed and approved by the U.S. Food and Drug Administration (FDA) before they go on the market. In 1938, the FD&C required that the drug manufacturers file a new drug application (NDA) with the FDA for each newly introduced drug and submit evidence to support safety of the drug product. Subsequently, the act has been amended on a number of occasions. A few landmark amendments that influenced the way the FDA approves drugs include the 1951 Durham–Humphrey Amendments (firmly establishing the concept of "prescription" drugs), the 1962 Kefauver–Harris Amendment, the Orphan Drug Act of 1983, and the Drug Price Competition and Patent Term Restoration Act of 1984 (also known as the "Hatch–Waxman Amendments").

Of the various amendments to the FD&C Act, the two that are most relevant to the approval of generic drugs are the 1962 and 1984 amendments. In 1962, the Kefauver–Harris Amendment required all drug manufacturers to establish effectiveness of their products for the claimed indications, in addition to demonstration of safety. Consequently, in 1966 the FDA, through National Academy of Sciences/National Research Council, conducted the Drug Efficacy Study Implementation (DESI) review—a retrospective evaluation for effectiveness of drugs approved between 1938 and 1962. The drugs that were determined to be effective through DESI could be marketed with NDA approval. Under this program more than 3400 prescription products and more than 500 over-the-counter products were reviewed and approximately 900 were removed from the market. As a result of the DESI review, the FDA established a procedure whereby manufacturers of copies of pre-1962 drugs could obtain the Agency approval of their products by submitting Abbreviated New Drug Applications (ANDAs), instead of repeating extensive clinical trials required of pioneer products. The Agency would approve a product if the manufacturer could prove that the generic product was the same as the innovator product and that it was properly manufactured and labeled and show acceptable relative bioavailability.

---

* This article was written during the author's affiliation with the U.S. FDA. It represents the personal opinions of the author and does not necessarily represent the views or policies of the U.S. FDA. Present address for corresspondence: 22386 Amber Eve Drive, Corona, California. Email:gur.jp.singh@gmail.com.

However, the abbreviated procedure was not available for generic versions of drugs approved after 1962.

The Hatch–Waxman Amendments to the FD&C Act extended the ANDA process to include generic versions of all drugs approved after 1962, as well as pre-1962 drugs. These amendments created Section 505(j) of the Act [21 USC 355 (j)] to establish the current ANDA approval process. On the basis of these amendments, manufacturers of generic drugs were required to demonstrate that generic drug products were, among other things, bioequivalent to the pioneer drug products.

The FDA enforces the law by promulgation of regulations, which are legally enforceable standards implementing the legislation. Proposed regulations are first published in the *Federal Register*, inviting comments from interested parties. Any comments received by the Agency are addressed in the preamble to the final rule when it is published in the *Federal Register*. Regulations issued by the Agency are incorporated annually in Title 21 of the Code of Federal Regulations (CFR). Agency regulations pertaining to approval of generic drugs are principally covered by 21 CFR parts 314 & 320. Among the various requisites for ANDAs, the applicants must meet the requirements for documentation of bioequivalence [21 CFR 314.94 (a) (7)] of the proposed generic drug product to the designated reference product, comparative product labeling [21 CFR 314.94 (a) (8)], and data related to chemistry, manufacturing, and controls [21 CFR 314.94 (a) (9)]. Regulations stating requirements for documentation of BE are included in 21 CFR 320.

The FD&C Act is intended to assure the consumer that, among other things, drugs are safe and effective for their intended use. The safety and effectiveness of generic drugs is based on documentation of their bioequivalence (BE) to the corresponding reference listed drugs (RLDs). This chapter provides a discussion on documentation of BE of locally acting nasal and inhalation drug products.

BIOEQUIVALENCE: DEFINITION AND APPROCHES

Based on section 505(j)(8) of the FD&C Act "a drug shall be considered to be bioequivalent to a listed drug if (*i*) the rate and extent of absorption of the drug do not show a significant difference from the rate and extent of absorption of the listed drug when administered at the same molar dose of the therapeutic ingredient under similar experimental conditions in either a single dose or multiple doses; or (*ii*) the extent of absorption of the drug does not show a significant difference from the extent of absorption of the listed drug when administered at the same molar dose of the therapeutic ingredient under similar experimental conditions in either a single dose or multiple doses and the difference from the listed drug in the rate of absorption of the drug is intentional, is reflected in its proposed labeling, is not essential to the attainment of effective body drug concentrations on chronic use, and is considered medically insignificant for the drug." The law also states that "for a drug that is not intended to be absorbed into the bloodstream, the Secretary may establish alternative, scientifically valid methods to show bioequivalence if the alternative methods are expected to detect a significant difference between the drug and the listed drug in safety and therapeutic effect." The types of evidence to establish BE stated in 21 CFR 320.24(b) include (*i*) in vivo studies in humans comparing drug/metabolite concentrations in an accessible biological fluid, (*ii*) in vivo testing in humans of an acute pharmacological effect, (*iii*) controlled clinical trials in humans to establish safety and efficacy, (*iv*) in vitro methods, and (*v*) any other approach deemed appropriate by FDA.

BE regulations (21 CFR 320.22) provide allowance for waiver of in vivo testing for certain drug products for which BA or BE may be self-evident. Thus, pursuant to the regulations [21 CFR 320.22(b)(3)] requirements for evidence of in vivo BE may be waived for drug products that are, among other things, "a solution for application to the skin, an oral solution, elixir, syrup, tincture, a solution for aerosolization or nebulization, a nasal solution, or similar other solubilized form." However, nasal solutions in metered dose devices intended for local delivery represent drug-device combination products. For these products, generic sponsors are requested to provide evidence for equivalent performance of the test and reference products based on a number of in vitro studies outlined below.

Equivalent performance of generic drug products and the corresponding RLDs are generally based on one or more of the methods listed under 21 CFR 320.24(b). The choice of in vivo method(s) is primarily based on the mode of drug delivery to the site(s) of action, whereas selection of the in vitro methods depends upon the type of drug product. Thus, for a majority of orally administered drugs, which are delivered to the site(s) of action through the systemic circulation, equivalent performance of the generic and reference products is based on in vivo BE studies, measuring drug concentrations in accessible biological fluids (blood or urine) and comparative in vitro dissolution testing (1,2).

BE studies that utilize pharmacokinetic (PK) metrics based on the measurement of drug concentrations in blood support comparative effectiveness and safety of most orally administered drug products, because these drugs reach the relevant site(s) of action in the body via systemic circulations. However, this approach by itself is generally not sufficient for documentation of BE of locally acting metered nasal and inhalation drugs, which may not depend upon the systemic circulation for delivery to the intended target sites. Determination of comparative performance of such products is complicated due to the (*i*) intended delivery of drugs directly to the local site(s) of action independent of absorption into the systemic circulation, (*ii*) entry of drugs into the circulation at multiple sites, and (*iii*) dependence of drug delivery on metering devices that are integral parts of the drug products. Therefore, the scientific paradigms for establishment of comparative performance of multisource locally acting nasal and inhalation products are unique, and they are different from the conventional PK methods used for establishment of BE of oral and some topical products designed for drug delivery to the systemic circulation. The evaluation of comparative in vivo performance of these products requires separate assessments of equivalence of local action and the systemic exposure. In addition, the approval of generic nasal and inhalation drug products requires evidence to support equivalent performance of metering devices, which includes a number of in vitro tests that characterize the drug delivery and spray plumes. A brief discussion of each of these approaches is provided below.

The FDA program for documentation of BE of locally acting nasal sprays and pressurized metered dose inhalers (MDIs) which includes a combination of the studies for establishment of equivalence of local action, systemic exposure and in vitro performance of devices, is based on the assumption that the formulations of the generic products to be qualitatively (Q_1) and quantitatively (Q_2) the same as those of the corresponding RLD products. The Q_1 (Qualitative) sameness means that the generic product contains the same inactive ingredients as the RLD, and the Q_2 (Quantitative) sameness means that the generic product contains all inactive ingredients at concentrations within ±5% of the concentrations in the RLD. In addition, the Agency recommends that certain BE evaluations should be based on more than one batch for the test product. Based on the Agency guidance (3), the in vitro BE studies should be performed on samples from three batches of the generic and RLD products. One of the test product batches should be large enough to be used for all in vitro and in vivo BE studies outlined below. The three batches of the test products should be prepared using three batches of the device, drug, and the excipients. Reference product samples should be from three different batches available in the marketplace. Test product samples should be from the primary stability batches used to establish the expiration dating period. The manufacturing process of these batches should simulate that of large-scale production batches for marketing. The test product batches should be manufactured under Current Good Manufacturing Practices (CGMP) conditions in the regular production facility using equipment of the same design and operating principles as will be used for commercial batches. The operations should be performed by appropriately qualified individuals and the batches fully packaged.

EQUIVALENCE OF LOCAL ACTION
The evidence for equivalence of local action is generally based on clinically relevant pharmacodynamic (PD) endpoints or clinical measurements. PD studies are preferred over clinical measurements [21 CFR 320.24(b)] because they provide greater sensitivity to detect potential differences in dose delivery to the local site(s) of action. However, for documentation of equivalence of local action in the absence of clinically relevant PD endpoints, the Office of Generic Drugs (OGD) relies on measurements based on clinical endpoints.

Nasal Drug Products
As stated above, the in vivo BE testing may not be required to support equivalence of locally acting solution nasal sprays. However, for the suspension-based nasal sprays, sponsors are requested to conduct in vivo BE studies to demonstrate equivalence of local action. The disparity in recommendations for solution- and suspension-based products is primarily due to the lack of suitable method(s) for determination of size of the drug particles in the suspension formulations in the presence of similar-sized excipient particles (3). Equivalence of local drug delivery of nasal suspension sprays intended for treatment of symptoms of rhinitis is based upon a clinical (rhinitis) study. Generally, treatment evaluations based on clinical measurements do not provide definable dose response over the labeled dose range. Likewise, the rhinitis studies are relatively insensitive in showing a dose–response relationship within the recommended dosing regimens. Therefore, these studies are intended only to confirm the lack of important clinical differences between generic and RLD formulations (Advisory Committee for Pharmaceutical Science, 2001), provided sameness of formulations, device performance, and equivalence of systemic exposure are established (3).

For establishment of equivalence of local action of suspension-based nasal sprays, the Agency Guidance (3) recommends rhinitis studies of two-week duration. Before dosing, the test and reference products should be primed according to labeling instructions. In vivo BE should be studied in patients with seasonal allergic rhinitis in a randomized, double-blind, placebo-controlled, parallel group study of two-week duration, preceded by a one-week placebo run-in period to establish a baseline and to identify placebo responders (4). The FDA recommends exclusion of placebo responders from the study to increase the ability to show a significant difference between active and placebo treatments (efficacy analysis) as well as to increase sensitivity to detect potential differences between test and reference products (equivalence analysis). Clinical evaluation of the placebo and active treatments is based on the Total Nasal Symptom Scores (TNSS). Treatment effectiveness is evaluated in terms of "reflective" scores to reflect the 12 hours immediately before dosing and "instantaneous" scores to determine the patient's condition at the time of evaluation. The FDA guidance (3) recommends that clinical evaluations should be made twice daily (AM and PM, 12 hours apart at the same times daily) throughout the 7-day placebo run-in period and the 14-day randomized treatment period. The readers are referred to the Agency guidance (3) for further details of the rhinitis study including clinical endpoints, scoring, and equivalence analysis.

Inhalation Drug Products

For determination of equivalence of local action of inhalation aerosols, the FDA relies on PD BE studies. Scientific considerations in the design and evaluation of the PD BE studies include (but are not limited to) the selection of the in vivo PD model and PD measures (endpoints), enrollment of appropriate subject/patient population(s), measurable within-study dose response, and statistical procedure for determination of BE. Scientific and regulatory aspects of each of these considerations are discussed below.

In Vivo PD Models

The mechanisms of action of inhaled respiratory drugs vary by class. Consequently, the choice of in vivo PD model for documentation of equivalence of local action may depend upon the class of the drug product and the biological model used for assessment of drug effect(s). The following provides a summary of the FDA recommendations of PD models for BE studies on some classes of the locally acting inhalation drug products.

Beta-agonists

Inhalation aerosols containing short-acting beta-agonist drugs like albuterol are indicated for both prevention and treatment of bronchospasm. Therefore, PD effects of such drugs may be measured in terms bronchodilatation or prevention of experimentally induced bronchoconstriction. For determination of in vivo BE of albuterol MDIs, the FDA guidances issued in 1989 (5) and 1994 (6) recommended the use of bronchodilatation and bronchoprovocation models, respectively. The Agency currently accepts either bronchodilatation or bronchoprovocation studies for documentation of BE of multisource formulations of short-acting beta-agonists.

The most common measure of bronchodilatation produced by inhalation aerosols is the increase in forced expiratory volume in one second (FEV_1). Following administration of the generic and RLD products, the FEV_1 versus time profiles can be used to compute a variety of metrics (7), which can be used for product comparison. The OGD recommends area under the effect versus time curve (AUEC; based on the trapezoidal rule calculations) and peak effect (FEV_{1max}). Determination of BE using an appropriate statistical analysis (see below) should be based on the AUEC and FEV_{1max} data adjusted for baseline (predose FEV_1) (8).

The duration of the observed increase in FEV_1 following administration of bronchodilators may vary by drug. Therefore, in the BE studies evaluating product performance using FEV_1, the duration of FEV_1 measurements should be relevant to the recommended clinical dosing regimens. For example, in the case of albuterol MDIs measurement of FEV_1 up to six hours after dosing is recommended. Furthermore, BE evaluations are based on comparison of $AUEC_{0-4}$ and $AUEC_{0-6}$, because the RLD labeling indicates two inhalations repeated every four to six hours.

Depending upon the severity of asthma and objective spirometric evaluation, the patients may be classified as mild, moderate, or severe asthmatics (9). Demonstration of dose response in bronchodilatation studies partly depends upon the severity of the disease, which determines the window of response for manifestation of bronchodilatory effects of the study drug. Results of bronchodilatation studies are highly dependent on the subject population to be studied, and it is relatively easy to enroll study populations that may not exhibit a dose response (10). The asthmatic populations most appropriate for bronchodilatation BE studies may be determined based on results of pilot studies. Based on the available information, mild asthmatics are generally not suitable for dose–response studies using multiple actuations of the marketed products. However, depending upon the study drug, populations of moderate-to-severe asthmatics may provide measurable dose response.

Inhibition of chemically induced bronchospasm by beta-agonists provides opportunities for discrimination of the PD effect of these drugs over a broad range of doses. This method is capable of showing beta agonist effectiveness where simple bronchodilation is not (11). Bronchoprovocation with methacholine or histamine following administration of varying doses of beta agonist aerosols has proven useful in studying their dose–response relationships (12) and documentation of BE of the generic and reference listed drugs (13). When single and multiple doses of the approved beta-agonists are used, dose–response relationships based on the bronchoprovocation models are generally steeper than those observed in bronchodilatation studies (14). The observed steepness in dose–response relationships imparts greater sensitivity to the bioassay to potential differences in dose delivery from multisource drug products (8).

Bronchoprovocation with either histamine or methacholine challenge may be used to quantitate and compare protective effects of short-acting beta-agonists. In such studies, the measure of the drug activity is PD_{20}-FEV_1 or PC_{20}-FEV_1, which refers to a provocative dose or concentration, respectively, of the challenge agent that reduces the drug-induced FEV_1 to a level 20% below an FEV_1 after administration of saline aerosol (8). The American Thoracic Society (ATS) guidelines (15) provide detailed recommendations regarding conduct of the methacholine change studies. The same recommendations are also applicable to studies using histamine as the challenge agent.

Bronchoprovocation studies are generally conducted in mild asthmatics. However, prestudy screening of mild asthmatics for responsiveness to the challenge agent can enrich the study population and provide improvement in dose response (12). In these studies, administration of the challenge agent (methacholine or histamine) can be achieved by the two-minute tidal breathing method or the five-breath dosimeter method (15). Based on the OGD recommendation in the guidance issued previously (6) either method could be used for the challenge studies. However, comparisons of the two methods have revealed limitations associated with the dosimeter method (16,17), which may influence determinations of dose response and product comparisons in BE studies. Hence for administration of the challenge agent in BE studies based on the bronchoprovocation model, the tidal breathing method may be preferable compared to the dosimeter method. The delivery of the challenge agent may also depend upon the nebulizer (18) used for generation and administration of the aerosol. The Agency prefers that the nebulizers used in a BE study should be of the same brand and model, and same set of nebulizers and operating conditions should be used for the test, reference, and placebo arms.

The FDA recommends that BE studies based on either in vivo model should be based on a single dose-separate day treatment schedule with sufficient washout between the treatments. The BE studies should be designed to provide high-quality data, which is determined by the accuracy and reproducibility of the observations. Accuracy of the observed data is imparted by regular calibration and validation of the instruments/equipment used. Reproducibility of observations is enhanced by proper use of the equipment and strict compliance with the protocols related to all aspects of the study. Standard Operating Procedures (SOPs) should include specifications for acceptable variability in equipment performance and spirometric observations. For example, the Agency previously recommended that in the bronchodilatation studies, the baseline FEV_1 on the treatment day should be within 88% to 112% of the qualifying day baseline values (5). For the challenge studies, it was recommended that the saline FEV_1 may fall no more than 10% from baseline FEV_1, and the baseline PC_{20} on each day should remain within 50% to 200% of the value measured on the qualifying day (6). The FDA recommends that all protocols and SOPs should be designed a priori by taking into consideration relevant recommendations from the ATS (15,19–21).

The FDA recommendations made previously (5,6) pertain to establishment of BE of the short-acting beta-agonists. The inhalation drugs approved by the Agency also include long-acting beta-agonists (LABA) such as formoterol and salmeterol. Compared with the short-acting beta-agonists, the LABA have prolonged duration of action determined principally by their physicochemical interaction with the membrane lipid bilayers (22). However, the bronchodilation or bronchoprovocation models recommended for the short-acting beta-agonists have also been used to study effectiveness of LABA (23–28). Thus, the PD models recommended for short-acting beta-agonists like albuterol may also applicable to determination of BE of inhalation aerosols containing LABA. However, the study protocols should be modified to accommodate shift in time to peak effect and prolonged duration of response.

Anticholinergic Drugs

Anticholinergics like ipratropium and tiotropium are indicated for the treatment of Chronic Obstructive Pulmonary Disease (COPD) (29,30). These drugs are quaternary ammonium derivatives of atropine, and they are the bronchodilators of choice for treating COPD patients because cholinergic mechanisms are known to contribute to obstruction in this disease (31). Increasing doses of anticholinergics has been associated with increasing improvement in FEV_1 (32–34). Therefore, bronchodilatation following inhalation of these drugs by the patients may exhibit measurable dose response, and the bronchodilatation model recommended for beta-agonists may also be useful for determination of BE of inhalation aerosol containing anticholinergic drugs. These studies should be performed in COPD patients. The selection of patients and development of study protocols should take into consideration all relevant recommendations stated in the ATS guidelines (35), in addition to the available information regarding dosing regimen and duration of the action of the study drug.

The anticholinergic bronchodilators also provide protection against bronchospasm in asthmatics with airway hyperresponsiveness (36,37). By virtue of their anticholinergic activity, these drugs inhibit methacholine-induced bronchospasm in asthma patients in a dose-dependent manner (38). Therefore, the bronchoprovocation model may also be applicable to BE studies on inhaled anticholinergic drugs. However, the choice of the challenge agent may be limited to methacholine, because these drugs may not block effects of histamine on the bronchial smooth muscle (39).

Inhaled corticosteroids

Measurement of airway function over several days of treatment and responses to bronchoprovocation has been frequently used for determination of the efficacy of inhaled corticosteroids (ICS). However, these models have not been applied to determination of BE of ICS products principally due to the lack of a measurable and reproducible dose–response relationship with this class of therapy (40–42). Recently, Ahrens et al. (43,44) described an "Asthma Stability" model, in which the dose response was measured by the extent to which low and high doses of ICS were able to sustain asthma control furnished by a "burst" of high-dose oral prednisone. Following establishment of baseline after the prednisone burst, the patients received randomly assigned low or high doses of test drugs. Responses from each treatment were evaluated for its ability to sustain asthma control based on FEV_1 measurements over a three-week dosing period. At the end of each treatment, the control baseline was reestablished with a course of prednisone. Though the "Asthma Stability" model appears promising for provision of dose response and perhaps suitable for crossover BE studies, it has not been used to support BE of any ICS product. At this time the utility of this model for BE studies on ICS is limited by the lack of information on (*i*) measurable dose response within the labeled dose range, (*ii*) optimum length of each study period between crossover arms, and (*iii*) whether a high dose of the inhaled corticosteroid under study can replace the high dose "burst" of prednisone to establish the maximum response baseline.

ICS therapy reduces airway inflammation (45). The noted lack of the dose response with ICS based on the conventional measures of airway function may be because the lung function tests measure airway caliber, whereas reduction of airway inflammation by ICS is not always directly reflected in airway caliber. Therefore, PD end points based on serial lung function measurements and bronchoprovocation studies may have little use in the determination of ICS action on the lung. The appropriate PD endpoint would be more directly reflective of and responsive to the degree of anti-inflammatory activity of ICS. Determination of airway inflammation has generally been based on observations that required application of invasive techniques (e.g., bronchial biopsies or lavages). Therefore, considerable attention has been paid to noninvasive markers of inflammation such as nitric oxide concentrations in the exhaled air (46–48). Nitric oxide is synthesized by nitric oxide synthase (NOS) found in several cell types. NOS isoforms located in the respiratory system may cooperatively regulate airway smooth muscle tone and immunologic-inflammatory responses (49). NOS exists in constitutive (cNOS) and inducible (iNOS) forms. iNOS is regulated at the posttranscriptional level; it can be induced by proinflammatory cytokines, and increased expression of iNOS leads to increase in the exhaled nitric oxide (ENO) levels in asthmatics (50). ENO is associated with eosinophilic inflammation in asthma (51), and it is considered to be a noninvasive biomarker reflecting such inflammation (49).

ENO is formed in the respiratory tract (49). In asthma, it is reported to correlate with measures of disease control such as asthma symptoms within the previous two weeks, dyspnea scoring, frequency of use of rescue medications, but had a weak correlation with FEV_1 (52). ICS therapy decreases bronchial NO (53). Reduction in ENO levels is due to both the ameliorative effects of steroids on the underlying airways inflammation in asthma and their inhibitory effects on iNOS expression (50). ENO is considered to be a sensitive and rapid marker of the dose-dependent effect of ICS treatment and asthma deterioration upon corticosteroid withdrawal (54). It also reflects asthma control upon treatment with ICS (55). Its levels in asthmatics treated with ICS may be closely related to the level of the lung exposure to corticosteroids because, in addition to suppressing inflammation aspects variably associated ENO, corticosteroids directly inhibit enzymatic production of NO (56). Therefore, ENO measurements may serve as the first responding indicator to assess patient compliance with ICS therapy and steroid dose titration (57–59) particularly in the absence of changes in FEV_1 (60,61).

The ICS-mediated reduction in the ENO concentrations is reversed upon cessation of the ICS treatment and they return to normal levels after appropriate washout periods (62). ENO provides notable (63,64) and reproducible (65) dose response within clinically relevant ICS dosing regimens. Based on these considerations, the ENO bioassay utilized in patients with asthma appears promising for PD BE studies of multisource ICS products. Its use for determination of BE of ICS has been previously suggested (66). However, up to the time of preparation of this chapter, the FDA has not issued any recommendation for the use of the ENO bioassay for BE studies on multisource ICS products.

Dose–Response Considerations and Statistical Evaluation

PD BE studies are required to have the capability to distinguish between the test and the reference products (6,8). The capability is ascertained by the demonstration of a definable within-study dose response. Dose–response relationships of the approved inhalation aerosols are generally nonlinear and, sometimes, very shallow. Therefore, the conduct of PD BE studies warrants consideration

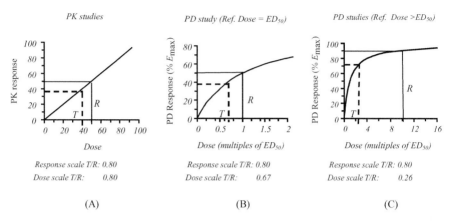

FIGURE 1 Comparison of difference in potencies (dose axis) of hypothetical generic (T) and RLD (R) products showing a 20% difference in pharmacokinetic (**A**) and pharmacodynamic (**B** and **C**) responses. The plots represent a linear (**A**) and nonlinear (**B** and **C**) dose–response relationships. Calculation of relative potencies (T/R) in parts B and C is based on the simple E_{max} model. *Source*: From Ref. 8.

of screening the appropriate patient population to provide a wider window for improvement of the underlying process (e.g., asthma) and enhance the likelihood of obtaining a measurable dose response. Demonstration of dose response also requires the use of more than one dose of the reference product. The basic PD BE study design may have two doses R_1 and R_2 (where 1 and 2 refer to one and two inhalations/actuations, respectively) of the reference product. Additional doses (i.e., R_1, R_2, and R_3) of the RLD can be used to enhance precision in dose response.

The dose–response relationships based on PD measures for inhalation drugs generally are not linear, and the extent of nonlinearity may also vary from one study to another due to a number of factors including the mode of action and potency of the study drug, the PD measure, and the study population and severity of the disease. The curvilinear nature of the dose response of inhalation drugs adds complexity to the statistical analysis of data to support BE of the test and reference products (8). If the dose–response relationship was linear as seen in most PK BE studies, a given change in dose on the *X*-axis would produce the same magnitude of change in the PD response on the *Y*-axis [Fig. 1(A)]. However, in the presence of nonlinear dose–response relationships, the observed difference in PD responses may not be a simple function of the difference in the dose delivered to the target site. The magnitude of difference in the responses may also vary with the positioning of the observed responses on the dose–response curve [Figs. 1(B) and 1(C)]. Therefore, for determination of BE based on PD measures, the statistical evaluation procedure should take into consideration the observed (i.e., within-study) nonlinearity in dose response.

Statistical analyses of the conventional BE studies based on pharmacokinetic measurements uses a two one-sided procedure (67). However, that procedure applied directly to the endpoint itself, as used for statistical analysis of most PK BE studies, is not appropriate for analysis of PD BE studies on inhalation aerosols, because it does not address the nonlinearity in dose–response relationships. To address this complexity, the FDA introduced a method known as the "Dose Scale" method (68). Based on this method, the assessment of BE is made in terms of relative bioavailability of the test and reference formulations at the site(s) of action in the lung. The relative bioavailability is determined in terms of "delivered" dose of the test formulation required to produce a PD response of the same magnitude as exhibited by the reference formulation, and its calculation takes into consideration the within-study dose response.

The development of the "Dose Scale" methodology was based on an assumption of a one-to-one relationship between "delivered dose" and the nominal (ex-actuator) dose of inhalations aerosols. The relationship between the ex-actuator dose (D) and the observed response (E) may be described as follows:

$$E = \phi(D).$$

If the dose–response data for multiple doses of the reference product are available, one can estimate a function ϕ_R, which relates the reference product dose D_R to its response E_R as follows:

$$E_R = \phi_R(D_R).$$

Application of the "Dose Scale" analysis to PD BE studies on a variety of inhalation aerosols warrants an a priori determination of the dose–response function ϕ_R. Based on simple theories of drug action, pharmacodynamic responses of drugs are mediated by the binding of drug molecules to receptors at the target site(s) (69). Given a finite number of drug receptors at the target site(s), PD responses generally exhibit nonlinear dose–response relationships; a maximum level of the drug effect may be achieved upon administration of dose sufficient to occupy all available receptors sites. Consistent with these pharmacological considerations, a model chosen to describe dose–response relationships should provide an estimate of the maximum achievable effect, as well as the establishment of no effect (above baseline response, if any) in the absence of the drug (70). Based on these considerations, the FDA selected a nonlinear function in the form of the following simple E_{max} model:

$$E = E_0 + \frac{E_{max} \times \text{Dose}}{ED_{50} + \text{Dose}}$$

where E is the response, Dose the Administered dose, E_0 the Baseline response (if applicable) in the absence of the drug, E_{max} the fitted maximum drug effect, ED_{50} the Dose required to produce 50% the fitted maximum effect.

When the dose–response relationship is described using this E_{max} model, then ϕ_R can be described as follows:

$$\phi_R = E_{0R} + \frac{E_{maxR} \times \text{Dose}_R}{ED_{50R} + \text{Dose}_R}.$$

For application of the "Dose Scale" method to determination of relative bioavailability, ϕ_R in the above equation can be fitted to the mean, or pooled, dose–response data for multiple (0, 1, 2, . . .) doses of the reference product. Mean responses may be computed as geometric mean or arithmetic mean depending on the distribution of the PD response data. If the data are normally distributed, arithmetic mean may be used, whereas geometric mean may be more appropriate for log-normally distributed data.

The relative bioavailability "F" of a dose of the test product relative to that of the reference product can be calculated by applying the inverse of ϕ_R to the mean of, or the pooled, response data of the test product as follows:

$$F = \phi_R^{-1}(E_T)/D_T,$$

where E_T is the mean of, or the pooled, response of the test product to the ex-actuator dose (D_T) of the test product.

The basic PD BE study designs require multiple doses (e.g., 0, 1, and 2 actuations) of the reference product, but one dose (1 actuation) of the test product may be sufficient. For such studies, the relative bioavailability of the test product can be estimated as follows:

$$F = \phi_R^{-1}(E_{T1})/D_{T1},$$

where E_{T1} is the mean, or pooled, response to the single ex-actuator dose (D_{T1}) of the test product.

The FDA may accept PD BE study designs using multiple doses of the reference product and only a single dose of the test product. However, the use of multiple doses of both test and reference products may enrich the study data and enhance precision of the estimated values. The Agency recognizes that conduct of crossover PD BE studies with additional treatment arms may be difficult for drug products that might take many days (or even weeks) to manifest measurable PD responses and require similarly long intervals for washout periods between the treatments. Nonetheless, it encourages (whenever possible) conducting PD BE studies based on multiple doses of both test and reference products (e.g., T_0, T_1, & T_2 and R_0, R_1 & R_2 representing 0, 1, & 2 doses of each product, respectively). For such studies, relative bioavailability "F" of the test product can be determined by simultaneously fitting the within-study dose–response data of both the test and reference products to the following modification of the above model:

$$y = E_0 + \frac{E_{\max} \times \text{Dose} \times F^i}{ED_{50} + \text{Dose} \times F^i},$$

where y is the response and i the treatment indicator ($0 = $ Ref., $1 = $ Test), with the understanding that $F^0 = 1$.

This modified model is based on assumption that both E_0 and $E_{\max}$ are the same for the test and reference products. ED_{50R} (for the Reference product) is ED_{50} itself, while ED_{50T} (for the Test product) is ED_{50}/F^1.

Determination of BE based on the "Dose Scale" method is a two-step procedure. In the first step, a within-study dose–response relationship is mathematically described by fitting the relevant version of the $E_{\max}$ model to the mean of, or pooled, dose–response data of the reference product (or both test and reference products, if available). In second step, a 90% confidence interval for "F" is estimated by a bootstrap procedure. Each bootstrap estimation includes calculation of "F" by fitting the above model to a "sample dose–response data set," which is generated by repetitive sampling with replacement. The FDA has used Efron's

BCa (bias corrected and accelerated) method (71) to compute a 90% confidence interval for "*F*."

The FDA has employed the "Dose Scale" method for regulatory evaluation of PD BE studies on albuterol MDIs. The studies were based on bronchoprovocation model (PD measure: Histamine PC_{20}) or the bronchodilatation model (PD Measures AUEC-FEV_1 and FEV_{1max}). The method has been applied to BE data from both models even though (*i*) the two models are clinically and physiologically distinct, and (*ii*) the nature (slope) of the dose response varied with the model as well as with the PD endpoint within the same model (72). Therefore, application of the "Dose Scale" method to PD BE studies is independent of the in vivo PD model and the PD endpoint.

For its acceptance of the PD BE studies on albuterol MDIs based on the "Dose Scale" analyses, the FDA used the interval between 67% and 150% as the bounds within which the 90% confidence interval (CI) must lie. However, it should be noted that the 67% to 150% acceptance limits are applicable *only* to the computed (based on the "Dose Scale" method) estimates of the relative dose delivered to the target site (lung); these CI limits are *not* applicable to BE evaluations based on measurements of PD/clinical responses as such and are also *not* applicable to determinations of BE using statistical procedures other than the "Dose Scale" method. This is because in the presence of nonlinear dose–response relationships, the 67% to 150% CI range on the dose axis generally represents a much narrower range on the corresponding response axis (8,72).

The selection of the 67% to 150% CI limits for BE studies on albuterol MDIs was based on the FDAs clinical judgment (73). The upper limit of 150 was based on (*i*) how the inhalation aerosols are used, (*ii*) the relatively higher degree of variability in effective drug delivery of MDIs compared to oral medications, and (*iii*) the lower reproducibility of the intraindividual PD response compared to PK determinations of systemically administered drugs. It was noted that, because of variability considerations, the point estimate for the test/reference comparison has to be close to unity for a product to meet the acceptance range of 67% to 150% (72,73).

Because the determinants of relative bioavailability based on the "Dose Scale" method are the observed difference in the test and reference product response, and the observed (within-study) dose–response relationship and variability in the PD response, the method should be applicable to analyses of a variety of PD BE studies regardless of the PD Model, PD measure, steepness of dose response, and the class of inhalation aerosols. Therefore, the "Dose Scale" method initially developed for PD BE studies on albuterol MDIs may be applicable to all such studies on a variety of respiratory drugs including ICS.

EQUIVALENCE OF SYSTEMIC EXPOSURE

Though the locally acting respiratory drugs are intended for delivery to the local sites (nose or lung), a portion of the aerosolized drug may be deposited in the oropharyngeal region, swallowed, and subsequently become available for absorption from the GI tract (Fig. 2). In addition, drug deposited in the lung may also enter the systemic circulation. Therefore, the FDA requests evidence for comparative systemic exposure from test and reference products to address

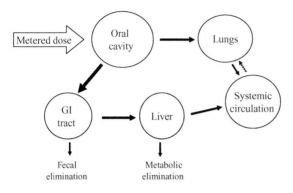

FIGURE 2 Schematic representation of disposition of a metered dose emitted from an inhalation aerosol. *Source*: From Ref. 8.

concerns regarding possible systemic side effects of locally acting nasal sprays (74) and inhalations respiratory drugs (75–82).

The evidence for equivalent systemic exposure preferably is based on PK (AUC and C_{max}) data. PD studies are generally less sensitive, as clinical side effects of inhaled drugs may be detected in the absence of measurable change in the PD measure (83,84). However, based on the Agency guidance (3), PD studies may be submitted if measurement of blood concentrations of the drug moiety(ies) is not feasible. Measurable plasma levels are those concentrations that are above the limit of quantitation and could consistently be quantified using a sensitive analytical method to the extent that the relevant PK parameters (AUC and C_{max}) can be calculated (3).

The systemic exposure studies should be conducted at the single maximum-labeled adult dose. This is particularly important in the case of the aqueous nasal sprays. For these products the administered sprays volume should be kept low to minimize the possibility of alteration of the drug deposition pattern within the nose that may occur at higher volumes when dosed above label claim. The deposition pattern may be altered due to loss of drug from the nasal cavity at higher volumes, due either to drainage into the nasopharynx or externally from the nasal cavity (3).

The recommendation for the use of the single the maximum-labeled dose is also applicable to studies of comparative systemic exposure from multisource MDIs. However, the concern regarding loss of drug due to drainage may not be applicable for MDIs due to the volatile nature of the carrier propellants used in these products. Therefore, if blood levels are not measurable following administration of the maximum labeled adult dose of MDIs, a higher dose may be used [per 21 CFR 320.29 (b)] as long as the administered dose is safe and in the linear pharmacokinetic range. Submission of a Bio-IND is required (per 21 CFR 320.31) if studies are conducted at doses exceeding the maximum labeled adult dose.

BE studies based on PK endpoints are generally conducted in adult healthy volunteers (85). The study populations should be representative of gender and race of the target populations. The studies are preferably based on randomized crossover (replicate or nonreplicate) designs enrolling adequate number of subjects.

Upon administration of the test and reference products, blood (plasma/serum) concentrations of parent drug and/or metabolite (if applicable) should be measured using a validated assay (86). The blood concentration versus time data should be used to compute bioavailability metrics AUC_{0-t}, AUC_{inf} (where applicable) and C_{max}.

For the oral products administered for drug delivery to the systemic circulation, BE is based solely on the PK measurements. Blood samples are generally taken for periods sufficient to describe the absorption, distribution, and elimination phases of the drug and the Agency guidance (85) recommends collection of blood samples for periods equal to at least three half lives of the drug. However, for locally acting drugs intended for inhalation and nasal delivery, feasibility of measurement of the blood concentrations over extended postdose periods may be limited by the labeled maximum dose, drug bioavailability, and sensitivity of the bioassay. Consequently, measurement of blood concentrations of the locally acting drugs for three half lives may not be feasible even with the use of very sensitive assays. The observed blood concentration profiles may be erratic and contain only a limited number of measurable blood concentrations. The Agency considers that, however limited, such blood concentration profiles are of value in determination of comparative systemic exposure from the locally acting generic and RLD products. The systemic exposure studies for these products represent only a part of the required BE testing paradigm. Thus only in the case of locally acting drugs delivered via inhalation and nasal routes, the FDA recommends that computation of the AUC_{0-t} may be based on at least four consecutive nonzero blood concentration values. The AUC computation should be terminated at the last quantifiable blood concentration before the first zero (i.e., below the limit of quantitation, BLOQ) value following these four or more values. The PK analysis should include only those subjects that meet this rule for both periods, that is, for both test product and RLD. The C_{max} should be selected as the maximum plasma concentration that occurs among the values used to compute the AUC_{0-t}. A second maximum concentration that may occur after the data points used in the computation of AUC_{0-t} [i.e., following the above-mentioned zero (BLOQ) value] may not be the C_{max} of interest.

Determination of equivalence of systemic exposure from the test and reference products is based on the natural log-transformed AUC and C_{max} data using the conventional two one-sided tests (ANOVA) procedure (87,88). Equivalence of systemic exposure from the test and reference products is established if the 90% confidence intervals for the AUC and C_{max} data fall within the acceptance limits of 80% to 125%. If the drug products contain more than one active drugs, documentation of BE requires demonstration of equivalent exposure for each drug.

EQUIVALENCE OF IN VITRO PERFORMANCE

Documentation of BE of generic drugs generally includes in vitro assessment of drug release from the drug products. For the solid dosage formulations intended for oral delivery, in vitro dissolution studies are generally adequate to characterize the drug product performance (89). However, drug delivery from metered inhalation aerosols and nasal sprays is complicated, as it is influenced by the formulation as well the metering device. The dimensions of the device components and operating principles/mechanisms are generally product specific. Variations in the design of the components of the drug delivery systems can influence the in vivo performance of the drug product (90–92). Therefore, the FDA recommends that the test product uses the same brand and model of devices (particularly the pump or metering valve and the actuator) as is used in the reference product (3). If this is not possible, the generic sponsors should use a valve or pump, and actuator, which are of designs resembling as close as possible in all critical dimensions to those of the reference product.

The inhalation aerosols and metered nasal spray products are generally designed to deliver multiple doses/actuations. Based on the labeled number of actuations, the product life can be divided into the Beginning, Middle, and End stages, where the Beginning stage represents the first actuation(s) following the labeled number of priming actuations, the Middle stage represents the actuation(s) corresponding to 50% of the labeled number of actuations, and the End stage corresponds to the label claim number of actuations (3). Because of the multidose nature of the products, testing for in vitro performance is performed at various stages of the product life, because potential changes in physicochemical properties of the formulation and wear and tear of the metering device over the use life of the drug product may influence drug delivery. In addition, some of the tests are performed at different time delays to determine equivalence at different stages of plume formation/dissipation, and different distances to impart sensitivity to the test to distinguish between the performance of the generic and RLD products (3).

The FDA has previously issued guidances that contain its recommendations for in vitro tests for nasal sprays and nasal aerosols (3) and inhalation aerosols (93). Based on these documents, comparative in vitro performance of metered nasal sprays and MDIs is based on a number of in vitro tests that characterize the drug delivery and spray plumes. The in vitro tests are designed to determine the priming and repriming (where applicable) characteristics, the potency (single actuation content), particle size and/or droplet size distributions, and plume shape (spray pattern and plume geometry). All comparative in vitro performance tests are now based on 30 units (10 units from each of the three lots) for each of the test and reference products (3).

The FDA recommends that, for all in vitro studies, the test and reference products should be actuated in a manner that removes potential operator bias, either by employing automatic actuation or by employing blinded procedures when manual actuation is used (3). The use of automated actuation systems is expected to decrease variability in drug delivery due to operator factors. Operational parameters for the automated actuation systems, which may include actuation force, velocity, acceleration, length of stroke, etc., can influence the results of in vitro evaluations (94). The Agency recommends that the settings used for operation of the automated actuators should be same for the test and reference products.

Priming and Repriming

Pursuant to 21 CFR 314.94(a)(8), the labeling for generic inhalation aerosols and metered nasal sprays is generally the same as that of the corresponding RLDs, except for specific changes described in the regulations (21 CFR 314.94(a)(8)(iv)). The labeling (or the patient package insert) of certain brand name products states the number of actuations that are wasted to prime the product for initial use and another set of actuations that are wasted for repriming following one or more periods of nonuse. For these products, the FDA solicits priming and repriming data.

Priming studies are noncomparative. However, they are necessary to document that each generic product delivers the labeled dose within the number of actuations to prime/reprime stated in the RLD labeling. Priming and repriming data for the test product in multiple orientations is provided in the Chemistry Manufacturing and Controls (CMC) portion of the ANDA submission. For BE evaluation, the studies are generally based on products stored in the valve upright position, with the exception of products for which RLD labeling recommends storage in the valve down position. For the latter products, priming data (and repriming data where applicable) should be provided following storage in the valve down position. Determination of product priming is based on the emitted dose of the single actuation at the beginning life stage immediately following the specified number of priming actuations in the RLD labeling. For generic products, priming is established if the geometric mean emitted dose (single actuation content—see below) of the 30 canisters at the labeled first primed actuation falls within 95% to 105% of label claim. Repriming should be similarly established based on a single actuation following the specified number of repriming actuations in the RLD labeling.

Single Actuation Content (SAC) Through Container Life

Generic inhalation aerosols and nasal sprays should demonstrate equivalence of SAC to the corresponding RLDs with respect to the amount of the aerosolized drug delivered per single actuation, based on a drug-specific chemical/chromatographic assay. The SAC determinations should be made at the Beginning, Middle, and End stages of aerosols, and Beginning and End stages of nasal sprays (3) using the Unites States Pharmacopeia (USP) recommended apparatus or another appropriate apparatus. The SAC determinations are based on the delivered (emitted ex-actuator) drug mass from primed products. The delivered mass of drug substance is expressed both as the actual amount and as a percentage of the label claim. Equivalence of the test and reference product data is based on the population bioequivalence methodology (PBE) stated below.

Particle Size Distribution (PSD) by Cascade Impaction

The PSD of the emitted dose (ex-actuatior) of inhalations aerosols is generally determined using cascade impactors (USP Apparatus 2 or Apparatus 5, with appropriate accessories) and/or other complimentary particle sizing methods, if applicable. The PSD determinations are made at the Beginning and End life stages of product use. The equipment and accessories are selected so that the majority of the dose is introduced into the impactor for fractionation. The cascade impactors should be operated at the compendial (if applicable) or the manufacturer recommended flow rates.

Multistage cascade impactors fractionate and collect particles of one or more drug components by aerodynamic diameter through serial multistage impactions. Measurable levels of drug below the top stage of the cascade impactor are a function of the specific drug product and the experimental setup and procedure, including the number of actuations and assay sensitivity. The Agency prefers the use of the fewest actuations justified by the sensitivity of the

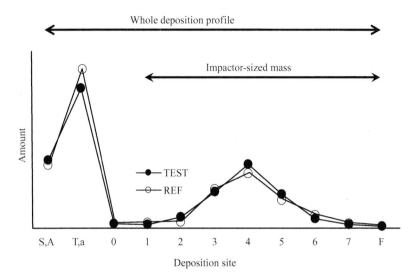

FIGURE 3 Schematic representation of the cascade impactor fractionation of emitted dose of inhalation aerosol and measures of PSD. *Abbreviations*: S, stem; A, actuator; T, throat; a, adaptor; F, filter.

assay. Thus, a validated and sensitive assay should be used for quantitation of drug deposition at the various sites.

The FDA recommends that drug deposition on each accessory and all impaction stages of the cascade impactor should be reported in mass units. Mass balance accountability based on the emitted (ex-actuator) dose is requested to be included in regulatory submissions. Mass balance should be based on drug deposition on adapters, induction port and any other accessories, the top stage, and all lower stages to the filter. Based on the Agency Guidance (95), the total mass of drug collected on all stages and accessories is recommended to be between 85% and 115% of label claim on a per actuation basis.

The cascade impaction data for the test and reference inhalation aerosols should be compared in terms of (a) whole deposition profile representing drug deposition on the individual sites to include stem and actuator, throat (atomization chamber) and adaptor, and all stages of the cascade impactor including filter, and (b) the impactor-sized mass representing total drug deposited below the top stage (stage 0) of the cascade impactor (Fig. 3).

Cascade impaction studies are also recommended for comparison of nasal sprays products. The objective of these studies is not to compare deposition profiles of the drug products, because most of the emitted dose is deposited prior to or on the top stage of the cascade impactor. Instead these studies are conducted to determine a fraction of dose in small droplets that may be delivered beyond the nose. Even though lung deposition following nasal delivery may represent a very small fraction of the emitted dose (96), its determination is important to address concerns regarding possible adverse pulmonary effects due to the possible delivery of excipients/additives beyond the nose (3).

The compendial cascade impactor apparatuses and accessories are appropriate for sizing inhalation aerosols that contain a significant fraction of the emitted dose in small particles. However, certain accessories like USP throat (induction port) may not be ideal for sizing nasal sprays, which predominantly contain large droplets that do not enter the cascade impactor. Entry of the ex-actuator dose into the cascade impactor can be enhanced by using larger atomization chambers (97–99). Therefore, FDA guidance (3) recommends the use of a 2 L or larger induction port for cascade impactor studies of aqueous nasal sprays.

Evaluation of comparative PSD of the test and reference products may be based on comparison of the (*i*) whole deposition profile and (*ii*) the impactor-sized mass. The whole profile comparisons may be based on the chi-square method described in the June 1999 Draft of the Nasal BA/BE Guidance (http://www.fda.gov/cder/guidance/2070dft.pdf). A Product Quality Research Institute (PQRI) Working Group has been studying the performance of the chi-square method for profile comparisons. The recommendation may be subject to change with future updates of the method, if published by the Agency.

Comparison of the impactor-sized mass is based on the population bioequivalence (PBE) approach described in Statistical Information from the June 1999 Draft Guidance and Statistical Information for In Vitro Bioequivalence Data (Posted on the Agency website on 4/11/2003). The method uses a confidence interval approach with an acceptance range of 90% to 111%. Based on the PBE methodology, the BE acceptance limits depend upon the relative variability of the test and reference products observed in the study, ensuring that the acceptance limits are appropriate for the specific products being compared and are based on the characteristics of the approved RLD.

PBE methodology employs two regulatory constants, $Sigma_{T0}$ and Epsilon. Of these, $Sigma_{T0}$ represents the Scaling Variance ($Sigma_{T0}^2$) and it determines the value at which the acceptance limits are scaled to the within-study variability of the reference product. When the reference product variance is greater than the scaling variance, the acceptance limits may be widened. When this variance is less than the scaling variance, the limits are held constant. Epsilon represents the variance offset term. It allows some differences between the total variances of the test and reference products that may be inconsequential.

Fixed values of the regulatory constants are necessary to promote fairness and consistency in regulatory outcome of in vitro BE studies performed by various testing laboratories using the same reference product. They also promote the same level of rigor for acceptance of in vitro studies on a variety of drug products. The selection of values of the regulatory constants requires careful consideration that the chosen values are not too restrictive to fail nearly identical products. Similarly, the chosen values could not be too liberal to pass products that were potentially clinically different. The FDA selected a $Sigma_{T0}$ of 0.1 and Epsilon of 0.01 based on extensive simulation studies that were performed to address the above considerations. Effective August 2005, the FDA implemented these values for all PBE analysis of the nonprofile data from in vitro performance studies on nasal sprays and inhalation aerosols.

Droplet-Size Distribution by Laser Diffraction

The droplet size plays an important role in regional deposition of drugs in the nasal cavity (100). The aqueous nasal spray plumes predominantly contain course droplets of sizes 17 to 215 μM (94,101). The FDA requires documentation of comparative droplet-size distribution (DSD) in the plume emitted from metered nasal spray products using laser diffraction methodology (3). Laser diffraction analyses size aqueous droplets in spray plumes irrespective of the presence or absence of particles in the droplets, and irrespective of the type of particle (drug or excipient or a mixture of both). Laser diffraction instruments use a low-power laser transmitter. The laser beam passes through a filter and a collimating lens to produce a clean parallel beam. The laser beam is focused by a collector lens known a Fourier Transform lens and focused to a point on a radial array of silicon diode detectors (102). The instruments measure the intensity of scattered light at known distances from a central target through which the laser would pass in the absence of droplets (103). The droplets passing through the laser beam scatter light at angles (104). The basic principle of determination of droplet-size laser diffraction is that angles at which the laser light is scattered are inversely proportional to the droplet size (105). The Fourier transform lens has the property of imaging the scatter from a particle to the same part of the detector for the entire time of residence in the beam, regardless of the particle's speed. Consequently at any given time, there is light energy across the detector array, which directly corresponds to the particle size distribution of the droplets that are present in the laser beam at that moment.

DSD determinations are made at the beginning and end stages of product life. The actuation distance may influence DSD measurements (101). Therefore, the FDA recommends (3) determination of DSD at two distances between 2 and 7 cm (at least 3 cm apart) from the actuator tip. At both distances the DSD measurements are made from the fully formed plumes that are characterized by stable regions of obscuration/transmission of the light depicted in the time–history plots generated by the equipment used for laser diffraction analyses. These plots provide obscuration/transmission and derived volume diameters versus time profiles for each plume. DSD is determined in terms of droplets up to less than 10% (D_{10}), 50% (D_{50}), and 90% (D_{90}) of the total droplet volume. The Agency also requests determination of polydispersity of the DSD from the median diameter (D_{50}). It is measured in terms of SPAN $[(D_{90}-D_{10})/D_{50}]$. Equivalence of the test and reference products is based on the comparative D_{50} and SPAN data using the PBE analysis.

Spray Pattern and Plume Geometry

An evaluation of comparative spray pattern and plume geometry is important for evaluating the performances of the valve and the actuator (95). The spray pattern and plume geometry can be influenced by the nature of the formulation (95,100) and a number of factors including the size and shape of the actuator orifice, the design of the actuator, the size of the metering chamber, the size of the stem orifice of the valve, and the vapor pressure in the container (95). The FDA therefore requests comparative studies of spray pattern and plume geometry of the test and reference products to support in vitro equivalence of nasal sprays and metered dose inhalers.

Spray pattern studies characterize the spray either during the spray prior to impaction, or subsequent to impaction on an appropriate target such as a thin-layer chromatography plate. Spray patterns are determined at the beginning life stage, and at two distances from the actuator orifice that allow capability to discriminate between the test and reference products. The selected distances are recommended to be at least 3 cm apart within the range of 3 to 7 cm. The FDA recommendations regarding comparative studies based on impaction or nonimpaction methods described in its guidance for nasal drug products (3) are also applicable to spray pattern studies on inhalation aerosols. Plume geometry describes a side view of the aerosol cloud parallel to the axis of the plume. It may be evaluated by high-speed freeze-frame photography or laser light sheet technology.

Plume geometry is described in terms of the plume cone angle and plume width that are known to influence the nasal spray regional deposition in the nasal airway (100). The plume angle is based on the conical region of the plume extending from a vertex that occurs at or near the actuator tip. The Agency guidance (3) recommends that the criteria for defining the plume angle and plume width borders should included in the study report.

The recommended plume width is the width at a distance equal to the greater of the two distances selected for characterization of the spray pattern. Therefore, the plume width data may be complementary to the spray pattern data obtained at the same distance. However, plume angle based on spray pattern dimensions and distance from actuator tip to an impaction surface is not accepted for evaluation of equivalence.

Plume geometry measurements are reported at a single delay time while the fully developed plume is still in contact with the actuator tip. The selection of the delay time should be based on method validation studies. The applicants are requested to provide documentation that the plume is fully developed at the selected delay time. Additional details regarding submission of data, such as representative photographs, etc., can be found in the Agency Guidance for nasal drug products (3).

WAIVER OF IN VIVO BE TESTING FOR HIGHER/LOWER POTENCY PRODUCTS

For orally administered drug products that exhibit linearity in the in vivo response (blood levels) over the relevant dose range, the OGD generally grants waiver of in vivo BE testing for the lower strengths based on (a) acceptable in vivo BE study(ies) on the highest strength or lower strength (where the use of higher strengths may pose safety issues), (b) formulation proportionality between the lower and the higher strengths, and (c) comparative in vitro dissolution of the lower and the higher strengths of the test products.

A direct application of the above approach for consideration of waiver of *all* in vivo testing for the lower/higher strengths of the inhalation aerosols is complicated by (*i*) the lack of linearity/dose proportionality in the PD response over the clinically relevant dose range, and (*ii*) the probable difference in the in vitro performance of the lower and higher strength products. Therefore, waiver of all in vivo testing for the lower/higher strength of aerosols may not be appropriate. However, if all strengths of the test and reference products are used in the PD study, the FDA may consider requests for waiver of in vivo testing for the demonstration of equivalent systemic exposure for the lower strength(s) of the test products in the presence of (a) acceptable PD study demonstrating equivalence in local action using all strengths, (b) acceptable comparative systemic exposure study on the highest strength, and (c) acceptable complete set of in vitro performance studies comparing each strength of the test and reference products.

If the RLD is marketed in more than one strength, and the PD BE studies use only one strength of the test product, the Agency may consider request for waiver of in vivo PD BE studies for the other strengths in the presence of (a) acceptable PD study comparing the lowest strength (or middle strength if its PD response is on the rising portion, that is, not on the plateau, of the dose–response curve) of the test and reference products, (b) complete set of in vitro performance studies comparing each strength of the test and reference products, and (c) two-way crossover or replicate design pharmacokinetic studies demonstrating equivalent systemic exposure from each strength of the test product and the corresponding strength of the reference product.

DRY POWDER INHLAERS

Following the introduction of the first dry powder inhaler (DPI) (106), these devices have gained increasing acceptance for inhalation drug delivery as they overcome limitations of the propellant-based MDIs with regard to (a) the amount of drug that can be administered (107); (b) coordination problems (108,109) frequently encountered with MDIs (110,111), which may contribute to reduced effectiveness of the treatment (112–114); and (c) environmental concerns from depletion of the ozone layer due to the CFC propellants used in MDIs (115–118).

The FDA has not issued its recommendations for determination of BE of DPIs. However, because the DPIs may contain the same drugs as in the MDIs and they may be designed to deliver the drug to the same target site(s) in the lung, it is logical to assume that the same three types of studies (i.e., in vitro studies to demonstrate equivalence in drug delivery, PK or PD studies to document equivalence of systemic exposure, and PD or clinical endpoint studies to document equivalence of action at the local site) would be necessary to establish BE of DPIs. Despite such similarity in the route of drug delivery, and similar target site(s) of action for respiratory drugs delivered by MDIs and DPIs, a direct application of BE testing paradigms recommended for MDIs to DPIs is complicated due to certain unique characteristics of drug formulations in the DPIs, mechanism of drug delivery from these products, and a variety in the design of devices used for drug delivery. Formulations of the currently approved DPIs are made up of either the active drug alone or the active drug loosely associated with a carrier, such as lactose. The marketed products differ with regard to the powder formulation and design of the delivery devices, and both formulation and device contribute to efficiency of aerosolization of the powder formulation (107,119), hence drug delivery to the patient.

The currently FDA-approved DPIs are breath-actuated. These products depend upon the patients' inspiratory effort to provide energy for entrainment, deagglomeration and aerosolization of the powder formulation (120), and its deposition in the respiratory tract (121). The inhalation effort determines the inspiratory flow rate through the device (122). In addition, the achievable flow rate through DPIs is also influenced by the flow resistance of the inhaler. Different DPIs may offer different levels of resistance to air flow (123–125), as the device resistance is dependent on the internal geometry of the inhaler and dimensions of the air channels. For a given inspiratory effort, the rate of air flow through inhalers may vary with the device (123,126), and the flow rate is considered to be an important determinant of in vitro drug delivery from DPIs (119–121). In addition, because of variation in the device resistance and achievable flow rates across the target patient populations (127–133), the flow rate may also influence in vivo performance of DPIs (134–139). Therefore, the development of recommendations for comparative performance of the generic and RLD DPIs warrants, among other things, scientific considerations regarding the powder formulation and its stability, device features, and the influence of flow rate on in vitro dose delivery and particle size distribution of the aerosolized drug. The currently available draft chemistry guidance (95) recommends consideration for flow rate in characterization of in vitro drug delivery (Dose content uniformity and particle size distribution) from DPIs. Information in that guidance pertains to chemistry, manufacturing, and controls, and up to the time of the preparation of this chapter no such guidance was issued for documentation of BE of DPIs. Recently however, the FDA has published (139) its view of scientific considerations for comparative assessment of in vitro performance of multisource DPIs. The fundamental considerations in the Agency publication are consistent with the views expressed above.

REFERENCES

1. Patnaik R, Lesko L, Chan K, et al. Bioequivalence assessment of generic drugs: An American point of view. Eur J Drug Metab Pharmacokinet 1996; 21:159–164.
2. Davit BM, Conner DP. Issues in bioequivalence and development of generic drugs. In: Sahajwala CG, ed. New Drug Development: Regulatory Paradigms for Clinical Pharmacology and Biopharmaceutics. New York: Marcel Dekker, 2004:399–416.
3. Guidance for Industry (Draft): Bioavailability and bioequivalence studies for nasal aerosols and nasal sprays for local action. Rockville, MD: Center for Drug Evaluation and Research, US FDA, 2003.
4. Guidance for Industry (Draft): Allergic Rhinitis: Clinical Development Programs for Drug Products. Rockville, MD: Center for Drug Evaluation and Research, Food and Drug Administration, 2000.
5. Guidance for in-vivo bioequivalence studies of metaproterenol sulfate and albuterol inhalation aerosols (Metered dose inhalers). Rockville, MD: Division of Bioequivalence, Food and Drug Administration, 1989.
6. Interim guidance for documentation of in vivo bioequivalence of albuterol inhalation aerosols (Metered dose inhalers). Rockville, MD: Office of Generic Drugs, US FDA, 1994.
7. Singh GJP. Bioequivalence methodology for albuterol metered dose inhalers: Results of pilot bronchodilator studies. Proceedings of the Second International Conference on the Pharmaceutical Aerosol: A Drug Delivery System in Transition 1994. Basel, Switzerland: Technomic Publishing AG.
8. Singh GJP, Adams WP. US regulatory and scientific considerations for approval of generic locally acting orally inhaled and nasal drug products. In: Dalby RN, Byron P, Peart J, Suman JD, eds. Respiratory Drug Delivery—Europe: A Regulatory and Analytical Symposium. Richmond, VA: Virginia Commonwealth University, 2005: 115–125.
9. Guidelines for the diagnosis and management of asthma. Component 1: Measures of assessment and monitoring—Initial assessment and diagnosis of asthma. National Institute of Health (National Heart Lung and Blood Institute), 1997.
10. Ahrens RC. On comparing inhaled beta adrenergic agonists. Ann Allergy 2001; 67:296–298.
11. Parameswaran K. Concepts of establishing clinical bioequivalence of chrofluorocarbon and hydrofluoroalkane [beta]-agonists. J Allergy Clin Immunol 1999; 104:S243–S245.

12. Creticos PS, Adams WP, Petty BG, et al. A methacholine challenge dose-response study for development of pharmacodynamic bioequivalence methodology for albuterol metered dose inhalers. J Allergy Clin Immunol 2002; 110:713–720.

13. Stewart BA, Ahrens RC, Carrier S, et al. Demonstration of in vivo bioequivalence of a generic albuterol metered-dose inhaler to Ventolin. Chest 2000; 117:714–721.

14. Parameswaran K, Inman MD, Elkholm BP, et al. Protection against methacholine bronchospasm to assess relative potency of inhaled ß2-agonist. Am J Respir Crit Care Med 1999; 160:354–357.

15. American Thoracic Society. Guidelines for methacholine and exercise challenge testing-1999. Am J Respir Crit Care Med 2000; 161:309–329.

16. Cockcroft DW, Davis BE, Todd DC, et al. Methacholine challenge: Comparison of two methods. Chest 2005; 127:839–844.

17. Allen ND, Davis BE, Hurst TS, et al. Difference between dosimeter and tidal breathing methacholine challenge: Contribution of dose and deep inspiration bronchoprotection. Chest 2005; 128:4018–4023.

18. Praml G, Scharrer E, de la Motte D, et al. The physical and biological doses of methacholine are different for Mefar MB3 and Jaeger APS Sidestream nebulizers. Chest 2005; 128:3585–3589.

19. American Thoracic Society. Standardization of spirometery—1994 update. Am J Respir Crit Care Med 1995; 152:1107–1136.

20. ATS/ERS Task Force. Standardisation of lung function testing: General considerations for lung function testing. Eur Respir J 2005; 26:153–161.

21. ATS/ERS Task Force. Standardisation of lung function testing: Standardisation of the measurement of lung volumes. Eur Respir J 2005; 25:511–522.

22. Anderson GP, Linden A, Rabe KF. Why are long-acting beta-adrenoceptor agonists long acting? Eur Respir J 1994; 7:569–578.

23. van der Woude HJ, Aalbers R. Long acting ß2 agonists: Comparative pharmacology and clinical outcomes. Am J Respir Med 2002; 1:55–74.

24. Palmqvist M, Ibsen T, Mellen A, et al. Comparison of the relative efficacy of formoterol and salmeterol in asthmatic patients. Am J Respir Crit Care Med 1999; 160:244–249.

25. Bisgaard H. Long acting ß2-agonists in management of childhood asthma: A critical review of the literature. Pediatr Pulmonol 2000; 29:221–234.

26. Kemp J, De Graff AC, Paerlman DS, et al; the Salmeterol Research Group. A 1-year study of salmeterol powder on pulmonary function and hyperresponsiveness to methacholine. J Allergy Clin Immunol 1999; 104:1189–1197.

27. van der Woude HJ, Postma DS, Politiek MJ, et al. Relief of dyspnoea by ß2-agonists after methacholine-induced bronchoconstriction. Respir Med 2004; 98:816–820.

28. Lemaigre V, van den Bergh O, Smets A, et al. Effect of long acting bronchodilators and placebo on histamine-induced asthma symptoms and mild bronchusobstruction. Respir Med 2006; 100:348–353.

29. Barnes PJ. The role of anticholinergics in chronic obstructive pulmonary disease. Am J Med 2004; 117(suppl 12A):24S–32S.

30. Gross NJ. Anticholinergic agents in asthma and COPD. Eur J Pharmacol 2006; 533:36–39.

31. Weder MM, Donohue JF. Role of bronchodilators in chronic obstructive pulmonary disease. Semin Respir Crit Care Med 2005; 26:221–234.

32. Gross NJ, Petty TL, Friedman M, et al. Dose response to ipratropium as a nebulized solution in patients with chronic obstructive pulmonary disease: A three center study. Am Rev Respir Dis 1989; 139:1188–1191.

33. Maesen FP, Smeets JJ, Costongs MA, et al. Ba 679 Br, a new long-acting antimuscarinic bronchodilator: A pilot dose response study in COPD. Eur Respir J 1993; 6:1031–1036.

34. Maesen PF, Smeets JJ. Sledsens TJ, et al; on behalf of the Dutch Study Group. Tiotropium bromide, a new long-acting antimuscarinic bronchodilator: A pharmacodynamic study in patients with chronic obstructive pulmonary disease (COPD). Eur Respir J 1995; 8:1506–1513.

35. Standards for the diagnosis and treatment of patients with COPD: A summary of the ATS/ERS position paper. Eur Respir J 2004; 23:932–946.
36. Terzano C, Petroianni A, Ricci A, et al. Early protective effects of tiotropium bromide in patients with airway hyperresponsiveness. Eur Rev Med Pharmacol Sci 2004; 8:259–264.
37. Sposato B, Mariotta S, Ricci A, et al. The influence of ipratropium bromide in the recovery phase of methacholine-induced bronchospasm. Eur Rev Med Pharmacol Sci 2005; 9:117–123.
38. O'Connor BJ, Towse LJ, Barnes PJ. Prolonged effect of tiotropium bromide on methacholine-induced bronchoconstriction in asthma. Am J Respir Crit Care Med 1996; 154:876–880.
39. Hiroshi K. Anticholinergic agents in asthma: Chronic bronchodilator therapy, relief of acute severe asthma, reduction of chronic viral inflammation and prevention of airway remodeling. Curr Opin Pulmon Med 2006; 12;60–67.
40. Wong BJ, Hargreave FE. Bioequivalence of metered-dose inhaled medications. J Allergy Clin Immunol 1993; 92:373–379.
41. Clark DJ, Lipworth BJ. Dose–response of inhaled drugs in asthma: An update. Clin Pharmacokinet 1997; 32:58–74.
42. Barnes PJ, Pedersen S, Busse WW. Efficacy and safety of inhaled corticosteroids: New developments. Am J Respir Crit Care Med 1998; 157:S1–53.
43. Ahrens RC, Teresi ME, Han S-H, et al. Asthma stability after oral prednisone—A clinical model for comparing inhaled corticosteroid potency. Am J Respir Crit Care Med 2001; 164:1138–1145.
44. Ahrens RC, Hendeles L, Teresi ME, et al. Relative potency of beclomethasone dipropionate (BDP) delivered by HFA-MDI, and fluticasone propionate (FP) delivered by Diskus. Eur Respir J 2003; 22(suppl 45):P1576 (Abstract).
45. Barnes PJ. Inhaled glucocorticosteroid for asthma. New Eng J Med 1995; 332:868–875.
46. Kharitonov SA. Exhaled markers of inflammatory lung diseases: Ready for routine monitoring? Swiss Med Wkly 2004; 134:175–192.
47. Brightling CE, Green RH, Pavord ID. Biomarkers predicting response to corticosteroid therapy in asthma. Treat Respir Med 2005; 4:309–316.
48. Choi JC, Hoffman LA, Rodway GW, et al. Markers of lung disease in exhaled breath: Nitric oxide. Biol Res Nurs 2006; 7:241–255.
49. Ricciardolo FL, Sterk P. Gaston B, et al. Nitric oxide in health and disease of the respiratory system. Physiol Rev 2004; 84:731–765.
50. Barnes PJ. Nitric oxide and airway disease. Ann Med 1995; 27:389–393.
51. Pyne DN, Adcock IM, Wilson NM, et al. Relationship between exhaled nitric oxide and mucosal eosinophilic inflammation in children with difficult asthma after treatment with oral prednisone. Am J Respir Crit Care Med 2001; 164:1376–1381.
52. Sippel J, Holden WF, Tilles SA, et al. Exhaled nitric oxide levels correlate with measures of disease control in asthma. J Allergy Clin Immunol 2000; 106:645–650.
53. Lehtimaki L, Kankaanranta H, Saarlainen S, et al. Inhaled fluticasone decreases bronchial but not alveolar nitric oxide output in asthma. Eur Respir J 2001; 18:635–639.
54. Kharitonov SA, Barnes PJ. Clinical aspects of exhaled nitric oxide. Eur Respir J 2000; 16:781–792.
55. Kharitonov SA, Barnes PJ. Does exhaled nitric oxide reflect asthma control? Yes, it does! Am J Respir Crit Med 2001; 164:727–728.
56. Deykin A. Targeting biological markers in asthma—Is exhaled nitric oxide the bull's eye? New Eng J Med 2005; 352:2233–2235.
57. Yates DH, Kharitonov SA, Robbins RS, et al. Effect of a nitric oxide inhibitor and a glucocorticosteroid on exhaled nitric oxide. Am J Respir Care Med 1995; 152:892–896.
58. Beck-Ripp J, Griese M, Arenz S, et al. Changes in exhaled nitric oxide during steroid treatment of childhood asthma. Eur Respir J 2002; 19:1015–1019.
59. Bates C, Silkoff PE. Exhaled nitric oxide in asthma: From bench to bedside. J Allergy Clin Immunol 2003; 111:256–262.

60. Kharitonov SA, Yates DH, Barnes PJ. Inhaled glucocorticoids decrease nitric oxide in exhaled air of asthmatic patients. Am J Respir Crit Care Med 1996; 153:454–457.
61. Delago-Corcoran C, Kissoon N, Murphy S, et al. Exhaled nitric oxide reflects asthma severity and asthma control. Pediatr Crit Care Med 2004; 5:48–52.
62. Silkoff PE, Mclean PA, Slutsky AS, et al. Exhaled nitric oxide and bronchial reactivity during and after inhaled beclomethasone in mild asthma. J Asthma 1998; 35:473–479.
63. Jones SL, Herbison P, Cowan JO, et al. Exhaled NO and assessment of anti-inflammatory effects of inhaled steroid: Dose response relationship. Eur Respir J 2002; 20:601–608.
64. Kelly MM, Leigh R, Jayaram L, et al. Eosinophilic bronchitis in asthma: A model for establishing dose-response and relative potency of inhaled corticosteroids. J Allergy Clin Immunol 2006; 117:989–994.
65. Silkoff PE, McClean P, Spino M, et al. Dose–response relationship and reproducibility of the fall in exhaled nitric oxide after inhaled beclomethasone dipropionate therapy in asthma patients. Chest 2001; 119:1322–1328.
66. Parameswaran K, Leigh R, O'Byrne PM, et al. Clinical models to compare the safety and efficacy of inhaled corticosteroids in patients with asthma. Can Respir J 2003; 10:27–34.
67. Schuirmann DJ. A comparison of the two one-sided tests procedure and the power approach for assessing the equivalence of average bioavailability. J Pharmacokinet Biopharm 1987; 15:657–680.
68. Gillespie WR. Proceedings of the joint meeting of the Advisory Committee for Pharmaceutical Science and Pulmonary-Allergy Drugs Advisory Committee held at the Holiday Inn in Gaithersburg, August 1996, Maryland.
69. Ariens EJ, Simonis AM. A molecular basis for drug action. J Pharm Pharmacol 1964; 16:137–157.
70. Holford N, Scheiner LB. Understanding the dose–effect relationship: Clinical application of pharmacokinetic–pharmacodynamic models. Clin Pharmacokinet 1981; 6:429–453.
71. Efron B, Tibshirani RJ. An Introduction to the Bootstrap. Boca Raton, FL: Chapman & Hall/CRC, 1993.
72. Singh GJP. Proceedings of the joint meeting of the Advisory Committee for Pharmaceutical Science and Pulmonary-Allergy Drugs Advisory Committee held at the Holiday Inn in Gaithersburg, August 1996, Maryland.
73. Jenkins J. Proceedings of the joint meeting of the Advisory Committee for Pharmaceutical Science and Pulmonary-Allergy Drugs Advisory Committee held at the Holiday Inn in Gaithersburg, August 1996, Maryland.
74. Wilson AM, McFarlane LC, Lipworth BJ. Effects of repeated once daily dosing of three intranasal corticosteroids on basal and dynamic measures of hypothalamic–pituitary–adrenal-axis activity. J Allergy Clin Immunol 1998; 101:470–474.
75. Fardon TC, Lee DKC, Haggart K, et al. Adrenal suppression with dry powder formulations of fluticasone propionate and mometasone furoate. Am J Respir Crit Care Med 2004; 170:760–766.
76. Allen DB. Safety of inhaled corticosteroids in children. Pediatr Pulmonol 2002; 33: 208–220.
77. Wohl ME, Majzoub JA. Asthma, steroids, and growth. New Eng J Med 2000; 343: 1113–1114.
78. Todd GRG. Adrenal crisis due to inhaled steroids is underestimated. Arch Dis Child 2003; 88:554–555.
79. Robinson JD, Angelini BL, Krahnke JS, et al. Inhaled steroids and the risk of adrenal suppression in children. Expert Opin Drug Saf 2002; 1:237–244.
80. Hochhaus G, Schmidt E-W, Rominger KL, et al. Pharmacokinetic/dynamic correlation of pulmonary and cardiac effects of fenoterol in asthmatic patients after different routes of administration. Pharm Res 1992; 9:291–297.
81. Kemsford R, Handel M, Mehta R, et al. Comparison of the systemic pharmacodynamic effects and pharmacokinetics of salmeterol delivered by CFC propellant and

non-CFC propellant metered dose inhalers in healthy subjects. Respir Med 2005; 99(suppl A):S11–S19.

82. Rosekranz B, Rouzier R, Kruse M, et al. Saftety and tolerability of high-dose of formoterol (via Aerosolizer®) and salbutamol in patients with chronic obstructive pulmonary disease. Respir Med 2006; 100:666–672.

83. Wolthers OD, Pedersen S. Measures of systemic activity of inhaled corticosteroids in children: A comparison of urine cortisol excretion and knemometry. Respir Med 1996; 89:347–349.

84. Lipworth BJ, Seckl JR. Measures for detecting systemic bioactivity with inhaled and intranasal corticosteroids. Thorax 1997; 52:476–482.

85. Guidance for Industry: Bioavailability and Bioequivalence Studies for Orally Administered Drug Products—General Considerations. Rockville, MD: Center for Drug Evaluation and Research, 2003.

86. Guidance for Industry: Bioanalytical method validation. Rockville, MD: Center for Drug Evaluation and Research, 2001.

87. Guidance for Industry: Statistical Approaches to Establishing Bioequivalence. Rockville, MD: Center for Drug Evaluation and Research, 2001.

88. Patnaik RN. Bioequivalence assessment: Approaches, designs and statistical considerations. In: Sahajwala CG, ed. New Drug Development: Regulatory Paradigms for Clinical Pharmacology and Biopharmaceutics. New York: Mercel Dekker, Inc., 2004:561–586.

89. Sathe PM, Raw AS, Ouderkirk LA, et al. Drug product performance, in vitro. In: Shargel L, Kanfer I, eds. Generic Drug Product Development: Solid Dosage Forms. New York: Mercel Dekker, 2005:187–209.

90. Whelan AM, Hahn NW. Optimizing drug delivery from metered-dose inhalers. DICP Ann Pharmacother 1991; 25:638–645.

91. Kublik H, Vidgren MT. Nasal delivery systems and their effect on deposition and absorption. Adv Drug Del Rev 1998; 29:157–177.

92. Labiris NR, Dolovich MB. Pulmonary drug delivery. Part II: The role of inhalant delivery devices and formulations in therapeutic effectiveness of aerosolized medications. Br J Clin Pharmacol 2003; 56:600–612.

93. Guidance for in-vitro Bioequivalence Studies of Metaproterenol Sulfate and Albuterol Inhalation Aerosols (Metered dose inhalers). Rockville, MD: Division of Bioequivalence, Food and Drug Administration, 1989.

94. Dayal P, Shaik MS, Singh M. Evaluation of different parameters that affect droplet size distribution from nasal sprays using Malvern Spraytec®. J Pharm Sci 2004; 93: 1725–1742.

95. Draft Guidance for Industry: Metered dose inhaler (MDI) and dry powder inhaler (DPI) drug products: Chemistry, Manufacturing and Control Documentation. Rockville, MD: Center for Drug Evaluation and Research, 1998.

96. Newman SP, Pitcairn GR, Dalby RN. Drug delivery to the nasal cavity: In vitro and in vivo assessment. Crit Rev Ther Drug Carr Syst 2004; 21:21–66.

97. Suman JD, Laube BL, Dalby RN. Documenting nasal bioequivalence from in vitro characteristics to physiologic response. In: Dalby R, Byron P, Paert J, eds. Respiratory Drug Delivery VIII. Raleigh, NC: Davis Horwood International, 2002:691–693.

98. Mitchell JP, Nagel MW. Cascade impactor for size characterization of aerosols from medical inhalers: Their uses and limitations. J Aerosol Med 2003; 16:341–377.

99. Naini V, Chaudhry S, Berry J, et al. Entry port selection for detecting particle size differences in metered dose inhaler formulations using cascade impaction. Drug Dev Ind Pharm 2004; 30:75–82.

100. Cheng YS, Holmes BS, Gao BS, et al. Characterization of nasal spray pumps and deposition patterns in replica of the human nasal airway. J Aerosol Med 2001; 14: 267–280.

101. Eck CR, McGarth TF, Perlwitz AG. Droplet size distribution in solution nasal sprays. In: Dalby RN, Byron PR, Farr SJ, Paert J, eds. Respiratory Drug Delivery VII. Raleigh, NC: Serentec Press, 2000:475–478.

102. Kippax P, Krarup G, Suman JD. Application of droplet sizing: Manual versus automated actuation of nasal sprays. Pharmaceut Technol (Outsourcing Resources) 2004:30–39.

103. Ranucci J. Dynamic plume-particle size analysis using laser diffraction. Pharm Tech 1992; 16:108–114.

104. Hollingworth GW. Particle size analysis of therapeutic aerosols. In: Moren F, Dolovich MB, Newhouse MT, Newman SP, eds. Aerosols in Medicine. Principles, Diagnosis and Therapy. Amsterdam, The Netherlands: Elsevier Science Publishers B.V., 1993:351–374.

105. Kippax P. Measuring Particle Size by Laser Diffraction Techniques. Paint and Coating Indus, 2005.

106. Bell J, Hartley P, Cox J. Dry powder aerosols. I. A new powder inhalation device. J Pharm Sci 1971; 60:1559–1564.

107. Dunbar CA, Hickey AJ, Holzner P. Dispersion and characterization of pharmaceutical dry powder aerosols. KONA 1998; 16:7–14.

108. Whelan AM, Hahn NW. Optimizing drug delivery from metered-dose inhalers. DICP Ann Pharmacother 1991; 25:638–645.

109. Virchow JC. Guidelines versus clinical practice—which therapy and which device? Respir Med 2004; 98(suppl 2):S28–S34.

110. Crompton G. Problems patients have using pressurized aerosol inhalers. Eur J Respir Dis 1982; 119(suppl):101–104.

111. Larsen J, Hahn M, Elkholm B, et al. Evaluation of conventional press-and-breathe metered-dose inhaler techniques in 501 patients. J Asthma 1994; 31:193–199.

112. Lindgren S, Blake B, Larson S. Clinical consequences of inadequate inhalation technique in asthma therapy. Eur J Respir Dis 1987; 70:93–98.

113. Chinet T, Huchon G. Misuse of pressurized metered-dose aerosols in treatment of bronchial diseases: Incidences and clinical consequences. Ann Med Interne (Paris) 1994; 145:119–124.

114. Giraud V, Roche N. Misuse of corticosteroid metered-dose inhaler is associated with decreased asthma stability. Eur Respir J 2002; 29:246–251.

115. Noakes TJ. CFCs, their replacements, and the ozone layer. J Aerosol Med 1995; 8(suppl 1):S3–S7.

116. Boulet LP. The ozone layer and metered dose inhalers. Can Respir J 1998; 5:176–179.

117. McDonald KJ, Martin GP. Transition to CFC-free metered dose inhalers—into the new millennium. Int J Pharm 2000; 201:89–107.

118. Tsai WT. An overview of environmental hazards and exposure risk of hydrofluorocarbons (HFCs). Chemosphere 2005; 61:1539–1547.

119. Smith IJ, Parry-Billing M. The inhaler of the future? A review of dry powder inhalers on the market today. Pulmon Pharmacol Therap 2003; 16:79–95.

120. Ganderton D, Kassem NM. Dry powder inhalers. Adv Pharmaceut Sci 1992; 6:165–191.

121. Timsina MP, Martin GP, Marriot C, et al. Drug delivery to the respiratory tract using dry powder inhalers. Int J Pharmaceut 1994; 101:1–13.

122. Perosson G, Olsson B, Solliman S. The impact of inspiratory effort on inspiratory flow through Turbuhaler® in asthmatic patients. Eur Respir J 1997; 10:681–684.

123. Clark AR, Hollingworth AM. The relationship between powder inhaler resistance and peak inspiratory conditions in healthy volunteers: Implications for in vitro testing. J Aerosol Med 1993; 6:99–110.

124. Olsson B, Asking L. Critical aspects of the function of inspiratory flow driven inhalers. J Aerosol Med 1994; 7(suppl 1):S43–S47.

125. Srichana T, Martin GP, Marriott C. Dry powder inhalers: The influence of device resistance and powder formulation on drug and lactose deposition in vitro. Eur J Pharm Sci. 1998; 7(suppl 1):73–80.

126. Cegla UH. Pressure and inspiratory flow characteristics of dry powder inhalers. Respir Med 2004; 98(suppl A):S22–S28.

127. Tarsin W, Assi KH, Chrystyn H. In vitro intra- and inter-inhaler flow rate-dependent dosage emission from a combination of budesonide and eformoterol in a dry powder inhaler. J Aerosol Med 2004; 17:5–32.

128. Bronsky EA, Grossman J, Henis MJ, et al. Inspiratory flow rates and volumes with the Aerolizer dry powder inhaler in asthmatic children and adults. Curr Med Res Opin 2004; 20:131–137.

129. Nsour WM, Alldred A, Corrado OJ, et al. Measurement of peak inhalation rates with an In-check Meter® to identify an elderly patient's ability to use a Turbuhaler®. Respir Med 2001; 95:965–968.

130. Gauld LM, Briggs K, Robinson P. Peak respiratory flows in children with cystic fibrosis. J Paediatr Child Health 2003; 39:210–213.

131. Kamps AW, Brand PL, Roorda RJ. Variation in peak inspiratory flow through dry powder inhalers in children with stable and unstable asthma. Pediatr Pulmonol 2004; 37:65–70.

132. Bentur L, Mansour Y, Hamzani Y, et al. Measurement of inspiratory flow in children with acute asthma. Pediatr Pulmonol 2004; 38:304–307.

133. Broeders AEAC, Molema J, Vermue NA, et al. Peak respiratory flow rate and slope of the inhalation profiles in dry powder inhalers. Eur Respir J 2001; 18:780–783.

134. Richards R, Simpson SF, Renwick AG, et al. Inhalation rate of sodium cromoglycate determines plasma pharmacokinetics and protection against AMP-induced bronchoconstriction in asthma. Eur Respir J 1988; 1:896–901.

135. Engel T, Scharling B, Skovsted B, et al. Effects, side effects and plasma concentrations of terbutaline in adult asthmatics after inhaling from a dry powder inhaler device at different inhalation flows and volumes. Br J Clin Pharmacol 1992; 33:439–444.

136. Chege JK, Chrystyn H. The relative bioavailability of salbutamol to the lung using urinary excretion following inhalation from a novel dry powder inhaler: The effect of inhalation rate and formulation. Respir Med 2000; 94:51–56.

137. Pedersen S, Hansen OR, Funglsang G. Influence of inspiratory flow rate upon the effect of Turbohaler. Archiv Dis Child 1990; 65:308–319.

138. Munzel U, Marschall K, Fyrnys B, et al. Variability of fine particle dose and lung deposition of budesonide delivered through two multidose dry powder inhalers. Curr Med Res Opin 2005; 21:827–834.

139. Lee SL, Adams WP, Li BV, Conner DP, et al. In vitro considerations to support bioequivalence of locally acting drugs in dry powder inhalers for lung diseases. AAPS J 2009; 11(3):414–423.

Transdermal Dosage Forms

Mario A. González

P'Kinetics International, Inc., Pembroke Pines, Florida, U.S.A.

Gary W. Cleary

Corium International, Inc., Menlo Park, California, U.S.A.

INTRODUCTION

Transdermal drug delivery systems were developed and first marketed commercially in the early 1980s (1). Since then, a number of drugs have been formulated into transdermal delivery systems and a listing of these may be found in Table 1 along with their most common brand names and physiochemical properties. With the exception of ISDN Tape, all these products are marketed in the United States and therefore are candidates for generic development. Table 2 lists the generic transdermal products currently approved via the ANDA (abbreviated new drug application) route as well as their Reference Listed Drugs (RLD).

The popularity of transdermal delivery systems is due in part to the advantages these systems have over more traditional routes of drug delivery. Besides offering a dosage route capable of avoiding presystemic or first-pass metabolism, transdermal delivery systems offer the advantage of being well accepted by patients because of their convenience and simplicity. Patient acceptance leads to improved patient compliance and better monitoring of drug utilization.

The function of a transdermal delivery system is to provide a systemic blood level that is therapeutically efficacious. Additionally, it must not only be nonirritating and nonsensitizing, adhere to the skin for the delivery time, but also be capable of being manufactured. Several transdermal products have achieved these criteria and have reached the marketplace. They deliver drugs through the skin anywhere from one to seven days. Often, they are referred to as "membrane controlled," "reservoir," or "monolithic" types. These terms describe the drug release mechanism used and can be misleading when it comes to design and manufacturing considerations. One of the objectives of this chapter is to propose a more consistent terminology when referring to different formulations of transdermal products. Every transdermal delivery system, however, has to allow the drug to permeate through the skin at the specific rate necessary to provide the desired therapeutic results. The drug delivery must be such that it can overcome a series of barriers within the skin to reach the systemic circulation as illustrated in Figure 1. Additionally, the device must adhere well to the skin yet be easily removed, be cost-effective, and esthetically acceptable to the physician and patient.

TABLE 1 Physiochemical Properties of Drugs in Marketed Transdermal Delivery Systems

Drug	Brand	M.W. (Daltons)	pK	M.P. (°C)	log P (o/w)	Permeation coefficient (cm/hr × 10³)
Clonidine	Catapres-TTS	230	8.2	140	0.83	35
Estradiol	Vivelle-Dot	272	–	176	2.49	5.2
Fentanyl	Duragesic	337	8.4	83	2.93	10
Fentanyl (iontophoretic)	Ionsys	337	8.4	83	2.93	10
Isosorbide dinitrate	ISDN-Tape	236	–	70	1.55	–
Methylphenidate	Daytrana	234	–	–	–	–
Nicotine	NicoDerm	162	6.2/10.9	<-80	–	3
Nitroglycerin	Nitro-Dur	227	–	13.5	2.05	20
Norethindrone acetate	Combipatch (+estradiol)	340	–	161	–	–
Oxybutynin	Oxytrol	357	–	–	–	–
Scopolamine	TransdermScop	303	7.8	59	1.24	0.5
Testosterone	Androderm	288	–	153	3.31	400

SYSTEM DESIGNS FOR TRANSDERMAL DELIVERY

When beginning a development program for a transdermal system, it is usually necessary for scientists from several different disciplines to focus their energies to develop an optimal transdermal delivery system. This includes scientists in the areas of biological sciences such as skin physiology, pharmacokinetics, and biophysics as well as physicochemical sciences such as pharmaceutics, adhesive, material sciences, analytical chemistry, quality control, and engineering. These team needs to consider what the final transdermal product is to achieve. They need to arrive at the performance criteria for the delivery system to achieve the therapeutics goals of the candidate product. Whether it is the first time a drug is to be incorporated into a transdermal delivery system or a generically equivalent system is being developed, a product profile needs to be designed, which describes the following:

- physical characteristics of the system (e.g., size, shape, thickness);
- rate of release through the skin;
- rate of release from the system;
- adhesion to the skin including duration of adhesion;
- patient demography;

TABLE 2 Marketed Generic Transdermal Delivery Systems and Their Probable Reference Listed Drug (RLD)

Drug	Strength	RLD
Nitroglycerin	0.1, 0.2, 0.3, 0.4, 0.6, and 0.8 mg/hr	Nitro-Dur®
Nitroglycerin	0.1 mg/hr	Transderm-Nitro®
Estradiol (once weekly)	0.0375, 0.05, 0.06, and 0.1 mg/day	Climara®
Estradiol (twice weekly)	0.05, 0.075, and 0.1 mg/day	Vivelle®
Clonidine Hydrochloride	0.1, 0.2, and 0.3 mg/day	Catapres-TTS®
Nicotine	7, 14, and 21 mg/day	Habitrol®

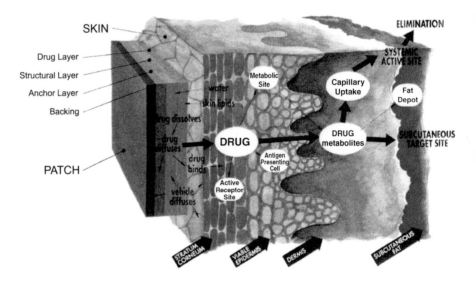

FIGURE 1 Schematic of drug permeation through the main skin barrier, the Stratum Corneum, to the vascularized viable epidermis and dermis. The disposition of the drug after absorption from the transdermal system is also illustrated.

- medical rationale and therapeutic goals, for example, plasma concentrations required;
- manufacturing efficiency;
- availability of raw materials;
- patent requirements; and
- regulatory requirements;

In the case of a generic transdermal product, the delivery target is easier to define, since the focus is to establish bioequivalence to a product already in the marketplace.

Irrespective of the development of a new product or a transdermal generic, information gained from studying the drug's pharmacokinetic properties is invaluable. Using parameters such as volume of distribution (Vd), area under the plasma concentration–time curve (AUC), total body clearance (Cl_T), desired plasma concentration (Cp), the steady-state plasma concentration (Css), and elimination half-life ($t1/2$), the designer of a transdermal system can get an estimate of the desired amount of drug that must permeate through the skin into the general blood circulation. Table 3 lists some of the current marketed transdermal products along with the pharmacokinetic parameters, which can be used to calculate the desired drug input rate or skin flux (J_{skin}). J_{skin} can be calculated as follows: (2)

$$k_o = J_{skin} = C_{ss} \times Cl_T, \tag{1}$$

where k_o is the input rate to achieve a steady-state plasma or blood level (C_{ss}). The calculation of J_{skin} may not appear to be necessary when formulating a generic transdermal, because the objective is to formulate a product that produces comparable plasma concentrations to the innovator's product, the

TABLE 3 Pharmacokinetic Parameters of Drugs in Marketed Transdermal Delivery Systems

Drug	Cl (L/hr)	V_d (L)	$t_{1/2}$ (hr)	Max. 24 hr dose (mg)	Oral Bioavailability (%)	Effective plasma level (ng/mL)
Clonidine	13	147	6–20	0.3	95	0.2–2
Estradiol	615–790	4.8	0.05	0.1	–	0.04–0.06
Fentanyl	27–75	280	3–12	2.4	–	1.0
Isosorbide Dinitrate	204	100–300	0.5–1.0	20	20	2–4
Methylphenidate	–	–	4–6	60	10–52	5–20
Nicotine	78	182	2	22	30	10–30
Nitroglycerin	966	231	0.04	15	<1	1.2–11
Norethindrone	–	–	–	0.25	–	0.5–1.0
Oxybutynin	PI	193	2	3.9	6	3–4
Scopolamine	67.2	98	2.9	0.17	27	0.04
Testosterone	–	–	–	6.0	<1	10–100

reference listed drug (RLD). However, calculating a desired J_{skin} and achieving this value in vitro will increase the chance of success in a bioequivalence study.

DIFFERENT MECHANISMS FOR CONTROLLING DRUG RELEASE

Up until now the transdermal systems that have reached the U.S. market-place have predominately delivered the drugs from the transdermal system and through the skin using passive diffusion of the drug. That is, with passive diffusion, the drug essentially follows Fick's laws of diffusion without any extra external forces such as electronics, ultrasound, heat, skin ablation, or mechanical poration to enhance drug permeation (3).

Different classifications have been used to describe commercially available transdermal systems. Some of the terminology used to describe these products has been unnecessarily confusing, and a classification system that focuses on the formulation of the products is preferable to one that relies on marketing claims. There are different types of transdermal systems marketed today, and it is useful to attempt to classify them in a consistent manner. They can be classified by the mechanism of drug release from the product or by the type of design of the product. One way is to consider the type of design of the transdermal without regard to the drug release mechanism itself. This allows a perception of the final product and its components in addition to how it might be formulated and fabricated. In the early stages of development, a transdermal system may be fabricated by hand on a laboratory bench or on pilot equipment. Final systems for a "Biobatch," however, have to be made on equipment that is capable of producing commercial quantities for the marketplace. Materials that have been selected to be part of a new transdermal system will have a major impact on how the product will be manufactured, what processes will be necessary, and how efficient and costly the manufacturing equipment will be. While transdermal delivery systems may employ different designs, they are all similar in that the delivery system contains a large drug concentration or a drug reservoir, which must be maintained essentially unchanged in order to produce ideally a drug delivery with zero-order

Backing layer with adhesive
Drug in solid gel-like reservoir
Release liner

FIGURE 2 Peripheral adhesive-laminate structure with the drug reservoir in a gel matrix.

kinetics. The different transdermal formulations or reservoir systems currently in use are discussed below.

Peripheral Adhesive Laminate Structure

One of the simplest transdermal delivery approaches is to create a drug reservoir by mixing the active ingredient with a polymer blend that solidifies into a solid gel. The drug is dispersed in the gel and diffuses from the gelatinous matrix into the stratum corneum. This type of system is illustrated in Figure 2. Examples of this type of formulation are among the earliest marketed in the United States, Nitro-Dur I from Key Pharmaceuticals and Nitro-Disc from Searle. The drug-loaded gel reservoir is typically attached to an adhesive foam pad or a paper tape to which an occlusive barrier, the baseplate, is attached as the outermost layer. The adhesive has to extend around the gel reservoir and thus this product tends to have a total surface area, which is much larger than the active surface.

The baseplate may be simply aluminum foil, which serves to produce occlusion as well as to contain the active ingredient in the gel reservoir. The gel matrix may also be attached to the baseplate instead of the adhesive tape, but adhesive is always found on the periphery of the foam pad in order to attach the transdermal system to the patient. This type of delivery system is easy to manufacture, and the initial costs for the manufacturing equipment are lower than for other systems. On the other hand, the product is usually less pharmaceutically elegant and results in larger transdermal units, which may preclude patient acceptance.

Liquid Form, Fill, and Seal Laminate

The fluid-filled reservoir system has proven to be effective for transdermal delivery of different drugs (Fig. 3). Transderm-Nitro® and Estraderm® originally marketed by Ciba-Geigy but developed by Alza are examples of this type of delivery system. Duragesic® a fentanyl transdermal system marketed by Ortho-McNeil, Inc., is a current example of this type of reservoir system. The active ingredient is usually contained in a cream or lotion-like vehicle and therefore must be contained by a heat-sealed plastic envelope. Contact with the skin is maintained by an outer adhesive layer, which is by necessity larger than the surface area of the reservoir because the heat sealing results in a border. Damage to the system may cause a rupture of the reservoir and is a disadvantage

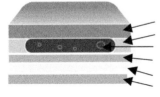

Backing layer
Heat seal layer
Drug in liquid reservoir
Membrane
Skin adhesive (add another layer, yellow)
Release liner (add another layer, light green)

FIGURE 3 Liquid form, fill, and seal laminate (liquid reservoir).

to this product design since drug may be put into contact directly with the skin with no absorption control. In the case of Duragesic®, transdermal fentanyl, this type of failure has resulted in accidental deaths and recalls of some lots of this product. Another disadvantage of this system is that the border formed by the heat-sealing process contains adhesive and drug will diffuse into the border. Yet, the border cannot be replenished rapidly from the reservoir during patient wear; therefore, the delivery from such a system will be greater during the first hours of wear. An advantage of this system is that large quantities of drug can be held in the fluid reservoir, and chemical enhancers such as ethanol in Estraderm®, the first transdermal Estradiol product, can be added to the reservoir. These chemicals can be used to alter the barrier properties of the stratum corneum, which increases the rate of drug delivery.

Solid-State Laminate

A more successful modification of the earlier transdermal formulations involves the development of adhesives in which the drug is dispersed. The drug and adhesive essentially form a monolith matrix from which the drug can be absorbed, yet the volatility of the drug is reduced through interaction with the adhesive thus resulting in a formulation that is easier to stabilize. This type of formulation is composed of an impermeable backing that is coated with a drug-loaded adhesive. The Impermeable backing serves as an occlusive barrier and also prevents evaporation of the drug from the system, while the transdermal system is on a patient. The adhesive polymers into which the drug is dispersed serve both as a reservoir and as an adhesive to hold the unit to the skin.

Release of the drug is controlled by slow diffusion from the adhesive-drug matrix to the stratum corneum, and drug release frequently occurs by apparent zero-order kinetics. A major advantage of this system over other designs is that it results in a more compact unit with greater flexibility that allows for improved patient convenience and acceptance. Figure 4 is schematic diagram of a solid-state laminate structure. This type of system has been used successfully for transdermal delivery of nitroglycerin by Key Pharmaceuticals/Schering-Plough (Nitro-Dur®) as well as the generic equivalent manufactured by Mylan Technologies and Minitran® from 3M Pharmaceuticals. While nitroglycerin is extremely volatile and can easily penetrate polymer barriers, the interaction between adhesive polymers and nitroglycerin lowers the volatility of the drug and stabilizes the nitroglycerin. Some other products utilizing this technology are Vivelle® and Vivelle Dot®, Estradiol transdermal systems from Noven Pharmaceuticals, as well as Transderm Scop® (scopolamine) from Novartis and Catapres-TTS® (Clonidine) from Boehringer Ingelheim.

Ointments, Creams, Lotions, or Gels Applied Directly to the Skin

While transdermal products are usually identified as drug delivery systems that are self-adhering, the following products are capable of delivering drugs through

— Backing layer
— Drug in skin adhesive
— Release liner

FIGURE 4   Schematic of a solid-state laminate (adhesive matrix).

the skin and achieving therapeutically effective concentrations: semisolid amorphous ointments, creams, lotions, and gels. The earliest transdermal delivery of nitroglycerin was accomplished by the ointment Nitro-Bid® marketed by Kremers-Urban. While this product had to be applied at least three times daily or every eight hours, it was still effective in controlling **angina**. Estradiol gels, progesterone gels, and testosterone gels have also been marketed for a number of years in Europe and recently have been approved for marketing in the United States. NSAID (nonsteroidal anti-inflammatory drugs) gels have been unable to produce systemically effective levels yet they have been successful in the treatment of localized pain. Given the adverse reactions from oral NSAIDs, it would not be surprising to see more of these products developed and approved.

The four types of transdermal formulations discussed above are not all inclusive since combinations of the different designs are always possible. New approaches to drug delivery such as the use of iontophoresis to enhance or accelerate drug permeation through the epidermis have been developed as illustrated by the fentanyl product Ionsys™ listed in Table 1 (4). Regardless of the formulation approach, all successful transdermal systems have the potential to deliver a drug to the skin surface so that the drug can migrate through the skin. By understanding these basic designs and their relative advantages and disadvantages, the system designer can incorporate the most suitable drug release mechanism for the application required. Using the appropriate plasticizers or vehicles, polymers, films, or membranes to match the diffusivity of the drug through the skin, the desired delivery rate and the optimum blood level can be effectively achieved. When designing a transdermal product to be approved as bioequivalent to an RLD, it behooves the system designer to utilize a mechanism for controlling the drug release, which is similar to that of the RLD. Although, the FDA does not require the use of the same mechanism of drug release, the approval of an ANDA will be facilitated if the total drug content in the generic product does not differ significantly from the RLD. This will typically be possible only when similar mechanisms of drug delivery are utilized. Irrespective of the mechanism used to control the drug delivery from a transdermal system, it will be critical that the Chemistry, Manufacturing, and Control (CMC) section of the ANDA clearly describe the delivery mechanism used in the formulation. In addition to the typical content of a CMC section, the following information should also be included:

- Formulation development history
- Description of the mechanism of drug release
- API source and drug master file information
- Characterization of the rate-controlling polymer including residual monomers
- Assay of residual solvents
- Quantification of drug migration into the packaging and release liner
- Stability testing of the final formulation

IN VITRO PERMEATION (FLUX) STUDIES

Diffusion studies can be used to determine if a drug candidate can transit the skin from a simple solution or a transdermal delivery system. There are different designs for diffusion or flux cells but they all have in common a donor and

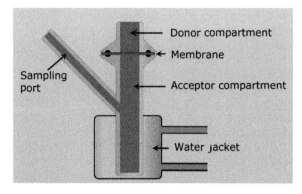

FIGURE 5 Design of an in vitro flux cell.

receptor compartment separated by a barrier membrane. A generic illustration of a flux cell may be seen in Figure 5. Studies involving drug diffusion through a barrier membrane such as animal or human epidermis are called flux studies and results in release rate data (J) with units of amount of drug diffused per unit surface area over time, for example, $\mu g/cm^2/hr$. The data collected from these diffusion studies can be presented graphically as J versus time or as cumulative amount released, Q_t ($\mu g/cm^2$), at sampling time t.

While flux data is not required by the FDA's Office of Generic Drugs for a generic transdermal, it is advisable to collect this type of data to aid in the design of a generic transdermal product. If the flux from the new generic is not statistically different from that of the RLD, the chances of demonstrating bioequivalence are much greater. The most predictive in vitro data comes from flux studies conducted with human cadaver skin in diffusion cells.

The best known of the different diffusion or flux cells is the Franz cell, which was developed by Dr. Tom Franz in the 1970s (5). When using this cell as illustrated in Figure 5, cadaver skin is the preferred barrier to drug diffusion from a transdermal product; however, polymer films or animal skin have also been used. On either side of the membrane to be tested is a cell compartment that serves either as a donor or a receptor. The barrier membrane with the attached transdermal product is actually the donor, and it will release its contents into the cell compartment serving as the receptor side.

An in vitro flux study can also be done with just a drug solution or suspension prior to the development of a transdermal delivery system. For this type of drug study, there is a liquid donor side and a liquid receptor side. The donor cell usually contains the drug either in suspension or in solution. The receptor cell contains a liquid that serves as a "sink" for the drug to diffuse into the receptor fluid. The receptor fluid can be water, normal saline, buffer, or some aqueous solution mixed with one or more solvents to solubilize drugs with poor solubility. The barrier membrane (or human cadaver skin) is placed between the donor cell and the receptor cell compartments. As the drug migrates across the test membrane or skin, samples are removed from the receptor cell and tested for content at various sampling times. When working with a transdermal product, the transdermal system is mounted at the opening of the donor cell with no liquid in the donor side. This is then attached to the cadaver skin, which will be on the opening of the receptor cell containing the receptor fluid.

Some cells are designed to have a continuous replenishing of the receptor solution. This allows for automated drug level monitoring in the receptor cell and facilitates the collection of multiple samples. In order to control temperature, cells may have water jackets surrounding the donor and receptor cell areas or the cells may be placed directly in a water bath. Diffusion studies are typically performed at 37°C, although some research groups prefer 32°C to 35°C as a closer simulation of skin temperature. An awareness of how a drug is applied to the skin and how this relates to the diffusion cell selected is important. The drug applied to the surface of the cadaver skin, the donor, may contain an "infinite" or a "finite" amount of drug. An "infinite" amount refers to the fact that there is more than enough drug present above its solubility on the donor side and far more available than will penetrate the skin during the lifetime of the flux study. For the most part there is an excess of drug present in a transdermal system to maintain a delivery with zero order kinetics during the transit through the skin for long periods of time. In this case, there is a constant amount of drug or an "infinite dose" available on the donor side to cross the membrane. Transdermal films or solutions can be placed on either vertical or horizontal cells with the drug diffusing in the direction of the receptor fluid only. Figure 5 illustrates a typical infinite dose diffusion chamber or cell. The graphical representation of the cumulative amount absorbed versus time may be seen in Figure 6. Ideally, sample in a flux study should be collected for a time period equal to the time that a patient will wear the transdermal product. This allows the system designer to evaluate the potential for success of the formulation. Obviously from the results seen in Figure 6, which were obtained from two transdermal formulations of the same drug, it would not be advisable to conduct a bioequivalence study with these two formulations. These profiles are typical of those seen for drugs studied using infinite dose techniques.

In the "finite dose" technique, a constant flux of the drug is not necessarily reached. Only a small amount of drug, usually 5 to 10 μL of drug suspension or solution (2–5 μg total drug content) is placed on a square centimeter skin sample that is mounted in the chamber. This type of test method is a better reflection of what occurs when a drug is delivered to skin with an ointment or lotion that is rubbed into the skin as a very thin film. The data from this type of experiment would differ from the results in Figure 6 in that the release profile decreases with time so that the cumulative amount diffused will reach a plateau.

An important consideration when conducting flux studies is to use cadaver skin from several donors. This is necessary because differences in the permeability of different skins can lead to misleading results. If a donor's skin happens to be one with a high permeability to the drug of interest, it can mislead the formulator into thinking that his product is delivering too much drug. The opposite is also true, that is, a skin with low permeability might make the formulator think that he needs to optimize the delivery of his transdermal system. Figure 7 illustrates the wide variability in flux from the same lot of transdermal nitroglycerin product. The different flux values were obtained using cadaver skin samples from different donors (6). A comparable variability in vivo has been reported by other authors (7). When comparing a generic product to an RLD, it is important to use a variety of skin samples from different donors to compare both products.

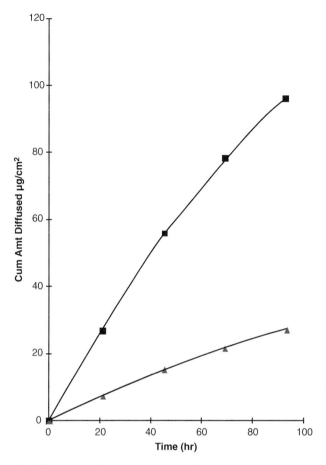

FIGURE 6 Plots of flux data from two different transdermal lots of the same drug.

Ideally, both products should be compared using skin samples from the same donors; this would be analogous to using a crossover design when conducting a bioavailability study. While flux studies are more difficult to conduct than dissolution studies, they are invaluable in optimizing the design of a transdermal system.

IN VITRO DISSOLUTION TESTING

As previously discussed, dissolution tests, in which in vitro drug release from a transdermal system is measured without the use of any barrier membrane, generate data which have a poor correlation with in vivo performance. These tests, however, are required by regulatory agencies including the FDA. As an example, if the in vitro dissolution data of nitroglycerin in water is compared for different products as illustrated in Figure 8, the release profiles are dramatically different as previously reported by FDA scientists (8). These results suggest that the differences seen in the in vitro dissolution profiles are obtained because different nitroglycerin systems have different mechanisms of release. As can be seen in

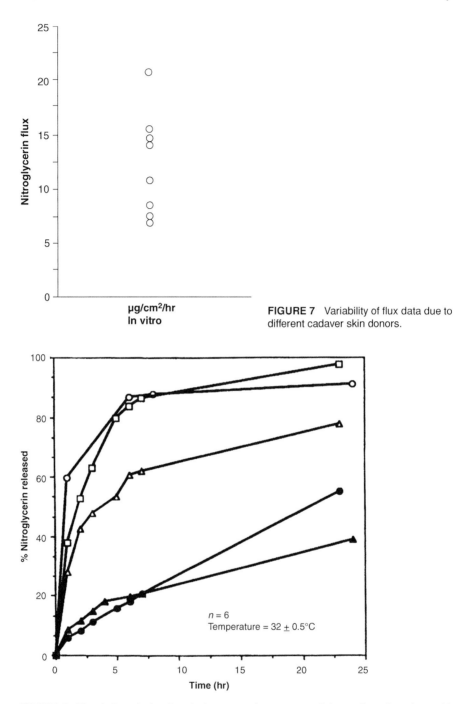

FIGURE 7 Variability of flux data due to different cadaver skin donors.

FIGURE 8 Dissolution of nitroglycerin into water from commercial transdermal products. (○) Ciba, (□) Bolar, (△) Searle, (●) Key Pharm, (▲) Wyeth.

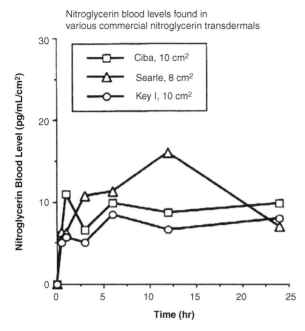

Nitroglycerin blood levels found in
various commercial nitroglycerin transdermals

FIGURE 9 GTN plasma levels from transdermal products manufactured by Ciba, Key, and Searle.

Figure 9, however, plasma levels of the Ciba (Transderm-Nitro®), Key (Nitro-Dur®), and Searle (Nitro-Disc®) are nearly the same.

While dissolution data may not correlate well with bioavailability results, it does serve a purpose as a quality control tool. As with other extended-release dosage forms, a transdermal product needs to be characterized by the ability of the delivery system to release the drug. In vitro dissolution allows this characterization, thus permitting changes in the drug release profile to be used for stability monitoring or for quality control from batch to batch. Obviously, for a dissolution method to be valid as a quality control tool, it must be able to detect formulation and manufacturing process changes and it should lend itself to automation.

Documentation similar to that of other drug delivery systems (e.g., batch records, formulation, weights and measurements, packaging) is also required for transdermal batch testing. Thus, testing of a transdermal delivery system for in vitro release is not unique or unexpected. As with other extended-release dosage forms, a transdermal system needs to be characterized by its ability to control the drug release. Dissolution differences may reflect differences in the design of the drug reservoirs used in different transdermal formulations. Differences in the drug release profile may also be seen during stability testing or batch-to-batch processing. Although there are many ways to study the release or dissolution of the drug from a transdermal system, regulatory agencies and official compendia list only a few methods. These methods are consistent in recommending that the dissolution profile include multiple time points, usually 3 or 4. This same requirement has been applied to extended-release oral formulations by the

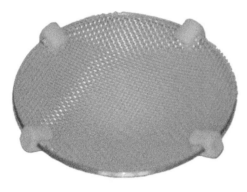

FIGURE 10 Disk assembly for Apparatus 5 consisting of a watchglass, a Teflon® screen to go over patch and clips to hold the assembly together.

FDA, so it is not surprising that the same would be necessary for the approval of a generic transdermal. At least 80% of the total drug content of the transdermal system should be released at the last sampling time and the amount release should exceed the amount absorbed in vivo.

The FDA preferred method uses a paddle in a dissolution vessel as described in the United States Pharmacopeia (USP) Apparatus 2, but the paddle is supported above a watchglass or stainless disk assembly that holds the transdermal system being evaluated, as illustrated in Figure 10. The transdermal system must be put on the watchglass with the adhesive side up followed by a mesh Teflon® screen to help secure the patch; clips are then used to hold the disk assembly together. Both the watchglass and stainless disk assembly are commercially available. This dissolution method is identified as Apparatus 5 in the USP (9). Apparatus 5 has been used successfully for many of the marketed products, but it may not be useful for all transdermal systems. As such the USP has also identified Apparatus 6, which uses the vessel assembly from USP Apparatus 1 but the rotating basket and shaft is replaced with a stainless steel cylinder, which is illustrated in Figure 11. The dosage unit is placed on the cylinder at the beginning of each test, and this method offers the advantage of being able to accommodate transdermal systems, which might be too large for Apparatus 5. The third method listed in the USP is Apparatus 7 that utilizes a reciprocating holder for the transdermal system. The assembly consists of a set of calibrated or solution containers made of glass or other suitable inert material, a motor and drive assembly to reciprocate the system vertically, and a set of suitable sample holders. Published results suggest that all of these methods can give reproducible results.

IN VIVO REQUIREMENTS: BIOAVAILABILITY/BIOEQUIVALENCE

As with oral extended-release generic products, in vivo bioequivalence studies with transdermal formulations are necessary to document that the performance of the formulation is equivalent between the generic and the RLD, and that it does not release the active drug substance at too rapid a rate (dose dumping). These objectives are met by conducting a single dose, nonreplicated, fasting study comparing the highest strength of the generic and reference listed drug product. At a minimum, the study should be a randomized, two-period, two-treatment, two-sequence crossover study comparing a generic or test

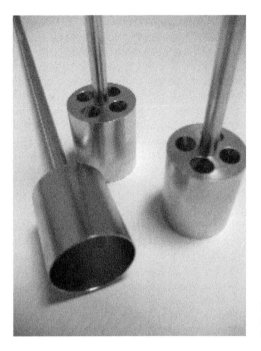

FIGURE 11 Rotating cylinder used to hold a transdermal system in USP Apparatus 6.

product to an RLD. To enhance the chances of success, the test product should have a flux rate that is not statistically different from that of the RLD. Although there is no regulatory requirement for this latter suggestion, it is highly recommended. Similarly, while a two-way crossover design is the only requirement, a replicate study design is recommended, whether the drug is considered highly variable or not. Whether a nonreplicate or replicate design is utilized in the bioequivalence study, the statistical analysis of the data should be based on average bioequivalence and not individual bioequivalence as is stated in the FDA's statistical guidance (10).

The bioequivalence study is usually conducted under fasting conditions, and this is a carryover from the design of studies to test oral extended-release formulations. While the fasting may not be necessary, it serves the purpose of yielding cleaner plasma samples at the earlier time points when drug concentrations are at their lowest levels. When designing a bioequivalence study, an adequate washout period must be utilized and should equal at least five elimination half-lives of the drug or metabolite being monitored. For a two-way crossover study, equal number of subjects should be randomly assigned to the two possible dosing sequences and the subjects must be crossed over such that all subjects receive both treatments. Any proposed protocol needs to be approved by an institutional review board (ethics committee) prior to initiation of the study.

For a transdermal product there is no need to evaluate the effect of dosing with food as is required with oral extended-release formulations. Also, the use of a multiple dose or steady-state crossover study is not necessary, since the FDA accepts the fact that a single dose study is more discriminating than one which utilizes a multiple-dose design. The 2002 guidance from the FDA states that the release of the drug from an extended-release delivery system into the systemic circulation can be best calculated from a single-dose study (11). The reader should be aware that while this guidance was prepared primarily for oral products, the introduction contains the following statement: "The guidance is also generally applicable to nonorally administered drug products where reliance on systemic exposure measures is suitable to document BA and BE (e.g., transdermal delivery systems . . .)." Applicants pursing a generic approval would be wise to submit a study protocol to the Division of Bioequivalence in the Office of Generic Drugs for review prior to the conduct of the study.

While "Pharmaceutical Equivalence" is required by CFR and the term is included in all Guidances dealing with generic approvals, all transdermal products can be considered to be pharmaceutically equivalent. That is, the type of drug-release mechanism is not important for the approval of a generic transdermal product. The drug content of the test product does not have to be equal to that of the RLD, unless there is a risk in having too much API remaining in the transdermal system after use by a patient. The same applies to the active surface area of a transdermal system. A generic product may be smaller or larger than the RLD. Furthermore, the use of a permeation enhancer is allowed even if it differs from the formulation used for the innovator product. The formulator of a generic transdermal, need only worry about demonstration of bioequivalence by traditional metrics, that is, the pharmacokinetic parameters achieved by the generic product must not be statistically different from those of the RLD. As with all ANDAs, this is the most critical requirement. A sample outline of a bioequivalence protocol for a transdermal nitroglycerin product may be seen below.

PROTOCOL SYNOPSIS

Objectives
Evaluate the bioavailability and wear properties of a generic transdermal nitroglycerin delivery system (Manufacturer, Address) relative to Nitro-Dur®, nitroglycerin Transdermal Infusion System (Key Pharmaceuticals, Inc; Kenilworth, NJ).

Study Design
This is a single-center, randomized, single-dose, two-way, crossover bioavailability study. A 24-hour washout period will be required between treatments.

Study Population
At least 48 healthy, nonsmoking, male and female volunteers, 21 to 45 years old will be enrolled in this study. A total of 44 volunteers must complete all the requirements of the protocol.

Study Treatments
Subjects will report to the clinical site by 17:00 hours on the evening before dosing. Each subject will be randomly assigned in accordance with a computer-generated randomization table to one of the treatments below:

Treatment A: Nitro-Dur 40 cm^2, 0.8 mg/hr
        (Key Pharmaceuticals, Inc.)

Treatment B: Nitroglycerin Transdermal System, 0.8 mg/hr
(Generic Manufacturer, Inc.)

Clinical and Laboratory Evaluations

A signed informed consent will be obtained from each subject. For screening purposes, the following procedures will be required of each volunteer and will be performed by the clinical site prior to initiation of the study: a complete physical examination including a medical history, vital signs, a 12-lead electrocardiogram (EKG), and clinical laboratory safety tests on blood and urine. During the treatments, blood pressure, heart rate, and respiratory rate will be monitored as a safety measure. A physical examination and clinical laboratory tests will be repeated at the completion of the study. Subjects will be observed and questioned throughout the study for the occurrence of any adverse events.

Sample Collections

A blood sample will be collected from each subject by venipuncture prior to each treatment (0 hour). Additional blood samples will also be collected at 1, 2, 3, 4, 5, 6, 8, 10, 12, 16, and 24 hours postapplication. Blood samples will also be collected at 0.5 and 1 hour following removal of the transdermal systems. Blood pressures and heart rates will also be monitored throughout the 25-hour period. Plasma harvested from these blood samples will be used for the assay of nitroglycerin plasma concentrations. The transdermal units will be checked for adhesion at each blood sampling time. After removal of the transdermal systems, the application sites will be examined by the investigator for any evidence of acute skin irritation and to assess the amount of adhesive residue on the remaining skin. The used transdermal units will be collected, individually wrapped, labeled, and stored frozen for subsequent analysis.

Pharmacokinetic and Statistical Analyses

The plasma concentration–time data will be used to determine the following pharmacokinetic parameters for each treatment: C_{max} (maximum nitroglycerin plasma concentration), T_{max} (time to C_{max}), AUC_{0-24} (area under the plasma concentration curve from time 0–24 hour, the time of removal of the transdermal units), and AUC_{0-25} (area under the plasma concentration curve from time 0 to the last sampling time, 25 hours). An analysis of variance (ANOVA) of the pharmacokinetic parameters calculated for this study as well as the two one-sided tests procedure will be conducted for ln-transformed C_{max} and AUC. The confidence interval for the ln-transformed parameters must fall within 80% to 125%.

WEAR, SKIN IRRITATION, AND SENSITIZATION STUDIES

Wear and Apparent Dose

Once a transdermal formulation has been optimized and good bioequivalence data have been obtained, it is important for the sponsor of the ANDA to properly define the wear and adhesion properties of the candidate transdermal system. The information gained from this type of study will help the Office of Generic Drugs decide that the generic transdermal is comparable to the RLD. Safety of the generic transdermal system as well as local skin irritation can also be evaluated in a wear study. This type of study can be run as an open label

TABLE 4 Adhesion Scoring Modified from the Expired FDA Guidance

Adhesion score	Definition
0	TDS adhered > 90%
	(Essentially no lift off the skin)
1	TDS adhered >75% to <90%
	(Only some of the edges lifting off the skin)
2	TDS adhered >50% to <75%
	(<50% of the system lifting off of the skin)
3	TDS adhered <50% but not detached
	(>50% lifting off of the skin without falling off)
4	TDS detached
	(Completely off the skin)

single dose, randomized, two-way crossover study, in which the two treatments are the RLD and the generic transdermal. The study should be conducted in 50 healthy male and female volunteers with at least a one-week washout between treatments.

The transdermal systems will be checked for adherence after application (0 hour) and at the various times after application. For a 24-hour product, the following time points would be recommended: 2, 4, 6, 8, 10, 12, 14, 22, and 24 hours postapplication. Adhesion scoring will be based on the scale contained in Table 4, which is a modification of the table found in a recalled FDA guidance (12). After the removal of each TDS, the application site will be examined for any evidence of local skin irritation using the scale listed in Table 5. Following removal of the TDS, the amount of adhesive remaining at the application site will be examined and graded as none (0), light (1), medium (2), and heavy (3). All scores should be recorded in a case report form by the clinical site conducting the study. It is also recommended that at each adhesion rating time point and when evaluating irritation, a digital photograph be taken of the transdermal system and the application site. These photographs should be valuable for future evaluation by the formulators and other transdermal scientists.

Although not necessary for a generic submission, a wear study can also be used to collect information regarding the dose delivered by a transdermal system. The used transdermal units can be collected, individually wrapped, labeled, and stored frozen for subsequent analysis. By comparing the drug remaining in these systems with the content of unused patches, the loss of drug from the transdermal system can by calculated, and this value is referred to as the Apparent

TABLE 5 Irritation Scoring Modified from the Expired FDA Guidance

0	No evidence of irritation
1	Minimal erythema (barely perceptible)
2	Definite erythema (readily visible) with minimal edema or papular response
3	Erythema with papules
4	Definite edema
5	Erythema, edema, and papules
6	Vesicular eruption
7	Strong reaction spreading beyond test site

Dose (13). It can be calculated for each used individual transdermal system from the following relationship:

Apparent Dose = Initial Potency − Residual Potency (2)

The Apparent Dose concept has been accepted by the FDA to define the dose delivered by a transdermal system for labeling purposes. For a generic product, since the labeling is based on the product being bioequivalent to the RLD, Apparent Dose calculations are supposedly not required but it seems that they are frequently requested by the regulatory agencies. The information gained from this type of study can help a formulator understand how the product is behaving in vivo without having to measure drug plasma concentrations.

Skin Irritation and Sensitization
Since the 1999 Irritation and Sensitization guidance has been recalled by the FDA, there is no official document to help guide the formulator of a generic transdermal (12). Indeed it could be argued that the irritation information gained from a wear study as described above and from the bioequivalence study should suffice. Health Canada organized a Scientific Advisory Panel to discuss this issue as part of bioequivalence requirements for transdermal fentanyl (14). The recommendation of the panel was that for generic products, skin irritation should be assessed as part of the bioequivalence study. The area of skin under the patch should be assessed and given an irritation score based on an acceptable standardized validated method such as described in Table 5. This position differs from that of the FDA's recalled guidance, which recommended a 200-subject study with repeated applications of the transdermal product for 21 days. The study was supposed to be a randomized, controlled, repeat patch test comparing the test patch to the innovator patch. Patches were to be applied to the same site for 21 days. Newly released guidances from the Office of Generic Drugs for lidocaine topical patches recommended a similar study design (15). As may be seen by any observer, this design was totally unrealistic in that no patient is ever advised to apply a transdermal system on the same site where a previous system has just been removed. Obviously, repeated applications to the same site result in stripping of the stratum corneum, thus allowing increased penetration of the active ingredient. In some cases this could result in toxic doses of drug being delivered to a healthy volunteer. While realistic or not, this cumulative irritation study design was still being advocated in the 2006 FDA Draft Guidance for lidocaine topical patches (15).

Sensitization
The Skin sensitization study design recommended by the recalled FDA Guidance was a Modified Draize Test. The study design was a randomized study in 200 healthy volunteers using three treatments: the generic transdermal system, the innovator product, and a placebo of the generic product, which contained all of the ingredients except for the API. The application sites were to be randomized among the volunteers and the study was divided into three sequential periods for a duration of six weeks. Based on the lidocaine topical patch guidance (15), the irritation and sensitization studies can be combined. After completion of the 21 day irritation study, subjects undergo a washout period of two weeks which is then followed by a "Challenge Phase." In this final period, the

transdermal systems are applied to new skin sites for 48 hours. Evaluations of skin reactions are to be made by a trained blinded observer at 30 minutes and at 24, 48, and 72 hours after patch removal. Obviously, if a study volunteer has been sensitized, a severe skin reaction will be observed during the challenge phase. From the Health Canada viewpoint, the Scientific Advisory Panel felt that it was not necessary to conduct sensitization studies on generic products. Individual excipients, however, should continue to be evaluated for their potential to cause sensitization. This is consistent with the logic that sensitization is most likely due to the API and not due to the polymers used in the transdermal formulation. A formulator skilled in the art is going to select polymers that have been well characterized as to their irritation and sensitization potential, and of course animal screening for irritation and sensitization has always been used in the preclinical phase and this practice will continue. The need for a sensitization study in humans for a generic product is questionable at best. These type of studies, however, will continue to be necessary as a measure of safety required by the regulatory agencies, especially the FDA. There are ongoing discussions between the FDA and the pharmaceutical industry, which hopefully will redefine and simplify the criteria for safe and effective generic transdermal products.

REFERENCES

1. González MA. Trends in Transdermal Drug Delivery, Chapter 8 in Topics in Pharmaceutical Sciences, 1991, Crommelin DJA, Midha KK eds. Stuttgart, Germany: Medpharm Scientific Publishers, 1992.
2. Franz TJ. Kinetics of cutaneous drug penetration. Int J Dermato 1983; 22:499–505.
3. Van Buskirk GA, González MA, Shah VP, et. al. Scale-up of adhesive transdermal drug delivery systems: AAPS/FDA Workshop Report. Pharm Research 1997; 14(7): 848–852.
4. Viscusi ER, Schechter LN. Patient-controlled analgesia: Finding a balance between cost and comfort. Am J Health-Syst Pharm 2006; 63(Suppl 1):S3–S13.
5. Franz TJ. Percutaneous absorption: On the relevance of in vitro data, J Invest Dermatol 1975; 64:190–195.
6. Noonan PK, González MA. Pharmacokinetics and the variability of percutaneous absorption. J Toxicol Cutaneous Ocular Toxicol 1989; 8(4):511–516.
7. Gerardin A, Gaudry D, Moppert J, et al. Glycerol trinitrate (nitroglycerin) plasma concentrations achieved after application of transdermal therapeutic systems to healthy volunteers. Arzneimittel Forschung 1985; 35:530–532.
8. Shah VP, Tymes NW, Skelly JP. Comparative in vitro release profiles of marketed nitroglycerin patches by different dissolution methods. J Controlled Release 1988; 7:79.
9. USP 29 NF 24. The United States Pharmacopeial Convention, Inc., 2006.
10. FDA Guidance Statistical Approaches to Establishing Bioequivalence, January 2001.
11. FDA Guidance. Bioavailability and Bioequivalence Studies for Orally Administered Drug Products—General, July 2002.
12. FDA Guidance. Skin Irritation and Sensitization Testing of Generic Transdermal Drug Products, December 1999.
13. Noonan PK, González MA, Ruggirello D, et al. Relative bioavailability of a new transdermal nitroglycerin delivery system. J Pharm Sci 1986; 75:688–691.
14. Health Canada web site dealing with transdermal fentanyl. http://www.hc-sc. gc.ca/dhp-mps/prodpharma/activit/sci-consult/ftds-sdtf/index_e.html. Accessed 2006.
15. FDA "Draft Guidance on Lidocaine" December, 2006.

9 Pharmaceutical Development of Modified-Release Parenteral Dosage Forms Using Bioequivalence (BE), Quality by Design (QBD), and In Vitro In Vivo Correlation (IVIVC) Principles

Siddhesh D. Patil

Millennium: The Takeda Oncology Company, Parenteral Formulation Sciences, Cambridge, Massachusetts, U.S.A.

Diane J. Burgess

Department of Pharmaceutical Sciences, University of Connecticut, Storrs, Connecticut, U.S.A.

INTRODUCTION

Successful clinical performance of pharmacotherapeutic regimens is related to efficient and safe delivery of medicinal agents at a targeted site of action. Modified-release parenterals (MR parenterals) are defined as novel or specialty dosage forms that achieve therapeutic objectives by maintaining blood levels through regulated and rate-controlled drug release from the formulation over extended periods of time (1). These formulations are designed by entrapment or encapsulation of the drug into inert polymeric or lipophilic matrices that slowly release the drug in vivo over durations as short as a week and up to several years (1). The polymers or lipophilic carriers used to deliver the drugs in MR parenterals either biodegrade in vivo or are nonbiodegradable. Nonerodible systems are removed at the end of the therapy.

Examples of MR parenteral dosage forms include microspheres, liposomes, drug implants, inserts, wafers, depots, pellets, drug-eluting stents, micro- and nanoparticles. These types of drug delivery systems represent some of the most clinically advanced products and are used for a wide range of therapeutic indications marketed worldwide. Selected examples of marketed MR parenterals are listed in Table 1. Because of their extended duration of action, these dosage forms are often classified and identified as "long acting," "sustained release," "controlled release," or "prolonged release" counterparts that differentiate them from immediate-release conventional dosage forms (1).

MR parenterals are most commonly administered by injection or implantation in the subcutaneous or intramuscular sites for systemic drug delivery. Examples of drugs delivered using these dosage forms include growth hormone analogue somatropin (2); incretin mimetic exenatide for the treatment of diabetes and obesity (3); gonadotropin analogues such as leuprolide (4) and goserelin (5) for the treatment of endometriosis and prostrate cancer, and synthetic progestagen levonorgestrel for use in hormonal contraception (6). MR parenterals can

TABLE 1 Selective Examples of Marketed Formulations of MR Parenterals

Type of MR parenteral	Commercial or trade name	Drug	Clinical indication
Microsphere	Lupron	Leuprolide acetate	Prostate cancer Endometriosis Fibroids
Microsphere	Sandostatin	Octreotide acetate	Acromegaly Carcinoid tumors Vasoactive intestinal peptide tumors
Microsphere	Nutropin	Somatropin (rDNA origin)	Growth hormone replacement
Microsphere	Risperdal Consta	Risperidone	Schizophrenia
Microsphere	Trelstar	Triptorelin pamoate	Prostate cancer
Microsphere	Retin-A Micro	Tretinoin	Acne
Microsphere	Vivitrol	Naltrexone	Alcohol dependence
Microsphere	Arestin	Minocycline hydrochloride	Periodontitis
Liposome	Doxil	Doxorubicin HCl	Ovarian cancer Kaposi's sarcoma Multiple myeloma
Liposome	AmBisome	Amphotericin B	Fungal infections Visceral leishmaniasis
Liposome	DaunoXome	Daunorubicin citrate	Kaposi's sarcoma
Liposome	DepoCyt	Cytarabine	Lymphomatous meningitis
Liposome	DepoDur	Morphine sulphate	Pain following major surgery
Wafer	Gliadel	Carmustine	Brain cancer
Depot	Suprefact	Buserelin acetate	Prostate cancer Endometriosis
Implant	Zoladex	Goserelin acetate	Prostate cancer
Implant	Implanon	Etonogestrel	Contraception
Implant	Retisert	Fluocinolone acetonide	Uveitis
Implant	Viadur	Leuprolide acetate	Prostate cancer
Implant	Supprelin	Histrelin acetate	Precocious puberty
Implant	Vitrasert	Ganciclovir	Cytomegalovirus retinitis

also be used for site-specific targeted and tissue-specific delivery for localized action of the drug. Examples of such MR parenterals include delivery of fluocinolone acetonide to the eye using an intravitreal implant for the treatment of inflammatory eye disease uveitis (7), delivery of minocycline hydrochloride to gum tissue using a microsphere delivery system for the treatment of periodontal disease (8), and the delivery of carmustine to brain tissue using a biodegradable wafer implant made up of the polyanhydride copolymer polifeprosan 20 as an adjunct to surgery and radiation in the treatment of brain cancer (9).

In addition to the use of inert and novel carrier-based systems, MR parenterals can attain sustained drug delivery using drug forms that have poor dissolution characteristics (1,10). These formulations achieve prolonged action due to slow dissolution of the chemical salt forms, drug-complexes, or crystalline polymorphs. Compared to their immediate-release rapid- or short-acting versions that are easily absorbed due to fast dissolution, the intermediate- and long-acting MR parenterals can maintain systemic blood levels over extended periods of time with fewer does of administration (10). Examples of slow-dissolving MR parenterals include insulin glargine and insulin zinc suspensions used in the treatment of diabetes (11,12). Insulin glargine upon administration can achieve long duration of action due to gradual precipitation of insulin micro-crystals with slow dissolution characteristics (11). Subcutaneous administration of the insulin glargine formulation causes insulin crystallization due to local-ized change in the formulation pH. Insulin zinc suspension contains microcrys-talline or amorphous complexes insulin and zinc ions that can achieve long action due to gradual dissolution. Varying combinations of drug polymorphs of insulin–zinc complexes can be used to obtain programmable and consis-tent sustained-release characteristics (12). Dexamethasone acetate and methyl-prednisolone acetate for symptomatic pain and inflammation management (13), penicillin G procaine and penicillin G benzathine as antibiotics (14), as well as estradiol cypionate and medroxyprogesterone acetate as contraceptives (15) are examples of poorly water soluble salts that can be used in MR parenterals for prolonged or sustained-release action.

PHARMACEUTICAL ADVANTAGES OF MR PARENTERALS

Some of the key drivers for the development of MR parenterals are their sig-nificant therapeutic and clinical advantages over conventional standard- or immediate-release dosage forms (16). As shown in Figure 1(A), immediate-release IV bolus infusions and subcutaneous injections achieve clinical efficacy by targeting the therapeutic window of the drug, that is, the systemic level of the drug required for optimal therapeutic benefit, which is defined by the min-imum effective level and the maximum tolerated level (16). Following admin-istration of the immediate-release dosage form, the systemic level of the drug increases as drug is absorbed in the circulation, achieves a maximum concentra-tion and then decreases as it is metabolized and eliminated (16). This classical pharmacokinetic behavior necessitates repeat administration of the immediate-release dosage form to maintain effective blood levels in the therapeutic window to achieve safe and effective clinical outcomes (16). The half-life, potency, and intended therapeutic levels of the drug determine the dosing frequency. On the other hand, as shown in Figure 1(B), MR parenterals, can achieve effective sys-temic concentration of the drug within the therapeutic window over extended periods of time through an optimized, consistent, and steady release of the drug after a single administration (16). MR parenterals are thus able to provide similar clinical benefit over the entire duration of the therapeutic regimen using fewer doses compared to the immediate-release dosage form (16).

   MR parenteral technologies can be used for drugs that have short half-lives, or limited bioavailability due to poor absorption or extensive degrada-tion when administered via the oral route. For example, leuprorelin acetate, a long-acting synthetic peptide analogue of gonadotropin-releasing hormone, used in the treatment of endocrine disorders such as prostrate cancer in men and polycystic ovarian syndrome in women, can be formulated as an immediate-release liquid or a sustained-release microsphere formulation administered par-enterally (17). As shown in Figure 2, in a preclinical rat model, subcutaneous or intravenous administration of a 1-mg dose of leuprorelin acetate, under-goes rapid clearance characterized by steady decrease in the serum levels of the

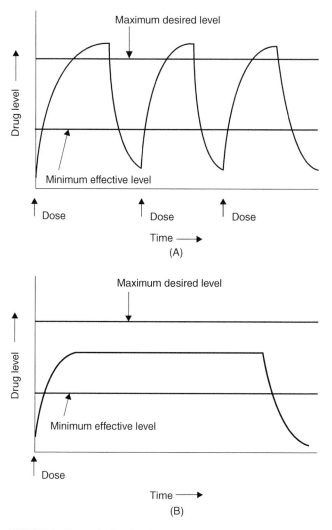

FIGURE 1 Systemic drug levels upon administration of (**A**) immediate-release dosage form and (**B**) modified-release parenteral. *Source*: From Ref. 16.

drug and very short elimination half-lives (17,18). On the other had, as shown in Figure 3, a similar dose (0.9 mg/rat) formulated in a microsphere delivery system and administered using a subcutaneous injection can achieve sustained levels of the drug over a one-month period (17,19). The sustained-release dose is well tolerated and clearly demonstrates the use of reduced dosing frequency to achieve comparable blood levels of the drug in vivo.

In a clinical setting, patient compliance is often superior for MR parenterals compared to immediate-release dosage forms, since the frequency of administration is significantly reduced for equivalent therapeutic efficacy (20,21). Using MR parenterals, the overall quality of life, drug adherence, drug satisfaction, and

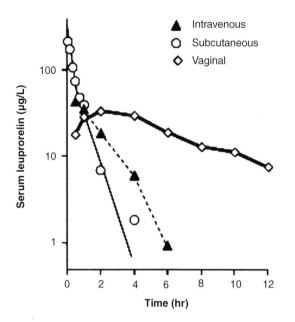

FIGURE 2 Serum levels of leuprorelin acetate upon administration of a 1-mg dose of an immediate-release formulation via subcutaneous, intravenous, or vaginal routes in a rat model. *Source*: From Ref. 18.

economics for treatment of patients with chronic conditions such as epilepsy (20), alcohol dependence (22,23), and prostrate cancer (24) are significantly improved due to dosing flexibility and better drug acceptability and tolerability due to reductions in fluctuations in the systemic levels that are commonly observed with immediate-release dosage forms (25).

From a commercial perspective, MR release parenterals offer a tremendous advantage for life-cycle management of new chemical entities as a pathway to

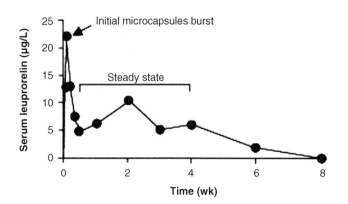

FIGURE 3 Serum levels of leuprorelin acetate upon administration of a 0.9-mg subcutaneous dose of a long-acting sustained-release microsphere formulation in a rat model. *Source*: From Ref. 19.

extend their commercial lifespan. As existing oral or parenteral formulations of pharmaceuticals face increasing marketplace competition by follow on molecules of similar chemical structure, MR parenterals serve as a line extension of the same molecule that is administered in a more patient compliant format. Since the toxicology impact, clinical efficacy, and safety of these molecules has already been evaluated and understood in large populations during use as immediate-release conventional dosage forms, the pharmaceutical development of MR parenterals can be relatively less risky than the effort of introducing a new chemical entity, as long as the stability of the molecule is ensured in the inert matrix.

In addition to their therapeutic and commercial advantages, MR parenteral products have been developed for specialized applications. Examples include the use of biodegradable microspheres to improve photostablity tretinoin (26) and decrease topical irritation (27) by slow release of the encapsulated drug. MR parenterals such as liposomes have been used for passive tumor targeting for anticancer applications (28). Liposomal products of anthracycline anticancer agents such as doxorubicin (Doxil) and daunorubicin (DaunoXome) utilize the enhanced permeability and retention effect to target tissues, due to which these vesicular particles are selectively accumulated in tumor tissue because of leaky and poorly defined vasculature (28,29). Localization of the liposomes in tumor tissues that is subsequently followed by degradation and release of the drug leads to targeted destruction of the tumor cells (30). Liposomal Amphotericin B (AmBisome) can be used to treat severe fungal infections due to efficient interaction of the drug-loaded liposome with fungal cell walls that leads to disruption of the bilayer assembly and selective transfer of the antibiotic into the cell (31). Drug delivery using these liposomal MR parenteral systems significantly reduces systemic exposure, and consequently reduces the toxicity associated with these high potency molecules (28–31). Liposomes also have been used to improve the circulation half-lives of drugs through stealth technology that involves PEGylation of the bilayer liposome systems to circumvent recognition by and activation of the cells of the immune system (28,32).

MR parenterals such as liposomes (28) and nanoparticles (33,34) have also been used for tissue-specific targeting through incorporation of specific chemical and structural moieties or ligands in the nondrug inert lipids bilayer assemblies and polymeric matrices of these dosage forms, respectively. Improved targeting can diminish nonspecific exposure, reduce the dose and dosing frequency, decrease potential side effects, and improve the therapeutic efficiency of molecules (35). MR parenteral systems can also be used for the sustained delivery of biologics such as antibodies, interleukins, growth factor proteins, and enzymes as well as nucleic acid therapeutics such as oligonucleotides and siRNA that have reduced oral bioavailability due to poor absorption and rapid degradation in the stomach (36–43). Microparticulate systems can also be used for long-term delivery as well as adjuvants for immunopotentiation of vaccine antigens (44). Several MR parenterals for advanced applications have shown tremendous promise in preclinical models and are currently undergoing clinical development (36–42).

CONSTRAINTS FOR GENERIC MR PARENTERAL PRODUCT DEVELOPMENT

Because of their tremendous pharmaceutical and therapeutic advantages, development of generic modified-release parenterals remains a field of high commercial value and rich opportunity, but one with little specific regulatory guidance and progress. To date there are no approved and marketed generic formulations of modified-release parenterals to be found in the U.S. FDA's Approved Drug Products with Therapeutic Equivalence Evaluations database (45). This list includes a wide range of approved generic and bioequivalent versions of MR oral dosage forms and immediate- or standard-release formulations such as tablets, capsules, injections, infusions, oral and topical suspensions, opthalmics, transdermal dosage forms, ointments, and gels; however, no examples of inert carrier-based generic MR parenterals can be found (45).

Some of the significant hurdles associated with generic MR parenteral development are the paucity of technical information about these products for unrestricted public use, the high costs and risks associated with the research and development of these complex products, and the lack of definitive regulatory pathways and strategies for product approval (46). Since much of the technical knowledge concerning these advanced dosage forms is patented and backed by stringent intellectual property protection, access to these technologies can be fee-based and/or highly restrictive (47). Strict enforcement of the terms and conditions regarding the use of the intellectual property around MR parenterals and world-wide laws that support and protect their scientific and commercial value effectively prevent unauthorized development of generic versions.

The overall complicated design and engineering of inert carrier-based MR parenteral dosage forms are in themselves strong deterrents to the development of generic versions. The carrier lipids or polymers used for entrapment, encapsulation, or complexation of drugs are highly specialized raw-materials and are generally patented for their functional abilities (48,49). In the case of polymer matrix–based MR parenterals, controlled-release of the drug is usually achieved by diffusion and/or erosion of biodegradable matrix and is highly specific to the physicochemical characteristics of the polymer and the drug. The innovator product sponsors that develop such advanced materials and programmable technologies patent them for wide-ranging applications and because of their knowledge lead in the field they often remain the sole manufacturers (48,49). As a result, procurement and use of carrier materials for developing generic versions and eventually establishing a supply chain for future commercial products necessitates close involvement and formal legal negotiation between the generic sponsor and the innovator sponsor.

Additionally, development of generic versions of MR parenterals can be particularly challenging due to the highly specialized nature of the unit processes used in the manufacture and development of these complex dosage forms. The processes, know-how, and equipment used to facilitate complexation or improve entrapment of the drug in the carrier matrices as well as those developed to modulate release are routinely patented and have similar restrictions on use as is the polymer or lipids used in the design MR parenterals. These specialized processes require large capital investment to establish commercial operations for the intended product. Information that can improve the manufacture and stability of the dosage forms as well as the stability of the entrapped drug in the dosage form are often classified as trade secrets. Using this strategy, innovator sponsors secure a competitive advantage over competing technologies by having basic patented information freely available and at the same time limiting complete technical information essential for the practical use of the technology. In effect, development of generic versions of such technologies is effectively prevented or requires collaboration with the innovator sponsor companies to support clinical development.

The difficulty in navigating the patent landscape poses additional design and engineering constraints in development of MR release parenterals due to significant technical similarities across a spectrum of controlled-release technologies. Thus, even if attempts are made to develop similar technologies based on scientific principles of innovator products without direct technical assistance from the innovator product sponsor, there is a high risk of unintended patent infringement. To alleviate risks and complications associated with intellectual property protection, sponsors of innovative MR parenteral technologies often commercially structure and market these dosage forms as drug development platforms for life-cycle management that can be tailored to develop commercial products for a wide range of pharmaceutical molecules using the assistance and technical know-how of the innovator sponsor. Because of the immense costs involved in developing commercial products from drug delivery technologies, it is routine to conduct feasibility studies or exploratory testing to identify if such platforms can be applicable to obtain a target product profile for the drug of interest. Should the feasibility study appear promising in a preclinical model, then future development is structured on the basis of toxicology of the drug using the platform technology, clinical studies, and competitive scientific, and business intelligence. Using such strategies the innovator sponsor can license the technology for pharmaceutical compounds of interest or can collaborate on the clinical development with other partners. Such an approach is very useful for sponsors who want to conduct feasibility studies for the drug of interest with potentially multiple MR technologies that can be used for life-cycle management of the molecule and eventually select the best approach that meets the target product profile. This strategy can also be beneficial in situations where the emphasis is to reduce the time to market the MR parenteral version of a drug where the pharmacology is fairly well-characterized via the immediate-release version and development time-lines are short. Drug development on these principles can also be less risky in terms of expenditure, especially in cases where it is eventually determined that the use of an initially promising MR parenteral technology to deliver the drug may not be clinically or commercially feasible.

In the absence of clear regulatory pathways for product approval, pharmaceutical development of generic MR parenterals can be extremely difficult to initiate, manage, and accomplish. Most of the regulatory guidance available in the United States, Europe, and Japan is based on standard- or immediate-release orals and parenterals and extended- or modified-release oral dosage forms. Application of guidance developed on products of relatively simpler technologies and extensive product development histories such as oral tablets or capsules or immediate-release injections or infusions to MR parenteral product development can be extremely challenging and in some cases irrelevant. As newer MR parenteral dosage forms overcome many of the formulation-based limitations associated with standard- or extended-release orals and immediate-release parenterals, these technologies also require advanced analytical control strategies and instrumentation to test for their quality and consistency.

Given the extraordinary diversity in the composition, manufacturing, critical quality attributes, mechanism of action, duration of therapeutic activity, overall performance, and quality systems for MR parenterals, it is nearly impossible to identify universal regulatory or scientific pathways that can be applicable for all MR dosage forms. This chapter focuses on the application of the basic principles of generic product development for oral and injectable immediate or standard-release dosage forms to MR parenterals. Current regulatory guidance and opinions from industry and academic organizations and scientific special interest focus groups on

(a) bioavailability considerations and bioequivalence (BE) determination,
(b) application of quality-by-design (QBD) application to drug development,
(c) importance of drug release and product performance in vitro,
(d) testing and instrumentation developed for in vitro drug release, and
(e) establishing in vitro–in vivo correlations (IVIVC) for MR parenterals are

discussed in detail. In addition, the limitations on the application of these principles on MR parenteral product development are identified as well as some of the technical advantages and opportunities that could be exploited for faster, efficient, and accurate outcomes are recognized.

APPLICATION OF BIOEQUIVALENCE PRINCIPLES FOR GENERIC MR PARENTERAL DEVELOPMENT

From a regulatory perspective, pharmaceutical development of MR parenteral generic drugs can be facilitated using guidances provided by the Office of Generic Drugs of the U.S. FDA that oversees the review and approval process for these dosage forms. A typical regulatory approval pathway involves submission of an Abbreviated New Drug Application (ANDA) to obtain approval for a generic drug product. The submission process and the regulatory interactions for immediate- and modified-release oral and immediate-release injectable dosage forms is routine as evidenced by the high number of generic approvals for these products listed in the Approved Drug Products with Therapeutic Equivalence Evaluations database (Orange book). On the other hand, there is a significant lack of information for MR parenterals in the formulary. Although much of the guidance available from the FDA for the approval of generics is directed toward oral and immediate-release injectables, elements from these source documents can be applied to the pharmaceutical development of generic MR parenterals.

The FDA guidance defines generic drug products as those that are considered comparable to innovator products on the basis of the overall total product profile of these formulations, which includes descriptions of the dosage form, strength, and route of administration and have equivalent quality, safety, and performance for their intended use. Since the preclinical and clinical safety of innovator drug products has already been established and approved by the regulatory agencies in the New Drug Application (NDA) of the innovator product, the major focus of the ANDA is to establish bioequivalence between the innovator drug and the generic drug.

The 21 CFR part 320 statutorily defines bioequivalence as the absence of statistically significant difference between two formulations in the rate and extent to which the active ingredient or moiety becomes available at the site of action when administered under similar conditions at the same molar dose (50). Bioequivalence is a measure of comparison between two products on the overall systemic exposure of the active drug resulting after release from the formulation upon administration (50). Since the systemic exposure of the drug can be directly correlated to the clinical effects and the resulting safety, efficacy, and toxicity, demonstrating bioequivalence serves as an important parameter for comparative analysis between innovator or reference products and the generic. Using the FDA's guidance documents, bioequivalence between generics and innovator products can be established using pharmacokinetic studies, pharmacodynamic evaluations, clinical trials, and in vitro studies (50).

Based on the classical definition of bioequivalence, pharmacokinetic studies comparing generic and innovator MR parenterals would be the most preferred approach to establish bioequivalence. Pharmacokinetic evaluation characterizes systemic exposure following administration of both formulations by measuring standard parameters such as area under the curve (AUC), peak concentration (C_{max}), and time to peak concentration (T_{max}) in blood, plasma, or serum. Sponsors are required to develop a sensitive and accurate assay to quantify the drug in biological fluids. Pharmacokinetic studies identify the absorption, distribution, elimination, and metabolism of the drug and thus are a direct indicator of the onset, duration, and intensity of the clinical effects of the drug. Pharmacokinetic studies typically involve crossover designs where both comparative formulations are administered to the same patient, separated by a washout period. These study designs can be complex and challenging to setup for MR parenterals of long duration of action or where several unit doses of the formulation are responsible for achieving steady state.

Pharmacokinetic evaluation of MR parenterals intended for systemic action typically involves standard plasma or blood sampling at appropriate time intervals upon injection or implantation of the MR parenteral. However, for those MR parenterals involving localized drug delivery, the measurement of systemic levels is irrelevant and it is impractical to obtain local drug levels. In case of MR parenterals where several doses are required to obtain steady state for the drug or those that are administered at low doses due to high potency, application of PK measurements on a single dose may not be accurate or representative of the intended therapeutic regimen. In such cases, use pharmacokinetic evaluation in representative preclinical models for formulation optimization can support human clinical trial data generated.

Pharmacokinetic evaluation provides the most direct and definitive evidence of the overall systemic exposure of a drug to support bioequivalence of a product (50). Pharmacodynamic responses can be a useful indicator to study the clinical effects of systemic exposure of the drug (50). However, evaluation of pharmacodynamic responses as a measure of bioequivalence is only recommended when pharmacokinetic studies are not feasible. Most clinical trials involve monitoring pharmacodynamic responses in addition to pharmacokinetic measurements during comparative analysis of the formulations. However, since pharmacodynamic responses can be statistically more variable than pharmacokinetic parameters, clinical trials based on these require a higher population of subjects compared to those that involve direct PK measurement. Pharmacodynamic responses may differ from the clinical effect of the drug and can complicate data analysis if they are transient in nature. The types of pharmacodynamic responses monitored in clinical trials depend on the pharmacology of the drug and can serve as clinical biomarkers to indicate potential therapeutic effects in patients. Examples of pharmacodynamic responses include indicators such as heart rate and blood pressure (51) for cardiovascular drugs; neutrophil and stem cell counts (52) for drugs used in the treatment of neutropenia; bactericidal effects for antibiotics (53); and plasma glucose and insulin levels for agents used in the treatment of diabetes (54).

Comparative clinical studies that use clinical efficacy endpoints as a measure for comparing products are some of the not preferred techniques for demonstrating bioequivalence (55,56). Clinical endpoints, just like pharmacodynamic responses, can be highly variable and may not be representative of the therapeutic success or failure of the drug in general populations. Therefore, comparative clinical studies for demonstrating bioequivalence require large numbers of patients to support statistical significance of the results (57,58). Sometimes, the number of patients required to demonstrate bioequivalence in these trials can be even larger than that required for demonstrating efficacy of the innovator product (57,58). Additionally, since a long-term drug regimen is typically required for establishing clinical efficacy, these trials can be of extremely long duration for long-term MR parenterals. Other challenges include expense associated with long-term trials and the potential to pose a greater risk for safety due to relatively high levels of exposure compared to PK trials. Due to use of clinical outcomes as endpoints, crossover designs for MR parenterals may be difficult to implement and therefore multiple studies would be necessary. Despite numerous challenges associated with the variability of comparative clinical trials, this approach for establishing bioequivalence can be useful for MR parenterals for localized action and for those delivering cytotoxic or high potency molecules. PK measurements are not feasible for MR parenterals involving localized action due to the low systemic drug levels and therefore bioequivalence using clinical efficacy is the most reliable and the only approach available. For cytotoxic molecules there are the ethical concerns of exposing healthy human volunteers to drugs with known toxic effects (59).

Demonstrating comparability between innovator and generic products using in vitro dissolution tests without the support of clinical studies is the least preferred approach for establishing bioequivalence for modified- and immediate-release orals and parenteral products over human clinical trial investigations involving PK parameters, PD responses, or clinical outcomes (60–63). This approach involves developing an in vitro test procedure that entails suspension of the MR parenteral in an external medium, which is periodically sampled to characterize and measure drug release from the inert carrier matrix (60–63). The basic premise of demonstrating bioequivalence using an in vitro test is that if the drug release profiles of the entrapped drug from generic and innovator products are similar and consistent under standardized and controlled conditions, then they would be expected to have similar in vivo release characteristics (60–63). Since, bioavailability, systemic exposure, and pharmacological effects are directly associated with the released drug, if their in vitro release profiles are comparable, then there is a high likelihood that comparator generic and innovator products can be considered bioequivalent, although given the variability and unpredictability associated in vivo and lack of robust method development for in vitro release, clinical studies remain the most direct line of evidence to demonstrate comparability to date.

Bioequivalence determination using in vitro release methodologies has the least expense compared to all the clinical approaches; however, the technique can be risky especially if the drug release characteristics in an in vitro medium may not be representative of the in vivo release profile (50). Considerable resources are required to develop an assay to quantify drug release and to identify appropriate test method conditions that are optimized to discriminate similar versus nonsimilar generics and innovator products. False positives demonstrating comparability between inequivalent products or false-negative test results excluding potentially therapeutically equivalent products can be detrimental to the pharmaceutical development program of generics.

In vitro release tests are most useful for establishing bioequivalence for highly soluble and highly permeable drugs classified under the biopharmaceutics classification system and for modified-release parenterals where in vitro drug release is one of the most critical attributes for performance of the product. In vitro drug release tests may then be used for such drug candidates as surrogates for traditional bioequivalence studies and for MR parenterals intended for localized applications where pharmacokinetic and pharmacodynamic measurements may not be feasible or clinical outcomes may require repeat administration. These tests can also present a shorter and more efficient substitute for clinical studies used to demonstrate bioequivalence for very long-term MR parenterals where drug release is programmed over several months or years, which can lead to very long and expensive clinical trials, or where the crossover design of the clinical trial is impractical.

Since no innovator and generic controlled-release microsphere product exists on the market, a head-to-head comparison of a clinical study demonstrating bioequivalence between the two could not be found. Clinical studies that have demonstrated safety and efficiency of immediate-release dosage form regimens or MR parenteral compared to placebos to obtain desired pharmacokinetic, pharmacodynamic, or therapeutic clinical outcomes are typical in pharmaceutical drug development. These also include studies that have also demonstrated distinct clinical advantages and short-term as well as long-term superiority of parenteral controlled-release dosage forms to obtain desirable clinical outcomes for chronic conditions such as schizophrenia (64), diabetes (3), alcohol dependence (22,23), and cancer (17,19).

QUALITY-BY-DESIGN AND MODIFIED-RELEASE PARENTERALS

Development of generic controlled-release products can be facilitated by the application of quality-by-design (QBD) principles stated in the ICH guideline for Pharmaceutical Development (Q8R1) (65–67). Typical pharmaceutical development of MR parenterals is an empirical approach where selection and impact of formulation components and manufacturing variables on product properties are investigated and optimized in isolation. The manufacturing procedures are fixed and unit operations are batch processes that are primarily controlled by use of in-process tests and analysis of intermediates off-line. The product quality and control strategy involve sampling and testing representative samples for compliance with predetermined specifications (65–67).

The QBD approach for pharmaceutical development of MR parenterals would include a comprehensive understanding of the impact of material components and their properties and manufacturing processes on product quality. The ICH guideline provides principles and practices that can lead to a systematic elucidation of all factors (formulation, process, and combinations of both) and their complex interrelationships on optimal product performance. This approach involves simultaneous evaluation of combinations of materials and processes on product performance and identifying the overall design space for the product that can ensure optimal performance and thus desirable clinical benefit. This mechanistic basis of product development can ensure that manufacturing procedures have a higher degree of flexibility that can be controlled and adjusted using tools involving process analytical technologies. Ideally, the goal is to use continuous processes that can be tracked and adjusted using inline feedback and feed-forward mechanisms. Although the product quality would be ensured using specifications, the overall control strategy is risk-based and focused on robustness evaluation with potentially reduced end-product testing (65–67). In effect, compared to the empirical approach for product development, QBD would employ a preventative basis for risk mitigation instead of a reactive setup.

Concepts based on QBD can help increase understanding of the key defining critical characteristics of the dosage form that result in successful clinical performance, identifying major characteristics that need to be controlled, as well as detecting those features that may lead to unfavorable results. Programmed in vivo release of the entrapped drug from controlled-release matrices and overall performance in MR parenterals are key factors that determine systemic or localized effective concentrations. Therefore, a systematic understanding of the factors that affect in vivo drug release using QBD and their cascading impact on pharmacokinetics and clinical performance can thus help identify or support a promising strategy for developing generics based on innovator controlled-release products (65–67). Since the overall drug release mechanism from MR parenterals is complex compared to immediate- or standard-release dosage forms and the factors that impact it are generally poorly understood, application of QBD principles can be challenging in a practical setting.

The QBD principles for pharmaceutical development as outlined under the ICH guideline Q8R1 involve the design a drug product that meets patient and physician needs with consistent, safe, and efficacious clinical performance by identifying and establishing a quality target product profile (65). Elements that help in characterizing the quality target product profile of an MR parenteral may include the route of administration, nature and strength of the dosage form, the clinical container closure systems, and attributes that impact the release of the entrapped drug (which in turn affects the pharmacokinetic profile), as well as other quality attributes that ensure sterility, potency, and stability of the drug product. An example of a quality target product profile for a microsphere product can include the design of a single use dual-chamber pen device that stores the microsphere powder in one compartment and a diluent for suspension of the microspheres in another. The two components of this delivery system are extemporaneously mixed in the pen to form a uniform suspension of the microspheres and then self-administered by the patient as a subcutaneous injection. The integrity of the pen device prevents contact of the two components prior to administration and during storage. This ensures stability and sterility of the product as well as consistent and easy delivery of the desired dose at the intended site. Upon injection, the microspheres degrade in vivo in the course of time and release the drug with a predetermined pharmacokinetic profile that corresponds to a defined and desirable clinical outcome.

An example of a quality target product profile for a liposome product for anticancer application for a lipophilic drug can include the design of a single use prefilled syringe that stores the liposomal suspension. The syringe maintains the sterility of the product and is used to administer the product using via dilution in an intravenous fluid in a physicians office. Upon infusion, the liposomes selectively accumulate in tumor tissues, degrade in vivo, release the entrapped drug thus achieving targeted delivery to the tumor with predetermined pharmacokinetic parameters associated with desirable pharmacodynamic and clinical outcomes.

Once the target profile for the MR parenteral is defined, QBD involves identifying the critical product attributes (CQAs) that ensure the product meets the quality standards and demonstrates consistent and reliable clinical performance. The ICH guideline defines critical quality attributes as properties, characteristics (physical, chemical, biological, or microbiological) that should be well controlled and within limits for any product to meet the target product profile. These properties can affect the drug purity, stability, potency, and stability. CQAs for MR parenterals can vary depending on the specific type of dosage form and have to be identified on a case-by-case basis. Examples for CQAs for poly-(lactic-co-glycolic) acid (PLGA) microspheres can include properties that can influence drug release and stability such as polymer composition, residual solvents, and moisture as well as properties that can impact syringeability (such as particle size and size distribution, composition, and viscosity of the diluent used to suspend the microspheres before injection) as well as sterility of the final product. For liposomal dosage forms, CQAs could include the residual amount of unentrapped drug, lipid composition, particle size and distribution, and sterility. In a practical setting, identifying CQAs can be an iterative process evolving processes that rely on the accumulative knowledge acquired during all phases of drug product development and testing. In the initial stages of product development, the critical product attributes can be assigned using empirical knowledge or prior experience with similar products that can be re-assigned as more data on product quality and impact on clinical performance become available.

QBD principles also include risk assessment that identifies formulation (composition and characteristics) and manufacturing process parameters that can influence the CQAs of a product. The relationships that define the impact of material characteristics and process parameters on the CQAs can be characterized in the design space of the formulation. The design space of the formulation identifies and establishes relationships between multivariable and multidimensional combinations of process inputs that ensure that the final product generated meets the predefined CQAs of the product. Essentially operating within a design space would ensure consistent quality and reliable performance with a high level of confidence. The design space mapping and evaluation for MR parenterals can involve robustness testing of the formulation components and process parameters, mechanistic investigations on the engineering of the dosage forms, scale-up studies, design-of-experiment approaches as well as mathematical modeling. In practice, the design space can also be used to define acceptable boundaries for achieving acceptable quality as well as those that indicate failure. The in vitro release from the MR parenteral can be helpful in mapping the design space of the formulation and as a critical performance indicator if the CQAs are met and the target product profile is achieved.

The final element of a QBD approach for product development involves designing a control strategy once the design space is mapped. An efficient control strategy ensures that the product with the desired quality attributes will be consistently produced when the product is manufactured within the design space. Control of the process and characteristics of the raw materials and drug substance along with diligent in process testing of intermediates at critical steps in the manufacturing process can ensure a product that meets the specifications generate based on the desired therapeutic benefit. Once a comprehensive understanding of the product performance and manufacturing process inputs (process and raw materials) is established, ideally a continuous adaptive manufacturing process can be developed. Such an adaptive process could rely on real-time–release testing instead of end-product testing and risk-based understanding to ensure generation of a final product within the design space of the formulation. In effect, quality is built in the product as it is assembled to meet the CQAs that ensure the target product profile with potentially less dependence on end product testing.

DRUG RELEASE FROM MODIFIED-RELEASE PARENTERALS

One of the most important determinants of the performance of MR parenterals is the release of the drug entrapped with the dosage form in a controlled, consistent, and reliable manner that leads to a safe and efficacious therapeutic effect (62,68). The application of QBD principles in the preparation of MR parenterals and the use of BE to demonstrate comparability between generics and innovator products can be enhanced through an understanding of the factors and conditions that influence drug release from MR parenterals (66). Since the in vitro release profile of MR parenterals is a key indicator of performance and the eventual therapeutic effect, it can be vital to focus product development for both innovator and generic products on the optimization of drug release characteristics.

The release of drugs from MR parenterals results occurs via: drug diffusion through the carrier matrix; degradation of the polymer matrix; or a combination thereof (69,70). The exact mechanism, extent, and rate of the drug release are determined by the physicochemical properties of the carrier matrices and of the entrapped drug (71,72). In general, polymers and/or lipophilic components undergo degradation or disassembly following implantation in vivo and these processes eventually lead to structural collapse of the dosage form and release of the drug (69,70,73). In addition, based on the individual chemistry and degradation pathways, various MR parenterals can have unique mechanisms of release (69,70,74).

Drug release from biodegradable particulate systems such as microspheres, nanoparticles, and depots using PLGA polymer derivatives is usually triphasic (75). The first phase of release, also known as initial release or burst release, is observed upon hydration of the polymer upon implantation, which facilitates diffusion and release of the loosely associated surface bound drug from the matrix (76). The second phase of release, known as the lag phase, is characterized by low amounts of drug release and degradation of the high-molecular-weight polymer into lower molecular weight monomers. This is followed by the third phase that involves zero-order drug release corresponding to erosion of the polymeric matrix. The polymer breakdown in PLGA microspheres is due to autocatalytic scission of the polymer chains resulting from hydrolytic degradation at low pH (77,78).

Factors affecting drug release from microspheres include the molecular weight of the PLGA polymer; the blend composition, and monomer compositions of the polymer, that is, the lactide-to-glycolide ratio of the polymer chain; the microsphere architecture; the particle size; the crystallinity of the drug in the microsphere; and the presence of excipients in the microsphere matrix such as porosigens, surfactants, and basic compounds that can modulate the degradation kinetics of the polymer by changing the pH (72,78–83). The manufacturing processes used for microspheres can impact the amount of surface-associated drug, which can affect the burst release characteristics (81,84).

Drug release from liposomal MR parenteral systems can be due to a combination of mechanisms that include passive diffusion of the drug from the membranes, disassembly of the liposomal matrix by interaction with cellular proteins, and enzymatic degradation of lipids (28,85). Liposomes upon in vivo exposure can be internalized into cells of the reticuloendothelial system by receptor-mediated endocytosis or phagocytosis into endosomes followed by endosomal fusion and formation of lysosomes (85,86). Liposomal recognition by phagocytic cells is facilitated by their coating by serum proteins known as opsonins (86,87). The acidic pH conditions within lysosomes contributes to hydrolytic degradation of the lipids and release of the entrapped drug intracellularly (85). Factors influencing drug release include liposome size (smaller liposomes release their contents faster), route of administration, composition of the lipid matrix, charge of the lipids, and the nature of interaction of the drug with the lipids (28,74,85,88).

IN VITRO RELEASE TESTING OF MR PARENTERALS
Since understanding the mechanistic basis of drug release and the factors that impact release can be critical in establishing the total product profile, development of appropriate in vitro release testing methodologies to support these activities is an important part of a comprehensive pharmaceutical product development program (62,89). In vitro drug release tests are also known as dissolution tests or drug elution tests (73,89). The most generalized methodology for in vitro drug release tests for MR parenterals involves monitoring the release profile of the drug from the dosage form in a medium of defined composition under controlled conditions in a standardized equipment (73,90). The medium into which drug release occurs is sampled using a predetermined sampling scheme and the amount of drug released is measured using a sensitive assay. Since the in vitro release profile is reflective of the overall performance of the MR parenteral, it can be the most useful test for formulation selection and optimization as well as for assessing process parameters and robustness of the manufacturing process (68,76,88). In vitro release tests are also used for batch release of a commercial product and play an important role in the overall control strategy (89).

In the absence of a detailed understanding of the performance of innovator products, development of generics can use the in vitro test as an important assay to identify formulation composition, evaluate and select component equivalents, optimize process development as well as scale-up and establish stability and shelf life. The in vitro test can be used to demonstrate comparable performance characteristics of drug release between the generic and the innovator product under controlled conditions prior to preclinical experimentation to evaluate in vivo drug release (89). An optimized lead candidate of a generic formulation based on an accurate and discriminative in vitro release test can mitigate some of the significant risks and costs associated with the possibility of selecting an unsuitable formulation in animal experimentation and clinical studies. Given the tremendous value and importance of the in vitro drug release test, it may remain in the best interest of a MR parenteral drug development program to dedicate resources as early as possible to develop an accurate and discriminative in vitro release test.

In vitro drug release method development can be an iterative process since the test has to have the appropriate level of discrimination ability to map the design space of the formulation. Formulations that have the most favorable clinical performance and/or in vivo release in representative preclinical models can be used to optimize the in vitro drug release test. In vitro drug release for the formulation can be optimized over composition, pH, and temperature of the drug release medium, sample preparation and sampling scheme, and the instrument parameters under which the test is conducted (68,89,91).

Most typically, aqueous media at physiological conditions are preferred, although hydroalcoholic media containing low levels of the organic content have also been used (90,92). The volume of the drug release medium in which the release test is performed is a critical parameter since large volumes can lead to significant dilution-induced errors in the measurement and adsorptive drug loss in the instrumentation. Very low volumes can lead to violation of sink conditions, which can generate artifacts in the release profile that are not representative of the intrinsic in vitro release characteristics of the MR parenteral. The pH of the medium is usually optimized to reflect physiological conditions where the MR parenteral is expected to reside in vivo and also to ensure drug solubility and drug stability upon release (91). In addition to the presence of buffering components in the medium to maintain the target pH, surfactants such as Tween 80 (42,93), Tween 20 (40), and sodium dodecyl sulfate (39) can also be included. Although nonphysiological in origin, surfactants and detergents can facilitate drug release from the formulation, solubilize the drug upon release, maintain sink conditions, and prevent adsorptive loses of the released drug within the instrumentation. The tonicity of the in vitro release medium can also be adjusted using salts to generate a more physiological environment (41).

The assay used to measure the total amount of drug released must be validated using standard parameters of validation including linearity, accuracy, precision, and stability over the range of drug concentrations measured (94). Most typically, a UV assay is used to measure the total amount of drug release, although HPLC (36) or fluorescence (37) can be used to improve accuracy and sensitivity. Surfactants and other agents in the release medium can interfere in the assay measurement and impact chromatographic properties and should be factored accordingly in the development of the assay. The in vitro test is also optimized to identify an accurate and representative sampling scheme to characterize release. The sampling scheme can be used to quantify and characterize the burst release as well as the sustained-release phases in MR parenterals and is eventually used to set specifications for the final product (95).

The instrumentation used in the in vitro release test is a critical component that is optimized to ensure a robust method (96). In vitro drug release from MR parenterals can be studied in compendial instruments such as USP type 1, type 2, or type 4 dissolution apparatus or using a noncompendial temperature and humidity controlled chamber that can be qualified to guarantee consistent operation conditions (96). The advantages of compendial apparatus include ease of use and maintenance since the installation, operation, and performance can be guaranteed based on predefined specifications and tests indicated in the pharmacopoeias (96).

Type 1 and type 2 apparatus are the most prevalent for in vitro release testing (97–99). These include suspension of the dosage form in a standard volume vessel-containing medium that is continuously mixed throughout the duration of the test using a basket or a paddle attached to a rotating shaft attached to the drive unit. Samples of media can be taken either manually or via using cannulas. These instruments are optimized for the speed of the paddles; prevention of nonspecific adsorptive losses to the vessels, paddles or baskets, shaft, and filters used in cannulas; and the temperature at which the instrument is operated (97–99). Adsorptive loses can be reduced by use of nonreactive coatings. Despite their advantages in ensuring consistent performance, type 1 and 2 USP dissolution apparatus may not be the most optimal instruments for MR parenterals. Particulate MR parenteral dosage forms such as microspheres, liposomes, and micelles may suspend in the medium during the test and interfere in the sampling procedure. There may also be problems with aggregation of these systems depending on their hydrophobicity and density. Low volumes of media may not be possible due to the instrumentation size limitations that can lead to pooling of multiple units of the dosage forms, which can lead to errors. These instruments may not be useful for evaluating release of biologics from MR parenterals that may be particularly sensitive to adsorptive loses and instability due to the rotation in the instrumentation.

Type 4 dissolution apparatus or the flow through apparatus can overcome some of the limitations associated with the type 1 and 2 apparatus by confinement of the samples in flow-through cells outfitted with filters that allow passage of the released drug but not the particulate dosage forms (34,97,100). Drug release medium is stored in an external chamber, preheated, if necessary, and circulated through the flow-through cell (100). The media reservoir is sampled periodically to measure the amount of drug released. The type 4 dissolution instrument is optimized for the speed of flow; and for prevention of nonspecific adsorptive loses to the cells, and filters, media reservoir, as well as the tubing and fittings used to connect these components; and the temperature at which the instrument is operated. Because of confinement of the dosage forms, contamination of the sampling procedure is eliminated. The volume of the media can be controlled in the external chamber and therefore issues related to dilution or violations of sink conditions can be easily prevented. The volume used can be adjusted to the sensitivity and precision of the assay and pooling of dosage forms can be effectively avoided. The cells can also be engineered to accommodate geometrical and architectural specifications of the dosage forms (100).

Drawbacks associated with the type 4 dissolution instrument are the cumbersome setup and sample preparation time. In addition, there may be problems associated with leakage if the tubing connections are not correct. There is also the possibility of adsorptive loses to the following components: plastic or polycarbonate vessels, polymeric filters, polypropylene tubing, and glass media reservoirs. This can be reduced or eliminated by judicious choice of component materials and more commonly by the use of surfactants in the in vitro drug release medium.

In addition to compendial apparatus, numerous noncompendial laboratory scale apparatus have been used to test in vitro drug release from MR parenterals (90,91,101). These include the use of dialysis membranes to separate the dosage forms and released drug, the use of rotating bottles apparatus, and evaluating release in static systems such as vials and bottles by the incubation of the dosage forms in temperature-controlled chambers (90,91,101). Additionally, the use of Franz diffusion cells or apparatus based on modifications to them (34,102,103) and the use of shaker platform systems (38,43,104) are also common for in vitro release testing due to their ability to simulate blood flow around the MR parenteral in a dynamic physiological environment (37,43). These noncompendial instruments require rigorous qualification programs to ensure reliable temperature maintenance and consistent performance for the various instrumentation parameters optimized in the test.

Although information on the setup and optimization of in vitro drug release tests is widely available, specific compositions and conditions that can discriminate and map the design space for innovator products are usually maintained as trade secrets. It is thus a challenge to identify conditions and tests that can demonstrate equivalent or inequivalent performance in vitro of generic and innovator products since in vitro performance may not necessarily have to be a meaningful and reliable predictor of the performance in vivo. No matter, how reliable an in vitro test is developed, the in vivo performance can never be guaranteed since a highly controlled in vitro environment can rarely be representative of the variables observed in the in vivo environment. Nevertheless, the in vitro drug release test remains an important part of the overall control strategy for generic MR parenteral products to make important decisions on formulation and process development. These tests are most commonly used for quality control applications to facilitate routine lot release and ensure consistent performance across the manufactured lots.

IN VITRO–IN VIVO CORRELATION (IVIVC) FOR MR PARENTERALS

In vitro–in vivo correlation (IVIVC) is a mathematical relationship that can be defined between the in vitro release from a dosage form and the in vivo release observed in a preclinical or human model (105). Although consistent in vitro performance of an MR parenteral can ensure robust engineering and design, application of the in vitro test to predict in vivo performance can only be meaningful when an accurate IVIVC has been established. An IVIVC can be useful in formulation optimization, design space evaluation, process optimization, scale-up, preclinical and clinical evaluation, as well as to support implementation of postapproval changes in formulation and process during commercial manufacturing (106). These applications of IVIVC help reduce product development time-lines and costs by circumventing additional human testing that would otherwise be required.

Currently, regulatory guidelines for developing an IVIVC are only available for extended-release oral dosage forms (61,107). However, elements of the scientific basis for developing an IVIVC for extended-release oral dosage forms can be applied to product development of MR parenterals (63,108). Using the guidance developed by the FDA for extended-release orals (61–63,108), best practices to develop methods for IVIVC, evaluate prediction accuracy, and set specifications for in vitro drug release tests for MR parenterals can be identified. Additionally, this guidance can be used to apply IVIVC as a surrogate for clinical bioequivalence studies and to obtain a regulatory biowaiver to support scale-up and postapproval changes such as modifications in the formulation, equipment, process, and site of manufacture.

IVIVC involves establishing a mathematical relationship between an in vitro release property of a dosage form and the in vivo performance (109). The FDA guidance identifies three levels of IVIVC that vary in the in vitro and pharmacokinetic factors used to model the mathematical data. Level A IVIVC is a linear model that represents a point-to-point relationship between in vitro and in vivo release. The fraction of drug released in vitro is correlated to the fraction of drug released in vivo. The in vivo release profile of the drug is obtained by deconvolution of the drug concentration profile in body fluids over the duration of release. In this level, the in vitro and in vivo release profiles are superimposable either directly or by the application of an appropriate scaling factor. Level A is the most preferred and comprehensive model, as it takes into consideration the entire duration of drug release. Level A IVIVC has been demonstrated for MR parenterals such as microspheres (81,83), hydrogels (90), and multireservoir systems (106).

Level B IVIVC is based on the application of statistical moment analysis on the in vitro and in vivo data and is the least preferred of all levels for regulatory considerations. In this IVIVC level, summary parameters that can profile in vitro release such as mean in vitro drug release time and the dissolution rate constant can be correlated to parameters that define release in vivo such as mean in vivo drug release time and absorption rate constant, respectively. Level B is not considered a point-to-point correlation and may not be the most representative to in vivo release characteristics, since similar mean in vivo dissolution times can be obtained by significantly dissimilar in vivo release profiles.

Level C IVIVC establishes correlations between a fixed summary characteristic of the in vitro release profile to a fixed pharmacokinetic parameter. An example of a typical in vitro release parameter is the time required for release of a certain percent of the drug (e.g., $t_{50\%}$ or $t_{90\%}$) or the amount of drug released at a particular set-point in the in vitro drug release profile, whereas pharmacokinetic factors such as AUC, C_{max}, or T_{max} can be used to characterize the in vivo release profile. Level C is not a point-to-point correlation since it does not factor in the complete duration of the in vitro and in vivo release profiles. Multiple level C IVIVCs that establish relationships between the various in vitro release characteristics and in vivo parameters can be useful for regulatory submissions, if level A IVIVC is not possible.

The guidance identifies a process of developing IVIVCs using formulations with varying drug release rates, one faster and one slower than the nominal target rate of the formulation (107). These nonrepresentative faster and slower rates are from formulations that are usually outside the design space of the MR parenteral, although for even tighter control on the routine manufacturing of the product, these can also be a subset of the overall design space. The in vitro and in vivo release profiles are determined using the drug release test and deconvolution, respectively. A mathematical model is built by comparing the pharmacokinetics in vivo to the drug release profile in vitro at levels described in the guidance. In practical terms, the relationship between the fraction of drug released in vivo can be expressed as a linear function of the fraction of drug released from the MR parenteral in vitro. Human clinical data are recommended for regulatory purposes of an IVIVC to support a biowaiver and the trials are conducted in a crossover study design recruiting up to 36 subjects in each treatment condition.

The application of an IVIVC model for the prediction of in vivo release profiles depends on the accuracy of the model generated. The magnitude of the prediction error can be understood using internal or external data. Internal predictability uses pharmacokinetic parameters from formulations used to establish the IVIVC, whereas external predictability uses formulations not considered in the development of the model. For successful regulatory submissions to obtain a biowaiver for bioequivalence clinical studies using an IVIVC, the validation criteria for either internal or external prediction should be met. For internal predictability, the average absolute percent prediction error (%PE) is 10% or less for C_{max} and AUC predicted using the model and the %PE for each formulation used in the model should not exceed 15%. For external predictability, the %PE average for the formulation should be 10% or less for C_{max} and AUC predicted using the model. Additionally, %PE between 10% and 20% and those above 20% generally indicate inconclusive and inadequate predictability, respectively, unless otherwise justified.

Although generic product development can be facilitated using an IVIVC, it may not be useful in situations where the release mechanism across the generic and the innovator is altered. However, once a generic approval is obtained, further development issues such as site change, scale-up, or change in unit manufacturing operation processes that are encountered postapproval may be resolved using an IVIVC (107,108). Since IVIVC can be used to map the design space using a systematic performance-based approach, it can be a useful tool in the control strategy of the product and to set specifications for the in vitro release test (95).

CONCLUSIONS

Due to their innovative nature, highly specialized design, as well as the use of advanced technologies and proprietary components, no generic versions of MR parenteral dosage forms exist or have been approved by regulatory agencies. The pharmaceutical development of generic versions of these dosage forms thus remains an extremely challenging endeavor. With the emerging trend of use of biologics in pharmacotherapy that require specialized handling during routine pharmaceutical manufacture and as more proprietary innovator drug molecules, polymers, and processes lose their patent exclusivity protections, generic MR parenteral development will assume even greater significance in providing competitive and cost-effective therapeutic alternatives. The application of scientific principles outlined in regulatory guidances for bioequivalence, QBD, and IVIVC can be used to facilitate the pharmaceutical development of these products.

REFERENCES

1. Allen LV, Popovich GN, Ansel HC. Novel Dosage Forms and Drug Delivery technologies. In: Popovich GN, Allen LV, Ansel HC, eds. Ansel's Pharmaceutical Dosage

Forms and Drug Delivery Systems. Baltimore, MD: Lippincott Williams & Wilkins, 2005:652–671.

2. Biermasz NR, Romijn JA, Pereira AM, et al. Current pharmacotherapy for acromegaly: A review. Expert Opin Pharmacother 2005; 6(14):2393–2405.

3. Halford JC. Obesity drugs in clinical development. Curr Opin Investig Drugs 2006; 7(4):312–318.

4. Perez-Marrero R, Tyler RC. A subcutaneous delivery system for the extended release of leuprolide acetate for the treatment of prostate cancer. Expert Opin Pharmacother 2004; 5(2):447–457.

5. Mitchell H. Goserelin ('Zoladex')—Offering patients more choice in early breast cancer. Eur J Oncol Nurs 2004; 8(suppl 2):95–103.

6. Cullins V. Injectable and implantable contraceptives. Curr Opin Obstet Gynecol 1992; 4(4):536–543.

7. Brumm MV, Nguyen QD. Fluocinolone acetonide intravitreal sustained release device—A new addition to the armamentarium of uveitic management. Int J Nanomed 2007; 2(1):55–64.

8. Hanes PJ, Purvis JP. Local anti-infective therapy: Pharmacological agents. A systematic review. Ann Periodontol 2003; 8(1):79–98.

9. Lin SH, Kleinberg LR. Carmustine wafers: localized delivery of chemotherapeutic agents in CNS malignancies. Expert Rev Anticancer Ther 2008; 8(3):343–359.

10. Allen LV, Popovich GN, Ansel HC. Parenterals. In: Popovich GN, Allen LV, Ansel HC, eds. Ansel's Pharmaceutical Dosage Forms and Drug Delivery Systems. Baltimore, MD: Lippincott Williams & Wilkins, 2005:443–506.

11. Elrishi MA, Jarvis J, Khunti K, et al. Insulin glargine and its role in glycaemic management of Type 2 diabetes. Expert Opin Drug Metab Toxicol 2008; 4(8):1099–1110.

12. Sadrzadeh N, Glembourtt MJ, Stevenson CL. Peptide drug delivery strategies for the treatment of diabetes. J Pharm Sci 2007; 96(8):1925–1954.

13. Wilkinson H. Intrathecal Depo-Medrol: A literature review. Clin J Pain 1992; 8(1): 49–56.

14. Floyd A, Jain S. Injectable Emulsions and Suspensions. In: Lieberman HA, Rieger MM, Banker GS, eds. Pharmaceutical Dosage Forms: Disperse Systems. London, U.K.: Informa Healthcare, 1996:261–318.

15. Linn ES. Progress in contraception: New technology. Int J Fertil Womens Med 2003; 48(4):182–191.

16. Brannon-Peppas, L. Polymers in Controlled Drug Delivery. Medical Plastics and Biomaterials Magazine, 1997, November.

17. Periti P, Mazzei T, Mini E. Clinical pharmacokinetics of depot leuprorelin. Clin Pharmacokinet 2002; 41(7):485–504.

18. Okada H, Yamazaki I, Yashiki T, et al. Vaginal absorption of a potent luteinizing hormone-releasing hormone analogue (leuprolide) in rats. IV: Evaluation of the vaginal absorption and gonadotropin responses by radioimmunoassay. J Pharm Sci 1984; 73(3):298–302.

19. Okada H, Heya T, Ogawa Y, et al. One-month release injectable microcapsules of a luteinizing hormone-releasing hormone agonist (leuprolide acetate) for treating experimental endometriosis in rats. J Pharmacol Exp Ther 1988; 244(2):744–750.

20. Genton P. Progress in pharmaceutical development presentation with improved pharmacokinetics: A new formulation for valproate. Acta Neurol Scand Suppl 2005; 182:26–32.

21. Steinijans V. Pharmacokinetic characterization of controlled-release formulations. Eur J Drug Metab Pharmacokinet 1990; 15(2):173–181.

22. Mannelli P, et al. Long-acting injectable naltrexone for the treatment of alcohol dependence. Expert Rev Neurother 2007; 7(10):1265–1277.

23. Pettinati HM, Rabinowitz AR. Choosing the right medication for the treatment of alcoholism. Curr Psychiatry Rep 2006; 8(5):383–388.

24. Persad R. Leuprorelin acetate in prostate cancer: A European update. Int J Clin Pract 2002; 56(5):389–396.

25. Richter A, Anton SE, Koch P, et al. The impact of reducing dose frequency on health outcomes. Clin Ther 2003; 25(8):2307–2335.
26. Gollnick H, Cunliffe W, Berson D, et al. Management of acne: A report from a global alliance to improve outcomes in acne. J Am Acad Dermatol 2003; 49(1):S1–S37.
27. Webster G. Topical tretinoin in acne therapy. J Am Acad Dermatol 1998; 39(2):S38–S44.
28. Patil SD, Burgess DJ. Liposomes: Design and Manufacturing. In: Burgess DJ, ed. Parenteral Dispersed Systems: Formulation, Processing and Performance. New York, NY: Marcel Dekker Inc., 2005:249–303.
29. Hofheinz RD, Gnad-Vogt SU, Beyer U, et al. Liposomal encapsulated anti-cancer drugs. Anticancer Drugs 2005; 16(7):691–707.
30. Allen TM, Martin FJ. Advantages of liposomal delivery systems for anthracyclines. Semin Oncol 2004; 31(6):5–15.
31. Adler-Moore J, Proffitt RT. AmBisome: Liposomal formulation, structure, mechanism of action and pre-clinical experience. J Antimicrob Chemother 2002; 49(1):21–30.
32. Immordino ML, Dosio F, Cattel L. Stealth liposomes: Review of the basic science, rationale, and clinical applications, existing and potential. Int J Nanomedicine 2006; 1(3):297–315.
33. Moghimi MM, Hunter AC, Murray JC. Long-circulating and target-specific nanoparticles: Theory to practice. Pharmacol Rev 2001; 53(2):283–318.
34. Venkateswarlu V, Manjunath K. Preparation, characterization and in vitro release kinetics of clozapine solid lipid nanoparticles. J Control Release 2004; 95(3):627–638.
35. Jiang W, Kim BYS, Rutka JT, et al. Advances and challenges of nanotechnology-based drug delivery systems. Expert Opin Drug Deliv 2007; 4(6):621–633.
36. Chen F, Zhao Y, Wu H, et al. Enhancement of periodontal tissue regeneration by locally controlled delivery of insulin-like growth factor-I from dextran-co-gelatin microspheres. J Control Release 2006; 114(2):209–222.
37. Leitner VM, Guggi D, Krauland AH, et al. Nasal delivery of human growth hormone: In vitro and in vivo evaluation of a thiomer/glutathione microparticulate delivery system. J Control Release 2004; 100(1):87–95.
38. Blum JS, Saltzman W. High loading efficiency and tunable release of plasmid DNA encapsulated in submicron particles fabricated from PLGA conjugated with poly-L-lysine. J Control Release 2008; 129(1):66–72.
39. Duncan G, Jess TJ, Mohamed F, et al. The influence of protein solubilisation, conformation and size on the burst release from poly(lactide-co-glycolide) microspheres. J Control Release 2005; 110(1):34–48.
40. Jain SK, Awasthi A, Jain NK, et al. Calcium silicate based microspheres of repaglinide for gastroretentive floating drug delivery: Preparation and in vitro characterization. J Control Release 2005; 107(2):300–309.
41. Yin Y, Chen D, Qiao M, et al. Preparation and evaluation of lectin-conjugated PLGA nanoparticles for oral delivery of thymopentin. J Control Release 2006; 116(3):337–345.
42. Varde NK, Pack DW. Influence of particle size and antacid on release and stability of plasmid DNA from uniform PLGA microspheres. J Control Release 2007; 124(3):172–180.
43. Patil SD, Papadmitrakopoulos F, Burgess DJ. Concurrent delivery of dexamethasone and VEGF for localized inflammation control and angiogenesis. J Control Release 2007; 117(1):68–79.
44. Perrie Y, Mohammed AR, Kirby DJ, et al. Vaccine adjuvant systems: Enhancing the efficacy of sub-unit protein antigens. Int J Pharm 2008; 364(2):272–280.
45. Approved Drug Products with Therapeutic Equivalence Evaluations, Center for Drug Evaluation and Research, Office of Pharmaceutical Science, Office of Generic Drugs, U.S. Department of Health and Human Sevices, Food and Drug Administration, 2009.
46. Genazzani A, Pattarino F. Difficulties in the production of identical drug products from a pharmaceutical technology viewpoint. Drugs 2008; 9(2):65–72.
47. Schellekens H, Ryff JC. 'Biogenerics': The off-patent biotech products. Trends Pharmacol Sci 2002; 23(3):119–121.

48. Fogueri LR, Singh S. Smart polymers for controlled delivery of proteins and peptides: A review of patents. Recent Pat Drug Deliv Formul 2009; 3(1):40–48.

49. Bikiaris D, Koutris E, Karavas E. New aspects in sustained drug release formulations. Recent Pat Drug Deliv Formul 2007; 1(3):201–213.

50. Guidance for Industry: Bioavailability and Bioequivalence studies for orally administered drug products—General considerations, U.S. Department of Health and Human Services, Food and Drug Administration, Center for Drug Evaluation and Research (CDER), 2003 Revision 1.

51. Aberg J, Abrahamsson B, Grind M, et al. Bioequivalence, pharmacokinetic and pharmacodynamic response to combined extended release formulations of felodipine and metoprolol in healthy volunteers. Eur J Clin Pharmacol 1997; 52:471–477.

52. Lubenau H, Bias P, Maly AK, et al. Pharmacokinetic and pharmacodynamic profile of new biosimilar filgrastim XM02 equivalent to marketed filgrastim Neupogen: single-blind, randomized, crossover trial. BioDrugs 2009; 23(1):43–51.

53. Lister PD. The role of pharmacodynamic research in the assessment and development of new antibacterial drugs. Biochem Pharmacol 2006; 71(7):1057–1065.

54. Serra D, He YL, Bullock J, et al. Evaluation of pharmacokinetic and pharmacodynamic interaction between the dipeptidyl peptidase IV inhibitor vildagliptin, glyburide and pioglitazone in patients with Type 2 diabetes. Int J Clin Pharmacol Ther 2008; 46(7):349–364.

55. Christensen E. Methodology of superiority vs. equivalence trials and non-inferiority trials. J Hepatol 2007; 46(5):947–954.

56. Gomberg-Maitland M, Frison L, Halperin JL. Active-control clinical trials to establish equivalence or noninferiority: Methodological and statistical concepts linked to quality. Am Heart J 2003; 146(3):398–403.

57. Ebbutt AF, Frith L. Practical issues in equivalence trials. Stat Med 1998; 17 (15–16):1691–1701.

58. Julious SA. Sample sizes for clinical trials with normal data. Stat Med 2004; 23(12):1921–1986.

59. Koyfman SA, Agrawal M, Garrett-Mayer E, et al. Gross CP Risks and benefits associated with novel phase 1 oncology trial designs. Cancer 2007; 110(5):1115–1124.

60. Siewert M. Perspectives of in vitro dissolution tests in establishing in vivo/in vitro correlations. Eur J Drug Metab Pharmacokinet 1993; 18(1):7–18.

61. Uppoor V. Regulatory perspectives on in vitro (dissolution)/in vivo (bioavailability) correlations. J Control Release 2001; 72(1–3):127–132.

62. Martinez M, Rathbone M, Burgess D, et al. In vitro and in vivo considerations associated with parenteral sustained release products: A review based upon information presented and points expressed at the 2007 Controlled Release Society Annual Meeting. J Control Release 2008; 129(2):79–87.

63. Hayes S, Dunne A, Smart T, et al. Interpretation and optimization of the dissolution specifications for a modified release product with an in vivo–in vitro correlation (IVIVC). J Pharm Sci 2004; 93(3):571–581.

64. Harrison TS, Goa KL. Long-acting risperidone: A review of its use in schizophrenia. CNS Drugs 2004; 18(2):113–132.

65. International Conference on Harmonization of Technical Requirements for Registration of Pharmaceuticals for Human Use, ICH Harmonized Tripartite Guideline, Pharmaceutical Development Q8 (R1). Federal Register 2006; 71(98).

66. Dickinson P, Lee WW, Stott PW, et al. Clinical relevance of dissolution testing in quality by design. AAPS J 2008; 10(2):380–390.

67. Yu LX. Pharmaceutical quality by design: Product and process development, understanding, and control. Pharm Res 2008; 25(4):781–791.

68. Burgess DJ, Crommelin DJA, Hussain AS, et al. Assuring quality and performance of sustained and controlled release parenterals: EUFEPS workshop report. AAPS J 2004; 6(1):100–111.

69. Siepmann J, Göpferich A. Mathematical modeling of bioerodible, polymeric drug delivery systems. Adv Drug Deliv Rev 2001; 48(2–3):229–247.

70. Kanjickal DG, Lopina ST. Modeling of drug release from polymeric delivery systems—A review. Crit Rev Ther Drug Carrier Syst 2004; 21(5):345–386.
71. Iyer SS, Barr WH, Karnes HT. Profiling in vitro drug release from subcutaneous implants: A review of current status and potential implications on drug product development. Biopharm Drug Dispos 2006; 27(4):157–170.
72. Desai KG, Mallery SR, Schwendeman SP. Formulation and characterization of injectable poly(DL-lactide-co-glycolide) implants loaded with N-acetylcysteine, a MMP inhibitor. Pharm Res 2008; 25(3):586–597.
73. Onishi H, Machida Y. In vitro and in vivo evaluation of microparticulate drug delivery systems composed of macromolecular prodrugs. Molecules 2008; 13(9): 2136–2155.
74. Drummond DC, Noble CO, Hayes ME, et al. Pharmacokinetics and in vivo drug release rates in liposomal nanocarrier development. J Pharm Sci 2008; 97(11):4696–4740.
75. Zolnik BS, Burgess DJ. Effect of acidic pH on PLGA microsphere degradation and release. J Control Release 2007; 122(3):338–344.
76. Allison SD. Analysis of initial burst in PLGA microparticles. Expert Opin Drug Deliv 2008; 5(6):615–628.
77. Vert M, Mauduit J, Li S. Biodegradation of PLA/GA polymers: Increasing complexity. Biopharm Drug Dispos 1994; 15(15):1209–1213.
78. Mauduit J, Pérouse E, Vert M. Hydrolytic degradation of films prepared from blends of high and low molecular weight poly(DL-lactic acid)s. J Biomed Mater Res 1996; 30(2):201–207.
79. Yeo Y, Park K. Control of encapsulation efficiency and initial burst in polymeric microparticle systems. Arch Pharm Res 2004; 27(1):1–12.
80. Desai KG, Mallery SR, Schwendeman SP. Effect of formulation parameters on 2-methoxyestradiol release from injectable cylindrical poly(DL-lactide-co-glycolide) implants. Eur J Pharm Biopharm 2008; 70(1):187–198.
81. Zolnik BS, Burgess DJ. Evaluation of in vivo–in vitro release of dexamethasone from PLGA microspheres. J Control Release 2008; 127(2):137–145.
82. Obeidat WM, Price JC. Viscosity of polymer solution phase and other factors controlling the dissolution of theophylline microspheres prepared by the emulsion solvent evaporation method. J Microencapsul 2003; 20(1):57–65.
83. Chu DF, Fu XQ, Liu WH, et al. Pharmacokinetics and in vitro and in vivo correlation of huperzine A loaded poly(lactic-co-glycolic acid) microspheres in dogs. Int J Pharm 2006; 325(1–2):116–123.
84. Jain R, Shah NH, Malick AW, et al. Controlled drug delivery by biodegradable poly(ester) devices: Different preparative approaches. Drug Dev Ind Pharm 1998; 24(8):703–727.
85. Patil SD, Rhodes DG, Burgess DJ. DNA-based therapeutics and DNA delivery systems: A comprehensive review. AAPS J 2005; 7(1):E61–E77.
86. Ishida T, Harashima H, Kiwada H. Liposome clearance. Biosci Rep 2002; 22(2): 197–224.
87. Yan X, Scherphof GL, Kamps JA. Liposome opsonization. J Liposome Res 2005; 15 (1–2):109–139.
88. Tardi PG, Boman NL, Cullis PR. Liposomal doxorubicin. J Drug Target 1996; 4(3): 129–140.
89. Burgess DJ, Hussain AS, Ingallinera TS, et al. Assuring quality and performance of sustained and controlled release parenterals: AAPS workshop report, co-sponsored by FDA and USP. Pharm Res 2002; 19(11):1761–1768.
90. Patil SD, Papadimitrakopoulos F, Burgess DJ. Dexamethasone-loaded poly(lactic-co-glycolic) acid microspheres/poly(vinyl alcohol) hydrogel composite coatings for inflammation control. Diabetes Technol Ther 2004; 6(6):887–897.
91. Jiang G, Woo BH, Kang F, et al. Assessment of protein release kinetics, stability and protein polymer interaction of lysozyme encapsulated poly(D,L-lactide-co-glycolide) microspheres. J Control Release 2002; 79(1–3):137–145.

92. Suzuki K, Price JC. Microencapsulation and dissolution properties of a neuroleptic in a biodegradable polymer, poly(D,L-lactide). J Pharm Sci 1985; 74(1):21–24.
93. Dharmala K, Yoo JW, Lee CH. Development of Chitosan–SLN Microparticles for chemotherapy: In vitro approach through efflux-transporter modulation. J Control Release 2008; 131(3):190–197.
94. Kostanski JW, DeLuca PP. A novel in vitro release technique for peptide containing biodegradable microspheres. AAPS PharmSciTech 2000; 1(1):E4.
95. Modi NB, Lam A, Lindemulder E, et al. Application of in vitro–in vivo correlations (IVIVC) in setting formulation release specifications. Biopharm Drug Dispos 2000; 21(8):321–326.
96. D'Souza SS, DeLuca PP. Methods to assess in vitro drug release from injectable polymeric particulate systems. Pharm Res 2006; 23(3):460–474.
97. Voisine JM, Zolnik Banu S, Burgess DJ. In situ fiber optic method for long-term in vitro release testing of microspheres. Int J Pharm 2008; 356(1–2):206–211.
98. He P, Davis SS, Illum L. Chitosan microspheres prepared by spray drying. Int J Pharm 1999; 187(1):53–65.
99. Adeyeye CM, JC Price. Development and evaluation of sustained-release ibuprofen—Wax microspheres. II. In Vitro Dissolution Studies. Pharm Res 1994; 11(4):575–579.
100. Kauffman JS. Qualification and validation of usp apparatus 4. Dissolution Technol 2005(May):41–43.
101. Prabhu S, Sullivan JL, Betageri GV. Comparative assessment of in vitro release kinetics of calcitonin polypeptide from biodegradable microspheres. Drug Deliv 2002; 9(3):195–198.
102. Salama RO, Traini D, Chan H, et al. Preparation and characterisation of controlled release co-spray dried drug–polymer microparticles for inhalation 2: Evaluation of in vitro release profiling methodologies for controlled release respiratory aerosols. Eur J Pharm Biopharma 2008; 70(1):145–152.
103. Morales ME, Gallardo-Lara V, Calpena AC, et al. Comparative study of morphine diffusion from sustained release polymeric suspensions. J Control Release 2004; 95(1):75–81.
104. Naraharisetti PK, Lew MD, Fu Y, et al. Gentamicin-loaded discs and microspheres and their modifications: Characterization and in vitro release. J Control Release 2005; 102(2):345–359.
105. Emami J. In vitro–in vivo correlation: from theory to applications. J Pharm Pharm Sci 2006; 9(2):169–189.
106. Prescott JH, Krieger TJ, Lipka S, et al. Dosage form development, in vitro release kinetics, and in vitro-in vivo correlation for leuprolide released from an implantable multi-reservoir array. Pharm Res 2007; 24(7):1252–1261.
107. Extended Release Oral Dosage Forms: Development, Evaluation, and Application of In Vitro/In Vivo Correlations, Center for Drug Evaluation and Research (CDER), U.S. Department of Health and Human Services, Food and Drug Administration, 1997.
108. SUPAC-MR: Modified Release Solid Oral Dosage Forms, Scale-Up and Postapproval Changes: Chemistry, Manufacturing, and Controls; In Vitro Dissolution Testing and In Vivo Bioequivalence Documentation, Center for Drug Evaluation and Research (CDER), Food and Drug Administration, U.S. Department of Health and Human Services, 1997.
109. Mauger DT, Chinchilli VM. Methods of establishing in vitro–in vivo relationships for modified release drug products. In: Sen PK, Rao CR, eds. Handbook of Statistics, Vol. 18: Bioenvironmental and Public Health Statistics. Amsterdam: Elsevier Science, 2000:977–1002.

Biosimilar Drug Products—Manufacture and Quality

Suzanne M. Sensabaugh

Hartmann Willner LLC, Columbia, Maryland, U.S.A.

INTRODUCTION

Biosimilar drug products are recombinant, cell-derived protein products whose safety, identity, purity, impurities, potency, and quality can be determined, monitored, and controlled. These products are comparable in quality, safety, and efficacy to their corresponding innovative biotechnology-derived product(s) (brand product) and are distributed after patent expiry of the brand. These products are also known as "biological generics," "generic biologics," and "follow-on protein products." As with chemical generics, biosimilars adhere to the same manufacturing standards and controls as the innovator. From a manufacture and quality perspective, approval of biosimilars is based on comprehensive chemistry, manufacturing, and controls data and information and review by regulatory authorities to include the Food and Drug Administration (1). Although manufacturing standards and quality of biosimilars are no different than those of brand products, biotechnology-derived drugs, in general, are different in many ways from chemically synthesized drugs.

Biosimilars are produced in living systems. Production in living systems, as compared to chemical synthesis, may be viewed by some as a more complicated means of manufacture. However, with state-of-the-art analytical methods, advances in the separation sciences, ability to monitor and control the manufacturing process, and modern concepts of quality management, biotechnology-derived products can be viewed more like conventional chemical drugs in regard to characterization, manufacture, and control. Biosimilars, like their innovative biotechnology-derived counterparts, are routinely manufactured through processes that are reproducible, consistent, and robust. Validation of all manufacturing steps, from propagation of the source material to preparation of the active pharmaceutical ingredient (API) to final filling, allows established specifications to become meaningful predictors of end product quality and consistency. Controls and specifications ensure identity, purity, potency, safety, and quality from batch to batch.

This chapter provides an introduction to those not familiar with the manufacture and quality of biotechnology-derived protein products, both innovative and biosimilar. As there are many ways to manufacture these products, this chapter describes the general process steps and is not meant to be all-inclusive. Many FDA, EMEA, and ICH guidance documents address in detail the manufacture and quality of biotechnology-derived products from a regulatory perspective. Interested parties should refer to these documents for more information (2–5).

EVOLUTION OF MANUFACTURE AND METHODS

Historically, biotechnology-derived products were complex mixtures that were difficult to characterize and which had the potential for transmission of infectious diseases due to the living cells in which they were manufactured. Limited analytical and biological methods were available to demonstrate identity, purity, quality, safety, and potency. The assumption made by manufacturers and regulatory bodies alike was that any changes to the manufacturing process could result in a change in the safety and efficacy profile, as methods were not available to detect and characterize the effect of these changes and their consequence on product quality. Thus, the manufacturing process was integral to the definition of the product and led to the "process = product" paradigm. Because of this equation of product and process, in order to demonstrate that no significant change had occurred in safety and efficacy following a manufacturing change, clinical trials and FDA approval were required prior to release of the product manufactured under the changed manufacturing process (6).

Since then, and over the past 20 years, analytical and biological methods have evolved allowing greater ability to characterize these products. In fact, some biologics are better characterized than some chemical drugs. These analytical methods are used during research and development, in-process control testing, and end product testing. The development of these state-of-the-art analytical and biological methods has allowed a database of knowledge to be established on the result of these manufacturing changes, as they relate to safety and efficacy. In addition, manufacturing technology has progressed, thereby increasing the homogeneity and purity of these complex mixtures. The principle of "process = product" is no longer accepted for most biotechnology-derived products. However, the importance of product and process design and control cannot be overlooked. In-process controls, validation, and testing are necessary to ensure consistent product quality, safety, potency, identity, and purity.

THE MANUFACTURING PROCESS

The manufacturing process for biotechnology-derived drugs is different than that for chemical drugs. Biotechnology-derived products, including biosimilars, are protein products that are produced in living organisms as opposed to chemical synthesis. Although recombinant products can be derived from many cellular systems, such as bacteria, yeast, and fungi, or cells from mammals and insects, the focus of this chapter will be on bacterial and mammalian cell-derived products, as these cell substrates are the most popular, and thus, have a strong regulatory track record. The selection of a host cell system is dependent on many factors, including production efficiency, biological activity of the expressed protein, the need for posttranslational modification(s), economics, and regulatory issues. Bacterial cell substrates include *Escherichia coli* and *Pseudomonas putida*. Advantages of bacterial cell substrates (as compared to mammalian cell substrates) include a better understanding of cell and molecular biology, vector construction that is more straightforward, more rapid cell growth in less-expensive media, higher levels of protein expression, intracellular secretion of protein, and uncomplicated cell bank characterization. Disadvantages include potential endotoxin production, need for protein refolding and separation of incorrectly folded protein, no posttranslational processing, and an N-terminal methionine that may require removal. Mammalian cell substrates include Chinese hamster ovary (CHO) and Baby hamster kidney (BHK). Advantages of mammalian cell culture over bacterial cell substrates include posttranslational modification(s), a higher likelihood of obtaining a properly folded protein, and extracellular secretion of the protein. Disadvantages include expense; slow growth and production; potential adventitious agent contamination, such as viruses and mycoplasma; and more extensive cell bank characterization.

The manufacturing process for a biosimilar, as for all biotechnology-derived products, should be robust, reproducible, validated, and designed to produce the API (drug substance) and drug product. The manufacturing process can be divided into the following broad areas: generation of the expression vector construct and transfection of the host cell, expansion of host cell in culture vessels, production in aerobic fermentation system(s) (bioreactor), purification of API (and intermediate, if produced), and formulation/fill of drug product.

To begin the process, a gene encoding the protein of interest is identified. The DNA of this gene is cloned into a vector, such as a plasmid (these are circular extrachromosomal single-stranded DNA elements). This modified plasmid is inserted into the host cell. The plasmid contains all the elements necessary for replication of the gene of interest by the host cell's machinery. Derivation of these host cell lines must be documented and characterized. Testing must be carried out to ensure the cell line's identity, purity, safety, and stability. For bacterial-derived cell lines, this includes documentation of the species, strain, genotypic and phenotypic characteristics, pathogenicity, toxin production, environmental hazards of the organism from which the cell substrate was derived, method of isolation, culture procedures, genetic manipulation or selection, and testing of endogenous and adventitious agents (7). The original DNA sequence must also be confirmed to establish that the correct coding sequence for the protein has been incorporated into the host cell and is maintained during production (8).

Once a pure culture is established, it is subcultured into a master cell bank (MCB). Production of a MCB reduces the risk of contamination or loss of the pure culture. Everyday working cell stocks or working cell banks (WCB) may be subcultured from the MCB. The pure culture, MCB, and WCB (if WCB is established) are tested to ensure that purity, productivity, viability, and identity have remained consistent. The MCB must be thoroughly characterized and qualified. The source of the cells, their history, and generation needs to be well documented. Characterization of the cell banks includes viability, culture purity, strain identity, genetic analysis, and stability. The testing program will vary according to the cell types, but may also include sterility, mycoplasma, and viral contaminants. The integrity of these cell banks must be ensured to have a consistent, safe, high-quality product.

Manufacture begins when a vial from the WCB (or MCB if only a single-tiered banking system is established) is used as an inoculum and expanded in a culture vessel. After a measurable parameter, such as cell density, surpasses a predetermined limit and in-process testing is completed, a small fermenter(s), then a larger commercial-scale fermenter(s), is(are) inoculated. These bioreactor vessels provide better control over physical and nutritional factors than a culture vessel, such as a flask. At each step, cell viability, productivity, and purity of inoculum are necessary to ensure the success of the fermentation process. Although the environment in the bioreactor is dynamic, within specified controlled parameters a reproducible product is manufactured. The commercial-scale bioreactor is where the cell is producing the protein (active ingredient) under optimal, efficient conditions. A desirable manufacturing process must balance productivity, cell concentration, and time in this environment.

Cultivation of cells through aerobic fermentation techniques can be classified into batch, fed-batch, and continuous operation. In batch processes, all nutrients required for growth and protein formation are contained in the medium prior to inoculation. A fed-batch operation is one where nutrients are added during culture growth after inoculation. For continuous operations, the feed medium containing all the nutrients is supplied at a constant rate and the cultured broth is simultaneously removed. Oxygen is necessary for the life of these cells and is supplied by aeration. Mixing with a propeller or sparging aeration from the bottom of the bioreactor serves to ensure access to nutrients and oxygenation. Mammalian cells grow either in suspension or require a surface for attachment and growth (anchorage dependent). For anchorage dependent mammalian cells, suitable adherence materials must be provided, such as glass, plastic, ceramic, or synthetic resins. Bioreactors, such as plastic bags, and microcarrier culture systems have been developed to expand the surface area for adherent cell cultures at commercial scale. Growth of mammalian cells can be more of a challenge than bacterial cells, as they are more sensitive to shear force of agitation or air sparging, and usually dependent on materials of animal origin (such as bovine serum), leading to regulatory issues that must be addressed, for example, transmissible spongiform encephalopathy (TSE) concerns. In addition, for cell lines of mammalian origin, the inactivation or removal of endogenous viral particles or potential adventitious viral agents is a critical step in the manufacturing process. All viral inactivation and/or removal of steps must be validated (9).

Measurements of bioreactor conditions are the key to understanding and controlling the fermentation process. Computer control and sensing technologies (using sensors located in the bioreactor) allow for constant monitoring and feedback of culture conditions. This includes monitoring of physical parameters necessary for growth (e.g., temperature), cellular metabolism (e.g., dissolved gaseous O_2 and CO_2 concentrations), foaming due to agitation, metabolic by-product concentration, carbon and nitrogen source concentrations, airflow rate and agitation speed, and medium feed rate. Exhaust gas analysis provides information about respiratory activity, which is closely related to cellular metabolism and growth. Biological parameters are also measured such as protein synthesis, cell growth rate, and cell viability. Fermentation conditions are very important and can affect the purity and structure of the desired protein. Monitoring, testing, and in-process controls ensure that the conditions remain consistent from batch to batch.

Once the fermentation process has been completed or while on-going, as in the case of continuous bioreactor systems, fermentation broth is harvested. This broth includes the protein of interest; intact cells; cellular DNA, proteins, and fragments; soluble and insoluble media components; metabolic by-products; and various impurities. The protein of interest may be stored in the cell or expressed by the cell. If the protein is stored in the cell, cell membranes must be disrupted prior to isolation/recovery. Some protein products, especially those produced in bacterial cells, require a refolding step to bring the protein back to its native conformation. Various analytical methods may be used, such as

SDS-PAGE (sodium dodecyl sulfate-polyacrylamide gel electrophoresis), IEF (isoelectric focusing), SEC-HPLC (size exclusion high-performance liquid chromatography), and RP-HPLC (reverse-phase high-performance liquid chromatography). This broth is then concentrated to remove solids from the liquid, concentrate the solids to avoid excess water, stabilize the protein by the removal of proteolytic enzymes, and isolate the protein of interest and/or cells. Centrifugation, filtration, ultrafiltration, reverse osmosis, and/or other means may be used to perform these production steps to result in material for further manufacture. This intermediate [material that undergoes further molecular change or purification before it becomes an API (10)] and/or API may be stored prior to further processing.

Chromatography is usually used to purify the broth and isolate the protein of interest. Purification is a stepwise process where chromatography columns are used in a logical sequence to maximize throughput, yield, and purity. Purity and yield are influenced by a number of parameters, including flow rate, sample load, and media particle size. Chromatographic purification techniques include gel filtration or size exclusion chromatography (SEC), affinity, ion exchange, reverse phase, and hydrophobic interaction. Gel filtration allows for the fractionation of molecules by size, such as the separation of macromolecules from low-molecular-weight substances and substances, such as aggregates, that are above the gel fractionation range; gel filtration is also often used for desalting. Affinity chromatography is based on the specific binding of a resin-bound ligand or protein to a target protein or class of proteins; an example of affinity chromatography is the use of a Protein A column to bind a monoclonal antibody. Affinity chromatography allows for the removal of materials that do not bind to the affinity matrix, such as host cell proteins (HCP), DNA, and other process-related impurities. Reverse-phase chromatography utilizes a nonpolar support and separates proteins based on their hydrophobic character—less hydrophobic materials elute more quickly from the column. Ion exchange separates molecules based on charge. Hydrophobic interaction chromatography (HIC) allows for the separation of protein based on their hydrophobic residue content—more hydrophobic proteins are retained on the column to a greater extent. Each step results in the reduction of impurities and contaminants thereby increasing the purity of the final product. A stepwise purification process for a monoclonal antibody may include, for example, Protein A, ion exchange, and HIC columns.

During purification, the protein of interest is concentrated and stabilized. During fermentation, isolation, and purification, the target protein may undergo a variety of transformations, including hydrolysis, deamidation, and oxidation. Specifications are established for these process-related impurities and testing is conducted to ensure that specifications are met. Process- and product-related impurities may be quantitated through various methods, such as RP-HPLC, SDS-PAGE, Western Blotting, immunoassays, SEC-HPLC, and IEF.

After purification is completed, the product may be formulated or filled directly into final containers. As the vast majority of biotechnology-derived drugs are administered via the parenteral route, this filling must take place under aseptic (sterile) conditions. Filling under aseptic conditions poses different challenges than compounding of nonsterile pills and tablets—the most common dosage form for chemically synthesized drugs (11).

Process validation in conjunction with a thorough understanding of the manufacturing process and final product characterization assures the safety, potency, and quality of the biosimilar as with all biotechnology-derived drugs. Process validation begins with a description of the system, equipment, and product specifications, and proceeds through installation, operational, and performance qualification, and finally product qualification. The manufacturing process should be robust, reproducible, validated, and designed to produce the active ingredient in a stable formulation. Process- and product-related impurities are inherent to the manufacture of all drugs, whether biotechnology-derived or chemically synthesized. Appropriate clearance studies using validated methods during process validation ensure that these impurities are removed or controlled within acceptable limits thus ensuring safety of the final product. Impurity levels can be used to indicate deviations in manufacturing process control, and thus, it is important they be monitored. Contaminants are exogenous items, such as viruses and other adventitious agents, endotoxin, and leached affinity ligands, not deliberately added to the culture. Appropriate clearance studies demonstrating their removal should be conducted during process validation.

Due to the nature of the production system for biotechnology-derived products, these products may be more heterogenous than chemically synthesized drugs. This heterogeneity may influence conformation, function, or antigenic properties. Heterogeneity of biotechnology-derived drugs is evaluated and characterized, specifications are set, controls are established to maintain these specifications, and in process and end product testing allow for this heterogeneity to be controlled within allowable limits.

For all biological and chemical drugs, the risk the manufacturing process poses to the safety of the drug product is mitigated through

- manufacture according to current Good Manufacturing Practices (cGMPs);
- a consistent manufacturing process;
- assignment of meaningful specifications;
- appropriate equipment operating parameters;
- monitoring of critical parameters and establishing in-process controls;
- conducting in-process and end product testing;
- monitoring intermediate, drug substance (API), and drug product stability; and
- proper validation of equipment, analytical and biological methods, the production facility, and the production process.

COMPARABILITY

The development program for an innovative biological product includes comparisons to an in-house reference standard. For a biosimilar, not only is characterization conducted against an in-house reference standard, but analytical and biological data comparing the biosimilar to the already marketed brand product is also required.

The use of a scientific exercise to determine comparability of protein products is not a new practice put in place solely for biosimilars. Well in use by industry and the Agency on a case-by-case basis before FDA issued Guidance on the topic in 1996 (6), this scientific exercise is conducted to demonstrate that a biological product is comparable pre- and postmanufacturing process changes. It is a step-by-step approach with an evaluation using first, in vitro analytical and biological methods, and then in vivo animal and clinical studies. At each step of this exercise, the comparability of the two products is evaluated. If the products are comparable, the exercise ends. For example, if in vitro analytical and biological methods indicate the two products—one produced premanufacturing changes, and the other postmanufacturing changes—are comparable, no studies in vivo would be necessary. If comparison using analytical and biological methods indicates that the two products are not comparable, animal and/or clinical studies would take place to further investigate the differences between the two products. In 2001, the European Medicines Association (EMEA) expanded the concept of comparability to biosimilars (12). Although there is no legal or regulatory definition of the term "comparable," scientific ability, understanding, and experience are used to make the decision on if comparability has or has not been established (13).

The analytical and biological methods used to characterize the biosimilar and to compare the biosimilar to the brand product are the same methods routinely used in the characterization of innovative biotechnology-derived products. This includes the methods used to conduct the comparability exercise evaluating the products pre- and postmanufacturing changes. These methods are vast in number and determined on a case-by-case basis. For example, primary sequence structure can be assessed and compared using such methods as peptide mapping, N-terminal sequence analysis, 2D gel electrophoresis, and HPLC gel filtration. Secondary structure can be assessed and compared using circular dichroism and Fourier-transform infrared spectroscopy (FT-IR). Three-dimensional structure can be assessed and compared using X-ray crystallography and NMR spectroscopy. Molecular weight can be assessed and compared using mass spectroscopy (to obtain an exact mass) as well as by laser light scattering and ultracentrifugation.

Folding and conformation (higher order structure) is essential to the mechanism of action and determination of the structure–function relationship. This higher order structure may be inferred from the biological activity of the protein. Biological activity is defined as the specific activity or capacity of the product to achieve a defined biological effect (14). Potency (biological activity) is measured by biological assays such as cell proliferation assays, in vivo assays, and ligand binding assays, such as the enzyme-linked immunosorbent assay (ELISA).

CONCLUSION

Manufacturing standards and quality for biosimilar drug products are the same as for biotechnology-derived innovative products. There are vast differences between the manufacturing process steps of biologics and chemically synthesized drugs, but the quality principles remain the same. As with chemical generics, biosimilars are comparable in quality, safety, and efficacy to the brand product(s).

REFERENCES

1. Woodcock J, Griffin J, Behrman R, et al. The FDA's Assessment of Follow-on Protein Products: A Historical Perspective. Nat Rev Drug Discov 2007; 6:437–442.

2. FDA Center for Drugs Evaluation and Research, Guidance Documents. http:// www.fda.gov/cder/guidance/index.htm. Accessed September 15, 2008.
3. FDA Center for Biologics Evaluation and Research, CBER Guidances/Guidelines/ Points to Consider. http://www.fda.gov/cber/guidelines.htm. Accessed September 15, 2008.
4. European Medicines Agency (EMEA), Human Medicines. http://www.emea. europa.eu/index/indexh1.htm. Accessed September 15, 2008.
5. International Conference on Harmonization, ICH Guidelines. http://www.ich. org/cache/compo/276–254-1.html. Accessed September 15, 2008.
6. FDA Guidance Concerning Demonstration of Comparability of Human Biological Products, Including Therapeutic Biotechnology-derived Products, April 1996. http://www.fda.gov/cber/gdlns/comptest.txt. Accessed September 15, 2008.
7. ICH Harmonized Tripartite Guideline, Derivation and Characterization of Cell Substrates Used for Production of Biotechnological/Biological Products Q5D, July 1997. http://www.ich.org/LOB/media/MEDIA429.pdf. Accessed September 15, 2008.
8. ICH Harmonized Tripartite Guideline, Quality of Biotechnological Products: Analysis of the Expression Construct in Cells Used for Production of R-DNA Derived Protein Products Q5B, November 1995. http://www.ich.org/LOB/media/ MEDIA426.pdf. Accessed September 15, 2008.
9. ICH Harmonized Tripartite Guideline, Viral Safety Evaluation of Biotechnology Products Derived from Cell Lines of Human or Animal Origin Q5A(R1), September 1999. http://www.ich.org/LOB/media/MEDIA425.pdf. Accessed September 15, 2008.
10. ICH Harmonized Tripartite Guideline, Good Manufacturing Practice Guide for Active Pharmaceutical Ingredients Q7, November 2000. http://www.ich. org/LOB/media/MEDIA433.pdf. Accessed September 15, 2008.
11. FDA Guidance for Industry: Sterile Drug Products Produced Through Aseptic Processing—Current Good Manufacturing Practice, September 2004. http://www. fda.gov/cder/guidance/5882fnl.htm. Accessed September 15, 2008.
12. EMEA Committee for Proprietary Medicinal Products Note for Guidance on Comparability of Medicinal Products Containing Biotechnology-Derived Proteins as Drug Substance (CPMP/BWP/3207/00), Adopted September 2001.
13. Sensabaugh SM. Biosimilars: A business case. J Gen Med 2007; 4:186–199.
14. ICH Harmonized Tripartite Guideline, Specifications: Test Procedures and Acceptance Criteria for Biotechnological/Biological Products Q6B, March 1999. http://www.ich.org/LOB/media/MEDIA432.pdf. Accessed September 15, 2008.

Index